Lecture Notes in Mathematics

A collection of informal reports and seminars

Edited by A. Dold, Heidelberg and B. Eckmann, Zürich

115

Klaus W. Roggenkamp

McGill University, Montreal

Verena Huber-Dyson

University of Illinois, Chicago Circle

Lattices over Orders I

Springer-Verlag

Berlin · Heidelberg · New York 1970

K. W. Roggenkamp
Dept. of Mathematics, McGill University
Montreal, P. Q./Canada

(During the preparation of these notes the first
author has enjoyed research positions at the University of Illinois,
Université de Montréal and at McGill University.)

V. Huber-Dyson
Dept. of Mathematics, University of Illinois
Chicago, IL 60680/USA

PREFACE *)

These notes constitute the preliminary version of a book on lattices over orders. This might explain why we have included elementary definitions and theorems, thus deviating from the original character of the LECTURE NOTES. The notes consist of two volumes, and to make the first one, which mainly develops the theory of maximal orders, self-contained we have included a list of references and an index. Because of the preliminary character of the notes we welcome every suggestion which might improve the later book.

This work is an attempt to close the gap between the producing research scientist and the consuming reader. Our own experience has convinced us that our field abounds with "mathematical folklore". The theory of orders is quite essentially based on this kind of common knowledge, especially from related fields, such as algebraic number and class field theory, homological algebra and the theory of algebras.

As a matter of fact, the rapid modern development since 1950 employs this type of results as well as many early papers both as tools and as guiding light. Moreover, it seems that this development has reached a certain culminating point with some of the most recent results, and we feel it is ready for presentation in a reasonably complete form. Thus we are undertaking the task of clarifying the theory of orders eo ipso, of systematizing the wealth of existing and non-existing literature, and last but not least, to bring into unified form and to modernize, where we feel it is necessary, this vast field of knowledge.

Apart from this subjective motivation, the book shall serve two - in our opinion equally important - purposes. On the one hand it shall

*) The authors wish to express their gratitude to all the institutions, that have generously supported this work. Special thanks to Mrs. Betty Kurrle for her gracious assistance with the typing!

allow a good student or the eager reader, who is not a specialist, to learn the subject or to develop a love for our field. To come down to specifics: The reader should at least have graduate level; this because we expect him to have a genuine interest in this topic and to be capable of developing mathematical maturity. Therefore, we have tried to make this book as self-contained as possible; however, to keep it down to a reasonable size, we assume that the reader is familiar with elementary set theory and with the theory of fields.
On the other hand we would like the book to serve as a reference for the specialist. In this respect, the need for it is borne out by the fact that, apart from the classical treatment of maximal orders in Deuring [1] (1935), and a rather specialized chapter in Curtis-Reiner [1] (1962), there are neither lecture notes nor books available on the subject. Finally, the wealth of open problems, with which the theory of modules over orders still abounds, should inspire and encourage the research mathematician.
As can be seen from the table of contents, we have tried to use modern tools, whenever possible, and to put some stress on progenerators and separability; in particular, for maximal orders we treat systematically the approach introduced by Auslander-Goldman [1] (1960). Moreover, we have also enclosed the beautiful papers of Hasse [2] (1931) and Chevalley [1] (1936), which occasionally seem to have been neglected.
Summa summarum we have attempted to give a self-contained treatment of the theory of lattices over orders, that leads up to and includes the most recent developments. For the presentation of the - in our opinion - most important and fruitful results we have introduced as much preliminary material as necessary for their proofs. We have also included exercises at the end of every section; some concern results in their own right and are sometimes referred to in the text.

CONTENT

Introduction

The theory of orders and the study of modules over orders have three main sources:

(i) Ideal theory and arithmetic can be developed in maximal orders as in Dedekind domains (cf. Ch. IV). From this point of view, the study of maximal orders can be considered as non-commutative number theory.

(ii) Orders and ideals in orders have been introduced by H. Brandt [1],[2] in his studies of quadratic forms (this justifies the name "orders" for the algebraic systems under consideration).

(iii) Orders and modules over orders generalize the theory of integral representations of finite groups.

These notes have been written under the third aspect, and we shall have a closer look at this last development. The theory of group representations has its origin in the study of permutation groups and matrix algebras. In the years 1896-1899, G. Frobenius [1],[2] introduced the concept of a group representation and of the character of a representation. During the years 1900-1911, the theory of representations over the field of complex numbers $\underset{=}{C}$ was brought to a climax by W. Burnside, G. Frobenius and I. Schur (Burnside [1],[2],[3]; Frobenius [3]-[7]; Frobenius-Schur [1],[2]; Schur [1]-[4]). In 1911 W. Burnside published the second edition of his book on group theory [4], with a systematic treatment of the representations of finite groups. There he obtained group theoretic results, using representation theory, some of which - even today - cannot be proved by purely group theoretic means. As a matter of fact, modern group theory seems to be impossible without representation theory (cf. e.g. Feit-Thompson [1], Feit [1]). For a list of results in group theory obtained with the help of representation theory we refer to Boerner [2], pp. 60-66.

For a commutative ring we denote by GL(n,S) the general linear group of invertible (n×n)-matrices over S. A <u>representation of degree</u> n <u>of the finite group</u> G <u>with coefficients in</u> S is a realization of G as a group of (n×n)-matrices over S; i.e., one passes from an abstract group G to a concrete group of matrices. To be more precise: A <u>representation of degree</u> n <u>of</u> G <u>in</u> S is a multiplicative homomorphism (G is written multiplicatively)

$$\varphi : G \longrightarrow GL(n,S).$$

Two representations of G, φ and φ', both of degree n, are said to be S-<u>equivalent</u>, notation $\varphi \underset{S}{\cong} \varphi'$, if there exists $\underline{\underline{U}} \in GL(n,S)$ such that

$$\varphi(g)\underline{\underline{U}} = \underline{\underline{U}}\varphi'(g), \text{ for every } g \in G.$$

A representation φ of G is called <u>reducible</u>, if

$$\varphi \underset{S}{\cong} \begin{pmatrix} \varphi_1 & * \\ 0 & \varphi_2 \end{pmatrix},$$

where φ_1 and φ_2 are representations of G, $\varphi_1, \varphi_2 \neq 0$.

φ is said to <u>decompose</u> if

$$\varphi \underset{S}{\sim} \begin{pmatrix} \varphi_1 & 0 \\ 0 & \varphi_2 \end{pmatrix},$$

for representations $\varphi_1, \varphi_2 \neq 0$.

The main problems that arise immediately are:

(i) The classification - up to equivalence - of all <u>irreducible representations</u>.

(ii) The classification - up to equivalence - of all <u>indecomposable representations</u>.

(iii) The <u>structure problem</u>: How is an arbitrary representation built from the indecomposable ones and from the irreducible ones?

None of these questions has been answered satisfactorily, except in the very special case where S = K is a field, the characteristic of which does not divide the order of G. The results in this direction have essentially been obtained by Burnside, Frobenius, Maschke and Schur:

Theorem 1: Every representation of G in a field K, the characteristic of which does not divide the order of G, decomposes uniquely - up to equivalence - into irreducible representations. The number of non-equivalent irreducible representations is finite.

It was Noether's genius that gave new inspiration to the theory of representations of finite groups, when in her lectures at Göttingen 1927/28 (cf. Noether [1]), she brought the theory of representations of finite groups into connection with the theory of finite dimensional algebras over fields, a step that stimulated not only the theory of representations of finite groups, but also the theory of semisimple algebras. Her idea was as follows (not quite in this generality):

If S is a commutative ring, we may form the group algebra SG of the finite group G, where $SG = \{\sum_{g \in G} s_g g : s_g \in S\}$ with componentwise addition and multiplication induced by the multiplication in G. The following basic theorem provides the link between the concrete theory of matrix representations and the more abstract theory of the so-called representation modules.

Theorem 2: There is a one-to-one correspondence between the non-equivalent representations of G in S of degree n and the non-isomorphic left SG-modules which, as S-modules, are free on n elements, the so-called representation modules. Indecomposable representations correspond to left SG-modules which cannot be decomposed into a direct sum of proper submodules, each of which is S-free. Irreducible representations correspond to left SG-modules with an S-basis of n elements, which do not have SG-submodules with an S-basis of $\leq n-1$ elements.

To be more explicit, let $\varphi: G \to GL(n,S)$ be a representation of G of degree n in S. A free S-module V with a fixed basis $\{v_i\}_{1\leq i\leq n}$ is made into an SG-representation module V_φ, defining for $g \in G$

$$g\cdot v_i = \varphi(g)v_i = \sum_{j=1}^{n} (\varphi(g))_{ji}v_j,\ 1\leq i\leq n,$$

and then extending this action S-linearly.

Conversely, given a left SG-module, which is S-free on n elements, we fix an S-basis $\{v_i\}_{1\leq i\leq n}$. For $g \in G$ we have

$$g\cdot v_i = \sum_{j=1}^{n} s_{ji}(g)v_j,\ 1\leq i\leq n,\ s_{ji} \in S,$$

and the module properties of V imply that

$$\varphi : g \longmapsto (s_{ij}(g))$$

is a representation $\varphi: G \longrightarrow GL(n,S)$. Equivalence of representations corresponds to SG-isomorphy of representation modules.

In 1908 Wedderburn proved his structure theorem for semi-simple algebras of finite dimension over a field K; i.e., for finite dimensional K-algebras, for which every indecomposable module is simple. If the characteristic of K does not divide the order of the group G, then KG is a semi-simple finite dimensional K-algebra, and all the above problems have completely satisfactory answers, as already stated in Theorem 1.

However, the natural question arises: What happens if the characteristic of K divides the order of G? In this situation one talks about the so-called modular representations; the theory of modular representations has been developed mainly by R. Brauer [1]-[5], and for the study of modular representations we refer the reader to Curtis-Reiner [1], Ch.XII. In the modular case, KG is a so-called Frobenius-algebra, and it is no longer semi-simple; there exist indecomposable reducible KG-representation modules. Still, the question (1) on the irreducible modules is partially answered, since the simple left KG-modules are precisely the simple left (KG/rad KG)-modules, where rad KG is the Jacobson radical and KG/rad KG is semi-simple. By Wedderburn's theorem, the number of irreducible KG-modules is finite; but an explicit description of all irreducible representation modules seems to be unknown except in some special cases (cf. Berman [4]). The problem (ii) on the indecomposable KG-representation modules has partially been solved by D. G. Higman:

Theorem 3 (Higman [1], Kasch-Kneser-Kupisch [1], Berman [4]): Let G be a finite group and K a field of characteristic $p > 0$. If the p-Sylow-subgroups of G are cyclic, then there are at most $|G|$ non-equivalent indecomposable representations of G in K. If G has a non-cyclic p-Sylow-subgroup, then G has indecomposable representations of arbitrarily large degree.

But here too, an explicit description of the indecomposable modules seems to be unknown in general (cf. Berman [4]). Since the Krull-Schmidt theorem is valid for left KG-modules, every KG-module has a unique decomposition into indecomposables; and by the Jordan-Hölder theorem, the composition factors of every KG-representation module are unique - up to isomorphism.

Digressing now from group representations we find it worthwhile to mention some recent results on the number of non-isomorphic modules over finite dimensional - not necessarily semi-simple - algebras.

Theorem 4 (Curtis-Jans [1]): Let K be an algebraically closed field and A a finite dimensional K-algebra. If the socle of every indecomposable A-module M (i.e., the sum of the simple A-submodules of M) contains each simple A-module with multiplicity at most one, then the number of non-isomorphic indecomposable left A-modules is finite.

We would like to mention here a result of Roiter which proves a conjecture of Brauer.

<u>Theorem 5</u> (Roiter [7]): Let K be a field and A a finite dimensional K-algebra. If A has infinitely many non-isomorphic indecomposable left modules, then it has indecomposable representations of arbitrarily high degree.

In 1940 F. E. Diederichsen [1] considered for the first time systematically the so-called <u>integral representations of a finite group G</u>; i.e., multiplicative homomorphisms

$$\varphi : G \longrightarrow GL(n,\mathbb{Z}),$$

where $\mathbb{Z}$ is the ring of rational integers. Already in this first approach, he encountered some difficulties.

We have a natural injection

$$\iota : GL(n,\mathbb{Z}) \longrightarrow GL(n,\mathbb{Q}),$$

where $\mathbb{Q}$ is the field of rational numbers, and thus we may associate with every integral representation

$$\varphi : G \longrightarrow GL(n,\mathbb{Z})$$

the $\mathbb{Q}$-representation $\iota\varphi$. As was already known to Diederichsen, $\mathbb{Q}$-equivalence does not imply $\mathbb{Z}$-equivalence, and a $\mathbb{Z}$-representation can be indecomposable, though reducible. Moreover, the Jordan-Hölder theorem is no longer applicable to $\mathbb{Z}$-representations; i.e., the "irreducible parts" of a $\mathbb{Z}$-representation need not be unique. Nor need the "indecomposable parts" of a $\mathbb{Z}$-representation be unique: examples have been constructed that have non-isomorphic indecomposable direct decompositions; i.e., there is no "Krull-Schmidt theorem" for integral representations.

However, there are also some encouraging results:

<u>Theorem 6</u>: (i) Given a representation

$$\varphi' : G \longrightarrow GL(n,\mathbb{Q}),$$

then there exists an integral representation

$$\varphi : G \longrightarrow GL(n,\mathbb{Z})$$

such that $\varphi' \sim_{\mathbb{Q}} \iota\varphi$.

(ii) An integral representation φ is irreducible if and only if $\iota\varphi$ is irreducible.

Using this and the Jordan-Zassenhaus theorem (cf. Zassenhaus [1]), which states that the number of non-equivalent $\mathbb{Z}$-representations, that are $\mathbb{Q}$-equivalent, is finite, one finds that the number of non-equivalent irreducible integral representations is finite; again there is little information on a concrete realization of the irreducible integral representations. As to the question on indecomposable representations: Diederichsen had already shown in 1940 that the number of non-equivalent integral representations of G is finite if G is cyclic of order p, where p is a rational prime number. However, the general problem on the finiteness of the number of non-equivalent indecomposable $\mathbb{Z}$-representations has been solved by A. Jones in 1962, [1], combining the results of Heller-Reiner [1]-[4] and his own.

Theorem 7: The number of non-equivalent indecomposable integral representations of the finite group G is finite if and only if, for every rational prime number p dividing $|G|$, the p-Sylow-subgroups are cyclic of order $\leq p^2$.

In the proof of this theorem essentially all indecomposable representations are constructed, if the number is finite. Recently, L. A. Nazarova [3] has tried to list all classes of inequivalent indecomposable representations, even if this number is infinite.

The structure problem is affected very much by the fact that the Krull-Schmidt theorem is not applicable.

A further extension of the concept of a representation was imminent: let K be an algebraic number field and R the ring of integers in K; then R is a Dedekind domain with quotient field K, and one has "integral representations"

$$\varphi : G \longrightarrow GL(n,R).$$

In an obvious way equivalence, reducibility and decomposability can be defined. However, since R is in general not a principal ideal domain, both points of Theorem 4 break down for these generalized integral representations, since there are not enough of them. In fact, as it turns out, what is missing — in terms of modules — are exactly those R-projective RG-modules of finite type, that are not R-free. However, since Theorem 4 had rendered itself so useful in the theory of $\underline{\underline{Z}}$-representations, a broader definition of a representation of G over R was required. Naturally, this definition had to coincide with the old one in case R was a principal ideal domain. In particular, many problems in the theory of $\underline{\underline{Z}}$-representations are solved by "localization". So, the new modules should at least localize suitably, and from this point of view, the proper generalization of the concept of an integral representation over R such that (4) remains valid, is the following:

Definition: A representation module of RG is a left RG-module, which is at the same time an R-lattice; i.e., an R-projective RG-module of finite type.

With this definition, (4) remains valid with the appropriate changes, and we call such a representation module an RG-lattice.

One further generalization now leads to the theory of orders: RG is a subring of the semi-simple K-algebra KG, and in many proofs in the theory of RG-lattices, it is necessary to consider subrings Λ of KG, which have properties similar to those of RG relative to KG. Thence we arrive at the category of R-orders which contains the category of group rings over R, just as the category of semi-simple K-algebras contains the category of group algebras over K:

<u>Definition</u>: Let A be a semi-simple K-algebra and Λ a subring of A with the same identity as A. Then Λ is called an <u>R-order in A</u>, if

(i) $K\Lambda = A$

(ii) Λ is a finitely generated R-module.

It is easily seen, that in case $A = KG$, $\Lambda = RG$ is an R-order in A. Instead of studying RG-lattices, we shall study Λ-lattices. The theory of Λ-lattices has been developed extensively in the years 1950 ff. However, one class of R-orders, the maximal R-orders have already been explored in the years 1930-1940 (cf. Brandt [2], Chevalley [1], Deuring [1], Eichler [1]-[4], Hasse [1]-[3], Zassenhaus [1]) and recently, many of these results have been unified and brought up to date by Auslander-Goldman [1].

These notes are dedicated to the study of lattices over orders; but instead of taking R to be the ring of algebraic integers in an algebraic number field, we choose R to be any Dedekind domain with quotient field K. In this case, one has to require A to be a separable K-algebra so as to ensure the existence of maximal orders. We give next a brief and informal sketch of the contents of this book, stressing what we consider to be some of its highlights.

The structure of maximal orders in separable algebras — these play a dominant rôle in our approach to lattices over orders — is clarified in Ch. IV. Here, the structure problem can be settled locally because of the local validity of the Krull-Schmidt theorem; however, globally, no satisfactory answer can be expected, since maximal orders are in general not Dedekind domains but only Dedekind rings. While for Dedekind domains the cancellation law for direct summands holds, this need not be true even for maximal orders. Jacobinski's cancellation theorem (cf. below) gives a complete answer to the cancellation problem in general. Decomposability and irreducibility coincide for maximal orders, since for such orders every lattice is a projective module, and an answer to the question on the number of irreducibles depends on the theory of genera; i.e., on problems between global and local equivalence. To be more specific, let M and N be lattices; then M and N are said to lie in the same <u>genus</u> if M and N are locally isomorphic; i.e., if $M_{\underline{\underline{p}}} \cong N_{\underline{\underline{p}}}$ for all prime ideals $\underline{\underline{p}}$ in R. The theory of genera (Ch. VII, VIII) is a purely arithmetic one. Because of this, we have given two approaches to maximal orders in separable algebras: a more structural one, combining generators, progenerators and Morita equivalences (Ch. III) with homological algebra (Ch. II), and, in Ch. IV, the approach of Hasse [1], using arithmetic in topologically complete algebras.

Thus, Chapters I-III, though of interest in themselves, contain only introductory material; in Ch. I we give a brief introduction to modules over rings, in particular over Dedekind domains. This section is tailored especially to fit our purposes, and we have included it since — except for 12 volumes of Bourbaki — there is no textbook available where this material can be found in unified form. Ch. II is a short introduction to the homological tools used extensively for maximal orders and in dealing with decomposability over commutative orders. The main purpose there is to prove the equivalence between $\mathrm{Ext}^{-}_{-}(-,-)$ as defined by projective resolutions and $\mathrm{Ext}^{-}_{-}(-,-)$ as defined in terms of short exact sequences. In Chapter III, the Morita theorems are derived and, later, applied to clarify the structure of separable algebras.

We turn now to the problems for an arbitrary R-order Λ in the semisimple K-algebra A, where R is the ring of integers in the algebraic number field K.

In Chapter V the Higman ideal and related ideals are treated - all these ideals play an important rôle in the theory of genera. The Jordan-Zassenhaus theorem guarantees that there are only finitely many non-isomorphic irreducible Λ-lattices. In case A is split by K, the number of genera of irreducible Λ-lattices is equal to the product of the number of maximal R-orders containing Λ and the number of simple components of A; each genus contains h isomorphism classes, where h is the ideal class number of K, (Ch. VII).

A necessary and sufficient condition for Λ to have only finitely many non-isomorphic indecomposable lattices is not yet known except for some special types of algebras. In this realm, the two most remarkable results to date are the following: Drozd-Roiter [1], Dade[1], Gudivok [1], Heller-Reiner [4], Jacobinski [2] and Jones [1] have settled the problem for group algebras by showing that the number of indecomposable RG-lattices is finite if and only if the group G has a very special metacyclic structure, namely: If for a rational prime p dividing the order of G, $pR = \prod_j P_j^{e_j}$ is the prime decomposition of the ideal pR, $e(p) = \max_j(e_j)$ and G_p denotes a p-Sylow group of G, then either G_p is cyclic of order p^2, and $e(p) = 1$, or G_p is cyclic of order p and either $e(p) \leq 2$ or $p \leq 3$ and $e(p) = 3$. In this case the indecomposable RG-lattices have been constructed explicitly, though laboriously. Recently Nazarova (cf. [3]) has tried to classify the indecomposable RG-lattices even in case their number is infinite.

For a commutative algebra A the solution has been given independently by Drozd-Roiter [1] and Jacobinski [2]: In this case, there is exactly one maximal R-order Γ in A containing Λ, and the conditions of Drozd-Roiter are: Λ has finitely many non-isomorphic indecomposable lattices if and only if Λ has at most index two in Γ (as abelian groups) and $\Gamma/\Lambda/\mathrm{rad}(\Gamma/\Lambda)$ is a cyclic Λ-module, where $\mathrm{rad}(\Gamma/\Lambda)$ is the intersection of the maximal Λ-submodules of Γ/Λ. The approach of

Drozd-Roiter seems to lend itself to generalization and some preliminary results have already been obtained (cf. Roggenkamp [8],[9]). In view of this theorem, the following results seem quite surprising:

(i) For any R-order Λ in A, there are only finitely many non-isomorphic indecomposable projective lattices (Jacobinski [4], Jones [1]), (Ch.VIII).

(ii) The number of non-isomorphic lattices in the genus of a lattice M is bounded, by the Jordan-Zassenhaus theorem, but, what is more, this bound is independent of M; i.e., it is an invariant of Λ, (Jacobinski [3], Roiter [2]), (Ch. VII).

The structure problem is very hard to handle because of the lack of a Krull-Schmidt type theorem. For the $\underline{\underline{p}}$-adic completion $\hat{\Lambda}_{\underline{\underline{p}}}$ of Λ, $\underline{\underline{p}}$ a prime ideal in R, the Krull-Schmidt theorem is valid for lattices (Reiner [6],[3]; Borevich-Faddev [1]; Swan [2]), and the problem is trivial — this stresses also the importance of the theory of genera. For the localization, the Krull-Schmidt theorem is, in general, not valid for lattices, but cancellation is still admissible; i.e., $M \oplus N \cong M' \oplus N \Longrightarrow M \cong M'$. In the global case, this cancellation law fails. However, Jacobinski [4] has given a condition on A under which cancellation can be applied to some modules:

If no simple component of A is a totally definite quaternion algebra, then for Λ-lattices M, M', N, such that N is a direct summand of $M^{(n)}$, the direct sum of n copies of M, then

$$M \oplus N \cong M' \oplus N \Longrightarrow M \cong M'.$$

In totally definite quaternion algebras, however, cancellation is not even possible for lattices over maximal orders (cf. Swan [4]), (Ch. VII).

The arithmetical background for Jacobinski's cancellation law is based on some deep results of Eichler [3]. Once these are established,

the results follow elegantly from an exact sequence of Grothendieck groups in algebraic K-theory.
As already mentioned in the preface, the path to these deep results must lead through much of the development of the theory of integral representations, from the late twenties to the present. As to the present, we shall develop as much of K-theory as is needed and devote a chapter to Grothendieck groups. Here, much will be based on the works of Bass, Heller, Reiner and Swan.

Though we have attempted to prove, as much as possible, there are still some deep results from algebraic number theory which we quote without proof.

As stated at the beginning, this introduction has been written under the aspect of orders as generalization of group rings. We shall treat integral representations of finite groups only as examples; and thus, much of this beautiful theory will not be presented here.

CHAPTER I

PRELIMINARIES ON RINGS AND MODULES

§1. Modules and homomorphisms

In this section the basic definitions and properties of modules and homomorphisms are given. Homomorphisms are written opposite to the scalars. Products and coproducts are defined.

1.1 Definitions: A ring R is a set R with two internal laws of composition, "+" and "·", such that $(R,+)$ is an abelian group, and $(R,\cdot)$ is an associative structure, which is two-sided distributive with respect to "+". In the future we shall always assume, that all rings under consideration possess a unit element, 1; i.e., a neutral element with respect to "·". A left R-module M is a set with an internal law "+" such that $(M,+)$ is an abelian group, and with an external law $R \times M \longrightarrow M$, $(r,m) \longmapsto rm$, which satisfies the following conditions

$$\begin{aligned} r(m+m') &= rm+rm', \quad r \in R,\ m,m' \in M, \\ (r+r')m &= rm+r'm, \quad r,r' \in R,\ m \in M, \\ (rr')m &= r(r'm), \quad r,r' \in R,\ m \in M, \qquad (1.2) \\ 1m &= m, \quad m \in M. \end{aligned}$$

(The last condition is sometimes expressed by saying, that M is a unitary R-module.) If the condition (1.2) is replaced by $(rr')m = r'(rm)$, one says that M is a right R-module, and we generally write the operators, $r \in R$, on the right. By ${}_{\underline{R}}\underline{M}$ we denote the class of left R-modules, and by $\underline{M}_{\underline{R}}$ the class of right R-modules. Let N be a subgroup of $(M,+)$, where $M \in {}_{\underline{R}}\underline{M}$. If N is also a left R-module, then N is called a submodule of the left R-module

M. If N is a submodule of the left R-module M, we can make the factor group (M/N,+) into a left R-module, be defining $r(m+N) = rm+N$, $r \in R$, $m \in M$. The factor group (M/N,+) with this structure is called the factor module of M with respect to N, and it is denoted by M/N. The opposite ring R^{op} of R is the set R with an additive structure "$+^{op}$", $x +^{op} y = x + y$ and a multiplicative structure "$\cdot^{op}$", $x \cdot^{op} y = yx$. These definitions make R^{op} into a ring.

1.3 Remark: M is a left R-module if and only if M is a right R^{op}-module.

Let R and R' be rings. A (unitary) ring-homomorphism from R to R' is a map $\varphi: R \longrightarrow R'$, which is multiplicative and additive (and satisfies $\varphi(1) = 1$). Then the external law of composition
$R \times R' \longrightarrow R'$, $(r,r') \longmapsto \varphi(r)r'$, $r \in R$, $r' \in R'$ (resp. $R' \times R \longrightarrow R'$, $(r',r) \longmapsto r'\varphi(r)$, $r \in R$, $r' \in R'$) makes R' into a left (resp. right) R-module; every $M \in {}_{R'}\underline{M}$ becomes a left R-module, if we define $rm = \varphi(r)m$, $r \in R$, $m \in M$. If φ is the identity homomorphism on R, then R becomes a left (resp. right) R-module, denoted by ${}_RR$ (resp. R_R), the regular R-module. The submodules of ${}_RR$ (resp. R_R) are called the left (resp. right) ideals of R. If $M,N \in {}_R\underline{M}$ then an R-homomorphism from M to N is an additive map $\varphi: M \longrightarrow N$, such that $(rm)\varphi = r(m\varphi)$, $r \in R$, $m \in M$. We define the image of φ, $\mathrm{Im}\varphi = \{n \in N : n = m\varphi$ for some $m \in M\} \in {}_R\underline{M}$, the kernel of φ, $\mathrm{Ker}\varphi = \{m \in M : m\varphi = 0\} \in {}_R\underline{M}$ and the cokernel of φ, $\mathrm{Coker}\varphi = N/\mathrm{Im}\varphi \in {}_R\underline{M}$. One says that φ is an R-epimorphism, if $\mathrm{Im}\varphi = N$ and an R-monomorphism if $\mathrm{Ker}\,\varphi = (0)$. An R-epimorphism which is at the same time an R-monomorphism is called an R-isomorphism.

1.4 Remark: We always write module-homomorphisms opposite of the operators. If R is commutative we write the homomorphisms on the left, unless otherwise stated.

If $M,N \in {}_{R}\underline{\underline{M}}$, we write $\underline{\mathrm{Hom}_R(M,N)}$ for the set of R-homomorphisms from M to N. Then $\mathrm{Hom}_R(M,N)$ is an abelian group, if we define $\varphi + \psi : M \longrightarrow N$, $(m)(\varphi+\psi) = (m)\varphi + (m)\psi$; $\varphi,\psi \in \mathrm{Hom}_R(M,N)$. If $M = N$, then $\mathrm{Hom}_R(M,M)$ is also a ring, $\underline{\mathrm{End}_R(M)}$, under $\varphi\psi : M \longrightarrow M$, $(m)(\varphi\psi) = ((m)\varphi)\psi$; $\varphi,\psi \in \mathrm{End}_R(M)$. Moreover, M is a right $\mathrm{End}_R(M)$-module. In addition, the structure of M as left R-module and the structure of M as right $\mathrm{End}_R(M)$-module are linked by the formula

1.5 $$(rm)\varphi = r(m\varphi),\ r \in R,\ m \in M,\ \varphi \in \mathrm{End}_R(M).$$

In this connection one says that M is an $\underline{(R,\mathrm{End}_R(M))\text{-bimodule}}$. If R,S are two rings, we denote by ${}_{R}\underline{\underline{M}}_{S}$ the class of (R,S)-bimodules. If $M, M', M'' \in {}_{R}\underline{\underline{M}}$, we have a law of composition

1.6 $$\mathrm{Hom}_R(M,M') \times \mathrm{Hom}_R(M',M'') \longrightarrow \mathrm{Hom}_R(M,M''),\ (\varphi,\psi) \longmapsto \sigma$$
$$\text{where } m\sigma = (m\varphi)\psi,\ m \in M;$$

σ is called the composite of φ and ψ. This law is two-sided distributive and, whenever the composite of three homomorphisms is defined, it is associative.

1.7 Proposition: Let $M \in {}_{R}\underline{\underline{M}}$; then $\mathrm{Hom}_R({}_RR,M) \in {}_{R}\underline{\underline{M}}$ and one has an isomorphism

$$\Phi_M : \mathrm{Hom}_R({}_RR,M) \longrightarrow M,$$
$$\Phi_M : \quad \varphi \quad \longmapsto \quad (1)\varphi$$

of left R-modules. Moreover, Φ_M is a natural homomorphism; i.e., if $\sigma \in \mathrm{Hom}_R(M,M')$, $M' \in {}_{R}\underline{\underline{M}}$, then the following diagram

$$\begin{array}{ccc} M & \xrightarrow{\quad\sigma\quad} & M' \\ \Phi_M \Big\uparrow & & \Big\uparrow \Phi_{M'} \\ \mathrm{Hom}_R({}_RR,M) & \overset{\sigma_*}{\dashrightarrow} & \mathrm{Hom}_R({}_RR,M') \end{array}$$

can be completed in one and only one way to a commutative diagram; i.e., $\Phi_M\sigma = \sigma_*\Phi_{M'}$. Thus, we obtain an isomorphism of $\underline{\underline{Z}}$ *)-modules

$$\Phi:\mathrm{Hom}_R(M,M') \xrightarrow{\sim} \mathrm{Hom}_R(\mathrm{Hom}_R({}_RR,M),\ \mathrm{Hom}_R({}_RR,M'))$$
$$\sigma \quad\longmapsto\quad \sigma_*,$$

where $(r)\varphi\sigma_* = ((r)\varphi)\sigma$, for $r \in R$, $\varphi \in \mathrm{Hom}_R({}_RR,M)$.
The *proof* is straight forward and is left as an exercise. #

1.8 *Definitions*: Let $M \in {}_{R}\underline{\underline{M}}$, and $\{N_i\}_{i\in I}$ a family of submodules of M. Then $N = \bigcap_{i\in I} N_i$ is an R-module in the obvious way, called *the intersection of the family* $\{N_i\}_{i\in I}$. If now S is a subset of M, then the intersection of all the submodules of M containing S is called the *submodule of M, generated by S*, and one says that S is a *set of generators for N*. If $M \in {}_{R}\underline{\underline{M}}$ has a finite set of generators, one says that M is a *finitely generated R-module*, or an *R-module of finite type*. By ${}_{R}\underline{\underline{M}}^{f}$ (resp. $\underline{\underline{M}}^{f}_{R}$, resp. ${}_{R}\underline{\underline{M}}^{f}_{S}$) we denote the *class of left R-modules* (resp. right R-modules, resp. (R,S)-bimodules) of finite type. It is easily checked, that the submodule N of a left R-module M, generated by the family $\{m_i\}_{i\in I}$, is the set of finite linear combination of the elements m_i with (left) coefficients in R. Let $\{M_i\}_{i\in I}$ be a family of submodules of $M \in {}_{R}\underline{\underline{M}}$. The *sum of the M_i*, $\sum_{i\in I} M_i$, is the submodule of M generated by the union of the M_i, $i \in I$. Let $\{M_i\}_{1\le i\le n}$ be a finite family of left R-modules; the *(external) direct sum of the*

*) $\underline{\underline{Z}}$ denotes the ring of rational integers.

$\underline{M_i}$, $\oplus_{i=1}^{n} M_i$, is the R-module X, where $X = \{(m_1, m_2, \ldots, m_n): m_i \in M_i\}$ with componentwise addition and $r(m_1, \ldots, m_n) \overset{\text{def.}}{=} (rm_1, \ldots, rm_n)$, $r \in R$, $(m_1, \ldots, m_n) \in X$. If $\{M_i\}_{1 \leq i \leq n}$ is a finite family of submodules of the left R-module M, then $\sum_{i=1}^{n} M_i$ is said to be the <u>(internal) direct sum of the M_i</u>, $\oplus_{i=1}^{n} M_i$ (we use the same notation for external and internal direct sums, since, in general, no confusion can arise), if the map $\oplus_{i=1}^{n} M_i \longrightarrow \sum_{i=1}^{n} M_i$, $(m_1, \ldots, m_n) \longmapsto \sum_{i=1}^{n} m_i$ is an R-isomorphism. Let $\{M_i\}_{1 \leq i \leq n}$ and $\{M_i'\}_{1 \leq i \leq n}$ be two families of left R-modules, and let $\varphi_i \in \mathrm{Hom}_R(M_i, N_i)$, $1 \leq i \leq n$, be given. Then we define an R-homomorphism

$$\oplus_{i=1}^{n} \varphi_i : \oplus_{i=1}^{n} M_i \longrightarrow \oplus_{i=1}^{n} N_i$$

$$\oplus_{i=1}^{n} \varphi_i : (m_1, \ldots, m_n) \longmapsto (m_1\varphi_1, \ldots, m_n\varphi_n).$$

It follows immediately that $\mathrm{Ker}(\oplus_{i=1}^{n} \varphi_i) = \oplus_{i=1}^{n} \mathrm{Ker}\, \varphi_i$ and $\mathrm{Im}(\oplus_{i=1}^{n} \varphi_i) = \oplus_{i=1}^{n} \mathrm{Im}\, \varphi_i$.

1.9 <u>Lemma</u>: Let $M \in {}_R\underline{M}$, and $\{M_i\}_{1 \leq i \leq n}$ a family of submodules of M. Then the following conditions are equivalent:

(i) $\sum_{i=1}^{n} M_i = \oplus_{i=1}^{n} M_i$,

(ii) the relation $\sum_{i=1}^{n} m_i = 0$, $m_i \in M_i$, implies $m_i = 0, 1 \leq i \leq n$,

(iii) for every $1 \leq j \leq n$, $M_j \cap (\sum_{\substack{i=1 \\ i \neq j}}^{n} M_i) = 0$.

<u>Proof</u>: (i) and (ii) are equivalent by definition, and (iii) is just another way of expressing (ii). #

1.10 <u>Remark</u>: Let $\{M_i\}_{1 \leq i \leq n}$ be a family of left R-modules. With

$\oplus_{i=1}^{n} M_i$ we may associate two families of R-homomorphisms

$\pi_i: \oplus_{j=1}^{n} M_j \longrightarrow M_i$, $i = 1,\dots,n$, the <u>projections</u>,

$\pi_i: (m_1,\dots,m_n) \longmapsto m_i$, and

$\iota_i: M_i \longrightarrow \oplus_{j=1}^{n} M_j$, $i = 1,\dots,n$, the <u>injections</u>,

$\iota_i: m_i \longmapsto (0,\dots,0,m_i,0,\dots,0)$, where m_i is at the i-th position.

These maps satisfy the following relations.

(i)
$$\iota_i \pi_j = \begin{cases} 1_{M_i}, & \text{if } i = j, \text{ where } 1_{M_i} \text{ is the identity homomorphism on } M_i. \\ 0, & \text{if } i \neq j. \end{cases}$$

(ii)
$$\sum_{i=1}^{n} \pi_i \iota_i = 1_{\oplus_{i=1}^{n} M_i}$$

Moreover, the ι_i are monomorphisms whereas the π_i are epimorphisms (cf. (i)).

1.11 <u>Proposition</u>: Let $\{M_i\}_{1\le i\le n}$, $\{N_j\}_{1\le j\le n'}$, be two families of left R-modules. The map

$$\Phi: \mathrm{Hom}_R(\oplus_{i=1}^{n} M_i, \oplus_{j=1}^{n'} N_j) \longrightarrow \oplus_{\substack{i=1\\ j=1}}^{n,n'} \mathrm{Hom}_R(M_i, N_j)$$

$$\Phi: \qquad \varphi \qquad \longmapsto (\iota_i(M)\varphi\pi_j(N))_{\substack{1\le i\le n\\ 1\le j\le n'}}$$

where $\iota_i(M)$ and $\pi_j(N)$ are the maps defined in (1.9), is a natural isomorphism of $\underline{Z}$-modules (cf. (1.7)), and its inverse is

$$\Psi: \oplus_{\substack{i=1\\ j=1}}^{n,n'} \mathrm{Hom}_R(M_i, N_j) \longrightarrow \mathrm{Hom}_R(\oplus_{i=1}^{n} M_i, \oplus_{j=1}^{n'} N_j)$$

$$\Psi: \quad (\varphi_{ij})_{\substack{1\le i\le n\\ 1\le j\le n'}} \longmapsto \sum_{i=1}^{n} \sum_{j=1}^{n'} \pi_i(M)\varphi_{ij}\iota_j(N).$$

Proof: Using the identities in (1.10) it follows immediately, that Φ and Ψ are inverse to each other. We leave it as an exercise to show, that Φ is natural. #

1.12 Definition: $M \in {}_R\underline{\underline{M}}^f$ is said to be a free left R-module with a basis of n elements, if there exists an R-isomorphism $\varphi: M \longrightarrow {}_RR \oplus {}_RR \oplus \ldots \oplus {}_RR = ({}_RR)^{(n)}$, where the sum consists of n copies of ${}_RR$. The elements $e_i = (0,\ldots,0,1,0,\ldots,0)\varphi^{-1}, 1 \leq i \leq n$, where 1 is at the i-th position, are called basis elements of M. Then every element in M can be expressed uniquely as a linear combination of the $\{e_i\}_{1 \leq i \leq n}$ with coefficients in R (cf. (1.9)).

Exercises §1:

1a.) Let Φ, Ψ: ${}_R\underline{\underline{M}} \longrightarrow {}_R\underline{\underline{M}}$, $M \longrightarrow M^{\Phi}$, M^{Ψ} be such that to every $\varphi \in \mathrm{Hom}_R(M,N)$, $M,N \in {}_R\underline{\underline{M}}$, there are unique $\varphi^{\Phi} \in \mathrm{Hom}_R(M^{\Phi}, N^{\Phi})$ and $\varphi^{\Psi} \in \mathrm{Hom}_R(M^{\Psi}, N^{\Psi})$. Define the concept of naturality for a family of homomorphisms $\{\chi_M: M^{\Phi} \longrightarrow M^{\Psi}\}_{M \in {}_R\underline{\underline{M}}}$.

b.) Prove that Φ of (1.7) is a natural isomorphism.

2a.) Let $\{M_i\}_{i \in I}$ be a family of left R-modules. A family $\{P \in {}_R\underline{\underline{M}}, \{\pi_i\}_{i \in I}; \pi_i \in \mathrm{Hom}_R(P, M_i)\}$ is called a product of the $\{M_i\}_{i \in I}$, if for every $X \in {}_R\underline{\underline{M}}$ and any family $\varphi_i \in \mathrm{Hom}_R(X, M_i)$, $i \in I$, there exists a unique $\psi \in \mathrm{Hom}_R(X,P)$ such that $\psi\pi_i = \varphi_i$, $i \in I$; i.e., the diagram

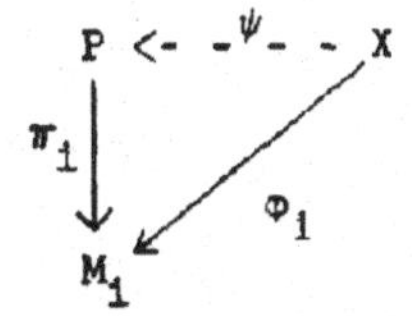

can be completed uniquely to a commutative diagram.

b.) Dualize this concept (i.e., reverse the arrows) to define a coproduct.

c.) Show that if a product or a coproduct exists, then it is unique up to isomorphism.

d.) Show that in $_R\underline{M}$ products and coproducts do exist, and describe them explicitly.

e.) If I is a finite set, show that products and coproducts coincide with direct sums (cf. (1.10)) in $_R\underline{M}$.

3.) Show that the homomorphism Φ in (1.11) is natural.

§2. Exact sequences

It is shown that $Hom_R(-,N)$ is left exact and contravariant, $Hom_R(M,-)$ is left exact and covariant, both are additive. Some properties of projective modules and dual modules are derived.

2.1 Definition: Let R be a ring, $M', M, M'' \in {}_R\underline{\underline{M}}$. and $\varphi \in Hom_R(M',M)$, $\psi \in Hom_R(M,M'')$. Then the sequence

$$M' \xrightarrow{\varphi} M \xrightarrow{\psi} M''$$

is said to be exact, if $Im\ \varphi = Ker\ \psi$. The sequence $0 \longrightarrow M' \xrightarrow{\varphi} M$ is exact if and only if φ is an R-monomorphism. The sequence $M \xrightarrow{\psi} M'' \longrightarrow 0$ is exact if and only if ψ is an R-epimorphism.

2.2 Remark: If

$$0 \longrightarrow M' \xrightarrow{\varphi} M \xrightarrow{\psi} M'' \longrightarrow 0$$

is an exact sequence of left R-modules and R-homomorphisms, then it follows from the first homomorphism theorem (one proves as for abelian groups: if $\sigma: M \longrightarrow N$ is an R-homomorphism between left R-modules, then $M/Ker\ \sigma \cong Im\ \sigma$ as left R-modules), that $M/Im\ \varphi \cong M''$. Conversely, if N is a submodule of the left R-module M, then the sequence

$$0 \longrightarrow N \xrightarrow{\kappa} M \xrightarrow{\lambda} M/N \longrightarrow 0$$

is exact, where $\kappa: N \longrightarrow M$, $\kappa: n \longmapsto n$, $n \in N$ and $\lambda: M \longrightarrow M/N$, $\lambda: m \longmapsto m+N$, $m \in M$, are the canonical homomorphisms.
The exact sequence

$$E: 0 \longrightarrow M' \xrightarrow{\varphi} M \xrightarrow{\psi} M'' \longrightarrow 0$$

of left R-modules is said to be <u>split exact</u> (or simply, it <u>splits</u>), if there exists $\sigma \in \mathrm{Hom}_R(M'',M)$ such that $\sigma\psi = 1_{M''}$. In this case σ is necessarily an R-monomorphism and $M = \mathrm{Im}\,\varphi + \mathrm{Im}\,\sigma$. If $x \in \mathrm{Im}\,\varphi \cap \mathrm{Im}\,\sigma$, then $x\psi = 0$ and, since σ is a monomorphism, $x = 0$; hence $M = \mathrm{Im}\,\varphi \oplus \mathrm{Im}\,\sigma$ by (1.8). Moreover, φ and σ are both monomorphisms; hence $M \cong M' \oplus M''$. We leave it as an exercise to show that the exact sequence E splits if and only if there exists $\tau \in \mathrm{Hom}_R(M,M')$ such that $\varphi\tau = 1_{M'}$.

2.3 <u>Proposition</u>: Let

$$0 \longrightarrow M' \xrightarrow{\varphi} M \xrightarrow{\psi} M'' \longrightarrow 0$$

be an exact sequence of left R-modules and R-homomorphisms. If M',M'' are of finite type, so is M. (The converse of this statement is not necessarily true; however, if $M \in {}_R\underline{\underline{M}}^f$, then $M'' \in {}_R\underline{\underline{M}}^f$).

<u>Proof</u>: Let S' and S'' respectively be finite systems of generators for M' and M'' respectively. If T is a finite subset of M such that $T\psi = S''(T\psi = \{t\psi : t \in T\})$, then $S = S'\varphi \cup T$ is a finite system of generators for M. In fact, the submodule M_o of M, generated by S contains $M'\varphi$; and since $S\psi = S''$, $M_o\psi = M''$. Hence $M_o = M$. #

2.4 <u>Lemma</u>: Let $M \in {}_R\underline{\underline{M}}$ and $\{N_i\}_{1\leq i\leq n}$ a family of submodules of M. Then the sequence

$$0 \longrightarrow \bigcap_{i=1}^{n} N_i \xrightarrow{\varphi} M \longrightarrow \bigoplus_{i=1}^{n} (M/N_i)$$

is exact. Here φ is the canonical injection, and

$\psi: M \longrightarrow \bigoplus_{i=1}^{n} (M/N_i)$, $m \longmapsto (m+N_1,\ldots,m+N_n)$, $m \in M$. (It should be observed, that ψ is in general not an epimorphism).

Proof: $\operatorname{Ker}\psi = \{m\in M: m+N_i \subset N_i,\ 1 \leq i \leq n\}$
$= \{m\in M: m \in \bigcap_{i=1}^{n} N_i\} = \bigcap_{i=1}^{n} N_i$. Hence the above sequence is exact by (2.2), since φ is a monomorphism. #

2.5 Definition: Let $M,M',N,N' \in {}_R\underline{\underline{M}}$. We define a map

$\hom: \operatorname{Hom}_R(M',M) \times \operatorname{Hom}_R(N,N') \longrightarrow \operatorname{Hom}_{\underline{\underline{Z}}}(\operatorname{Hom}_R(M,N),\operatorname{Hom}_R(M',N'))$

$\hom: \quad (\varphi,\psi) \longmapsto \hom(\varphi,\psi)$,

where for $\sigma \in \operatorname{Hom}_R(M,N)$, $\hom(\varphi,\psi)\sigma \overset{\text{def}}{=} \varphi\sigma\psi$. Moreover, hom satisfies the following identities

(i) $\hom(\varphi_1+\varphi_2,\psi) = \hom(\varphi_1,\psi) + \hom(\varphi_2,\psi)$,

(ii) $\hom(\varphi,\psi_1+\psi_2) = \hom(\varphi,\psi_1) + \hom(\varphi,\psi_2)$,

(iii) $\hom(0,\psi) = \hom(\varphi,0) = 0$, $\varphi,\varphi_1,\varphi_2 \in \operatorname{Hom}_R(M,N),\psi,\psi_1,\psi_2\in\operatorname{Hom}_R(M',N')$,

(iv) $\hom(1_M,1_N) = 1_{\operatorname{Hom}_R(M,N)}$ for $M = M'$, $N = N'$,

(v) $\hom(\varphi'\varphi,\psi\psi') = \hom(\varphi',\psi')\hom(\varphi,\psi)$ where $M'',N'' \in {}_R\underline{\underline{M}}$ and $\varphi \in \operatorname{Hom}_R(M',M),\varphi' \in \operatorname{Hom}_R(M'',M)$, $\psi \in \operatorname{Hom}_R(N,N')$, $\psi' \in \operatorname{Hom}_R(N',N'')$.

The verification of these identities is left as an exercise.

We remark shortly, what happens to right modules:

Let $M,M',N,N' \in \underline{\underline{M}}_R$ then,

$\hom: \operatorname{Hom}_R(M',M) \times \operatorname{Hom}_R(N,N') \longrightarrow \operatorname{Hom}_{\underline{\underline{Z}}}(\operatorname{Hom}_R(M,N),\operatorname{Hom}_R(M',N'))$

$\hom: \quad (\varphi,\psi) \longmapsto \hom(\varphi,\psi)$,

where for $\sigma \in \operatorname{Hom}_R(M,N)$,

$$\hom(\varphi,\psi)\sigma = \psi\sigma\varphi \qquad \text{(cf. (1.4))}.$$

The formulae (i)...(v) (even (v)) remain valid.

2.6 <u>Theorem</u>: Let $M', M, M'' \in {}_R\underline{\underline{M}}$. Then sequence

(i) $M' \xrightarrow{\varphi} M \xrightarrow{\psi} M'' \longrightarrow 0$ is exact if and only if, for every left R-module N, the sequence

(ii) $0 \longrightarrow Hom_R(M'',N) \xrightarrow{\psi^*} Hom_R(M,N) \xrightarrow{\varphi^*} Hom_R(M',N)$ is an exact sequence of $\underline{\underline{Z}}$-modules. Here $\psi^* = hom(\psi, 1_N)$, $\varphi^* = hom(\varphi, 1_N)$.

<u>Remark</u>: (i) The operation -* "reverses arrows"; i.e., $\varphi: M' \longrightarrow M$, implies $\varphi^*: Hom_R(M,N) \longrightarrow Hom_R(M',N)$ (cf. later: <u>contravariant functor</u>).

(ii) Since $Hom_R(M,N)$ is a $\underline{\underline{Z}}$-module, and since $\underline{\underline{Z}}$ is commutative, we write the homomorphisms on the left (cf. (1.4)).

<u>Proof</u>: Let the sequence (i) be exact. To prove, that (ii) is exact, it suffices to show, that (ii) is exact at $Hom_R(M,N)$. In fact, the exactness of (ii) at $Hom_R(M'',N)$ follows by applying the result to

$$M \xrightarrow{\psi} M'' \longrightarrow 0 \longrightarrow 0.$$

We have $\varphi^*\psi^* = hom(\psi, 1_N)\ hom(\varphi, 1_N) = hom(\varphi\psi, 1_N) = hom(0, 1_N) = 0$. Thus $Im\ \psi^* \subset Ker\ \varphi^*$. Now, let $\sigma \in Hom_R(M,N)$, such that $\varphi^*(\sigma) = 0$, then $\varphi\sigma = 0$ and so $(Ker\ \psi)\sigma = 0$. Hence we can complete the following diagram commutatively (cf. Exercise (2.3)), since $Ker\ \psi \subset Ker\ \sigma$ and, since ψ is an epimorphism:

D: 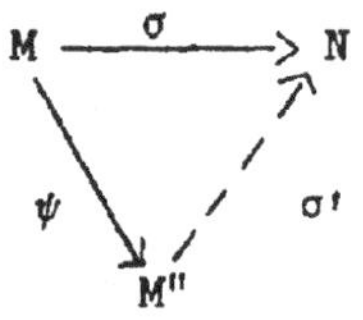

Thus, from the commutativity of D, we obtain $\psi\sigma = \sigma$; i.e., $\psi^*(\sigma') = \sigma$ and $Ker\ \varphi^* \subset Im\ \psi^*$.

<u>Conversely</u>, let the sequence (ii) be exact for every $N \in {}_R\underline{\underline{M}}$. To show that ψ is epic, let $N = M''/\mathrm{Im}\,\psi$ and let $\sigma: M'' \longrightarrow M''/\mathrm{Im}\,\psi$ the canonical homomorphism. Then $\psi^*(\sigma) = \psi\sigma = 0$. Since ψ^* is monic, $\sigma = 0$. To show exactness at M, we observe that $\varphi^*\psi^* = 0$; hence, in particular, for $N = M$, $\rho = 1_M$, we have $\varphi^*\psi^*(\rho) = \varphi\psi 1_M = \varphi\psi = 0$; i.e., $\mathrm{Im}\,\varphi \subset \mathrm{Ker}\,\psi$. Conversely, let $\sigma: M \longrightarrow M/\mathrm{Im}\,\varphi$ be the canonical homomorphism, and put $N = M/\mathrm{Im}\,\varphi$ in (ii). Since $\varphi^*(\sigma) = 0$, there exists $\sigma' \in \mathrm{Hom}_R(M'', M/\mathrm{Im}\,\varphi)$ such that $\psi^*(\sigma') = \sigma$; i.e.,

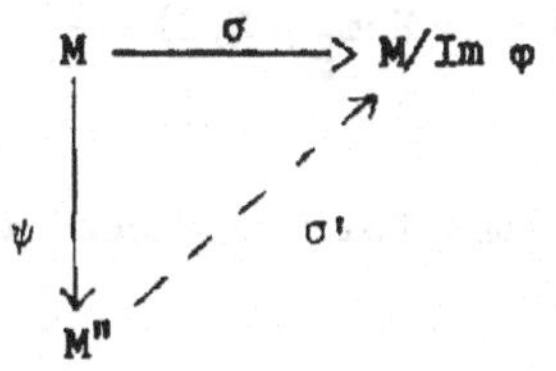

is a commutative diagram. Hence $\mathrm{Ker}\,\psi \subset \mathrm{Ker}\,\sigma = \mathrm{Im}\,\varphi$, and the sequence (i) is exact. #

2.7 <u>Theorem</u>: Let $N', N, N'' \in {}_R\underline{\underline{M}}$. Then the sequence

$$0 \longrightarrow N' \xrightarrow{\varphi} N \xrightarrow{\psi} N''$$

is exact if and only if for every $M \in {}_R\underline{\underline{M}}$ the sequence $0 \longrightarrow \mathrm{Hom}_R(M,N') \xrightarrow{\varphi_*} \mathrm{Hom}_R(M,N) \xrightarrow{\psi_*} \mathrm{Hom}_R(M,N'')$ is an exact sequence of $\underline{\underline{Z}}$-modules. Here $\varphi_* = \mathrm{hom}(1_M,\varphi)$ and $\psi_* = \mathrm{hom}(1_M,\psi)$. The <u>proof</u> is similar to the one of (2.6) and is left as an exercise. #

2.8 <u>Remark</u>: (i) The operation $-_*$ "preserves arrows"; i.e., $\varphi: N' \longrightarrow N$ implies $\varphi_*: \mathrm{Hom}_R(M,N') \longrightarrow \mathrm{Hom}_R(M,N)$ (cf. later: <u>covariant functor</u>).

(ii) If $0 \longrightarrow M' \xrightarrow{\varphi} M \xrightarrow{\psi} M'' \longrightarrow 0$ is an exact sequence of left R-modules and homomorphisms, then neither

$$0 \longrightarrow \mathrm{Hom}_R(N,M') \xrightarrow{\varphi_*} \mathrm{Hom}_R(N,M) \xrightarrow{\psi_*} \mathrm{Hom}_R(N,M'') \longrightarrow 0$$

nor

$$0 \longrightarrow \mathrm{Hom}_R(M'',N) \xrightarrow{\varphi^*} \mathrm{Hom}_R(M,N) \xrightarrow{\psi^*} \mathrm{Hom}_R(M',N) \longrightarrow 0$$

need be exact. As an example consider the exact sequence

$$0 \longrightarrow 2\mathbb{Z} \longrightarrow \mathbb{Z} \longrightarrow \mathbb{Z}/2\mathbb{Z} \longrightarrow 0,$$

with the canonical homomorphisms. Then

$$0 \longrightarrow \mathrm{Hom}_{\mathbb{Z}}(\mathbb{Z}/2\mathbb{Z},\mathbb{Z}) \longrightarrow \mathrm{Hom}_{\mathbb{Z}}(\mathbb{Z},\mathbb{Z}) \longrightarrow \mathrm{Hom}_{\mathbb{Z}}(2\mathbb{Z},\mathbb{Z}) \longrightarrow 0$$

is not exact (cf. Exercise (2.5)).

2.9 <u>Proposition</u>: For $P \in {}_R\mathfrak{M}^f$ the following conditions are equivalent:

(i) For every exact sequence

$$0 \longrightarrow M' \xrightarrow{\varphi} M \xrightarrow{\psi} M'' \longrightarrow 0$$

of left R-modules and homomorphisms, the sequence

$$0 \longrightarrow \mathrm{Hom}_R(P,M') \xrightarrow{\varphi_*} \mathrm{Hom}_R(P,M) \xrightarrow{\psi_*} \mathrm{Hom}_R(P,M'') \longrightarrow 0$$

is exact.

(ii) One can complete every diagram with an exact row

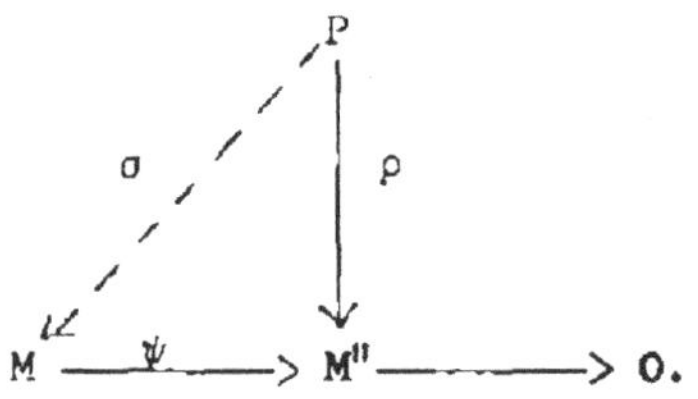

(iii) Every exact sequence

$$0 \longrightarrow M' \xrightarrow{\varphi} M \xrightarrow{\psi} P \longrightarrow 0$$

is split (cf. (2.2)).

(iv) There exists a free left R-module F with a finite basis, and a submodule X of F such that $F \cong X \oplus P$.

<u>Proof</u>:

<u>(i) ==> (ii)</u>: Given the diagram with exact row

$$\begin{array}{lccc} & & P & & \\ D & & \downarrow{\scriptstyle\rho} & & \\ & M & \xrightarrow{\psi} M'' & \longrightarrow 0 \,. \end{array}$$

One can complete the bottom sequence to the exact sequence

$$0 \longrightarrow \operatorname{Ker} \psi \xrightarrow{\iota} M \xrightarrow{\psi} M'' \longrightarrow 0,$$

where ι: Ker $\psi \longrightarrow M$ is the injection. Now (i) implies that there exists $\sigma \in \operatorname{Hom}_R(P,M)$ such that $\psi_*(\sigma) = \rho$; i.e., $\sigma\psi = \rho$. Consequently, σ completes D commutatively.

<u>(ii) ==> (iii)</u>. The diagram

$$\begin{array}{ccccc} & & & P & \\ & & & \downarrow{\scriptstyle 1_P} & \\ E: 0 \longrightarrow M & \xrightarrow{\varphi} & M & \xrightarrow{\psi} P & \longrightarrow 0 \end{array}$$

can be completed commutatively by (ii). Hence E splits (cf. (2.2)).

<u>(iii) ==> (iv)</u>. Since P is of finite type, there exists a free left R-mdoule F with a finite basis such that P is the epimorphic image of F; i.e., we have an exact sequence

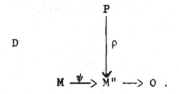

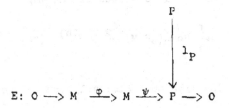

$$0 \longrightarrow \text{Ker}\,\varphi \longrightarrow F \xrightarrow{\varphi} P \longrightarrow 0,$$

which splits by (iii). Hence P is isomorphic to a direct summand of F (cf. (2.2)).

(iv) ==> (i). We show first that a free left R-module F with a finite basis satisfies (i). Because of (2.7) we only have to show, that ψ_* is an epimorphism. We have the following commutative diagram

$$\begin{array}{ccccc} \text{Hom}_R(F,M) & \xrightarrow{\psi_*(F)} & \text{Hom}_R(F,M'') & & \\ \Psi(M)\downarrow & & \downarrow\Psi(M'') & & \\ M^{(n)} & \xrightarrow{\psi^{(n)}} & M''^{(n)} & \longrightarrow & 0, \end{array}$$

where $X^{(n)}$ stands for the direct sum of n copies of X. Here n is the number of basis elements of F; $F \cong {}_RR^{(n)}$.

$$\overset{n}{\underset{1}{\oplus}} \psi = \psi^{(n)}: M^{(n)} \longrightarrow M''^{(n)} \quad \text{(cf. (1.8))},$$

$\Psi(M)$: $\text{Hom}_R(F,M) \longrightarrow M^{(n)}$ is composed of the maps

$$\text{Hom}_R(F,M) \xrightarrow{\Phi_1} \overset{n}{\underset{1}{\oplus}} \text{Hom}_R({}_RR,M) \xrightarrow{\Phi_2} M^{(n)} \quad \text{(cf. (1.7),(1.11),(1.12))}.$$

Then $\Psi(M)$ is an isomorphism. Similarly one defines the isomorphism $\Psi(M'')$. The equations

$$\Phi(M'')\psi_*(F)\sigma = \Psi(M'')\sigma\psi = ((1)\iota_i(F)\sigma\psi)_{1\leq i\leq n},$$
$$(\Psi(M)\sigma)\psi^{(n)} = [((1)\iota_i(F)\sigma)_{1\leq i\leq n}]\psi^{(n)} = ((1)\iota_i(F)\sigma\psi)_{1\leq i\leq n}$$

show, that the above diagram is commutative. Since $\psi^{(n)}$ is an epimorphism, so is $\psi_*(F)$; here $\iota_i(F)$: $R_i = R \longrightarrow F$ is the i-th injection. Now, for $P \in {}_R\underline{\underline{M}}^f$ we show: (iv) ==> (ii).

By (iv) there exists a free left R-module $F \in {}_{\underline{\underline{R}}}\underline{\underline{M}}^f$, such that $F \cong P \oplus X$. Given the diagram with an exact row

$$\begin{array}{ccccc} & & P & & \\ & & \downarrow{\scriptstyle\rho} & & \\ M & \xrightarrow{\psi} & M'' & \longrightarrow & 0. \end{array}$$

We can complete - by the above reasoning - the diagram

$$\begin{array}{ccccc} & & F & & \\ & \swarrow{\scriptstyle\sigma'} & \downarrow{\scriptstyle\pi} & & \\ & & P & & \\ & & \downarrow{\scriptstyle\rho} & & \\ M & \xrightarrow{\psi} & M'' & \longrightarrow & 0 \end{array}$$

commutatively; here $\pi\colon F \longrightarrow P$ is the projection (cf. (1.10)). Now we define $\sigma\colon P \longrightarrow M$ by $\sigma = \iota\sigma'$, where $\iota\colon P \longrightarrow F$ is the injection (cf. (1.10)). Then $\iota\sigma'\psi = \iota\pi\rho = 1_P\rho = \rho$ (cf. (1.10)); i.e., (ii) is satisfied, and (i) follows at once. #

2.10 Definition: A left R-module P of finite type, which satisfies the equivalent conditions of (2.9) is called a projective left R-module of finite type. By ${}_{\underline{\underline{R}}}\underline{\underline{P}}^f$ we denote the class of projective left R-modules of finite type. (Similarly, $\underline{\underline{P}}^f_{\underline{\underline{R}}}$.)

2.11 Definition: The dual of the left R-module M is defined as

$$M^* = \mathrm{Hom}_R(M, {}_RR).$$

Then M^* is a right R-module under the following action

$$m(\varphi r) = (m\varphi)r, \quad m \in M,\ \varphi \in M^*,\ r \in R.$$

Moreover, we have a homomorphism of left R-modules $\delta(M)\colon M \longrightarrow M^{**}$

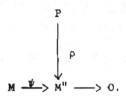

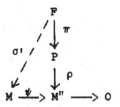

$\delta(M)\colon m \longmapsto m\delta(M)$, where $(m\delta(M))\varphi = m\varphi$, $m\in M$, $\varphi \in M^*$.

This $\delta(M)$ together with $\delta(\varphi) = \hom(\hom(\varphi, 1_R), 1_R)$ for $\varphi \in \mathrm{Hom}_R(M,M')$ is a <u>natural homomorphism</u> (cf. (1.7)). We leave the verification of the naturality of δ as an exercise.

2.12 <u>Lemma</u>: Let $P \in {}_R\underline{\underline{P}}^f$. Then $P^* \in \underline{\underline{P}}^f_R$, and $\delta(P)\colon P \longrightarrow P^{**}$ is a natural isomorphism.

<u>Proof</u>: Let F be a free left R-module of finite type with a basis $\{e_i\}_{1\leq i\leq n}$ (cf. (1.12)). We define elements $\{e_i^*\}_{1\leq i\leq n} \in F^*$ by $(e_j)e_i^* = \begin{cases} 1, & \text{if } i = j \\ 0, & \text{if } i \neq j \end{cases}$. Since every $\varphi \in F^*$ is uniquely determined by its values $\{(e_i)\varphi\}_{1\leq i\leq n}$, φ has a unique expression as $\varphi = \sum_{i=1}^{n} e_j^* r_j$, for some elements $r_j \in R$, $1 \leq j \leq n$. Hence F^* is a free right R-module with a basis of n elements. If now P is a projective left R-module of finite type, then there exists a free left R-module F of finite type such that $F \cong P \oplus X$ (cf. (2.9)). But $F^* \cong P^* \oplus X^*$ as $\underline{\underline{Z}}$-modules (cf. (1.11)); however, since F^*, P^* and X^* are right R-modules, the above isomorphism is also an isomorphism of right R-modules. Moreover, F^* is a free right R-module of finite type. Thus P^* is a projective right R-module of finite type (cf. (3.9)). From the above considerations it is clear, that F^{**} is a free left R-module on a basis $\{e_i^{**}\}_{1\leq i\leq n}$, where

$e_i^{**}(e_j^*) = \begin{cases} 1, & \text{if } i = j \\ 0, & \text{if } i \neq j \end{cases}$. On the other hand, $(e_i)\delta(F) = e_i^{**}$, $1 \leq i \leq n$. Hence $\delta(F)$ is an isomorphism. Since $\delta(M)$ is a natural homomorphism, we have the following commutative diagram

$$\begin{array}{ccccc}
P & \xrightarrow{\iota} & F & \xrightarrow{\pi} & P \\
\downarrow \delta(P) & & \wr\downarrow \delta(F) & & \downarrow \delta(P) \\
P^{**} & \xrightarrow{\iota^{**}} & F^{**} & \xrightarrow{\pi^{**}} & P^{**}
\end{array}$$

where $\iota: P \longrightarrow F$ and $\pi: F \longrightarrow P$ are the injection and projection resp. (cf. (1.10)).

$$\iota^{**} = \hom(\hom(\iota, 1_R), 1_R)$$

$$\pi^{**} = \hom(\hom(\pi, 1_R), 1_R) \quad \text{(cf. (2.5))}.$$

Then $\hom(\hom(\iota, 1_R), 1_R) \hom(\hom(\pi, 1_R), 1_R)) = \iota^{**}\pi^{**}$

$$= \hom(\hom(\iota\pi, 1_R)) = \hom(\hom(1_P, 1_R)) = 1_{P^{**}}.$$

(It should be observed, that $\mathrm{Hom}_R(M, {}_RR)$ is a right R-module; and thus, according to our convention(1.4), homomorphisms are written on the right; similarly for M^{**}). This implies that ι^{**} is a monomorphism and π^{**} is epic. Diagram chasing shows that $\delta(P)$ is an isomorphism. #

Exercises §2:

1a.) Let

$$E: \quad 0 \longrightarrow M' \xrightarrow{\varphi} M \xrightarrow{\psi} M'' \longrightarrow 0$$

$$0 \longrightarrow N' \xrightarrow{\sigma} N \xrightarrow{\tau} N'' \longrightarrow 0$$

be two exact sequences of left R-modules and homomorphisms. Show, that the sequence

$$0 \longrightarrow M' \oplus N' \xrightarrow{\varphi \oplus \sigma} M \oplus N \xrightarrow{\psi \oplus \tau} M'' \oplus N'' \longrightarrow 0$$

is exact (cf. (1.8)).

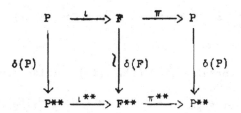

b.) Show, that E is split $\Longleftrightarrow \exists \rho \in \mathrm{Hom}_R(M,M')$ such that $\varphi\rho = 1_{M'}$.

2.) Verify the formulae (2.2,(i)...(iv)) (also for right modules).

3.) Let $M_1, M_2, M_3 \in {}_{R}\underline{\underline{M}}$, and let $\varphi \in \mathrm{Hom}_R(M_1,M_2)$, $\psi \in \mathrm{Hom}_R(M_1,M_3)$. If $\mathrm{Ker}\,\psi \subset \mathrm{Ker}\,\varphi$, show that there exists exactly one $\sigma \in \mathrm{Hom}_R(\mathrm{Im}\,\psi, M_3)$ such that the diagram

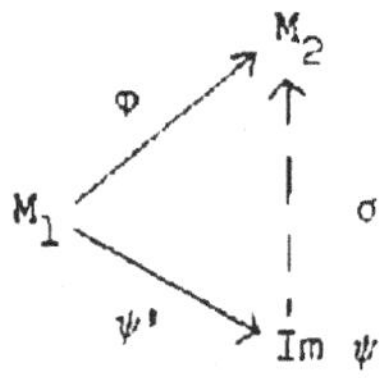

is commutative, where $\psi': M_1 \longrightarrow \mathrm{Im}\,\psi$, $\psi': m_1 \longmapsto m_1\psi$, $m_1 \in M_1$.

4.) Prove (2.7).

5.) Show that the sequence

$$E^*: 0 \longrightarrow \mathrm{Hom}_{\underline{\underline{Z}}}(\underline{\underline{Z}}/2\underline{\underline{Z}},\underline{\underline{Z}}) \xrightarrow{\kappa^*} \mathrm{Hom}_{\underline{\underline{Z}}}(\underline{\underline{Z}},\underline{\underline{Z}}) \xrightarrow{\iota^*} \mathrm{Hom}_{\underline{\underline{Z}}}(2\underline{\underline{Z}},\underline{\underline{Z}}) \longrightarrow 0$$

is not exact, where E^* is derived from the sequence

$$0 \longrightarrow 2\underline{\underline{Z}} \xrightarrow{\iota} \underline{\underline{Z}} \xrightarrow{\kappa} \underline{\underline{Z}}/2\underline{\underline{Z}} \longrightarrow 0$$

with the canonical homomorphisms.

6.) Show that δ (cf. (2.11)) is a natural homomorphism.

§3. Tensor products

The tensor product is covariant, additive and right exact. Projective modules are flat. The natural map $\mu : \mathrm{Hom}_R(M,R) \otimes_R N \longrightarrow \mathrm{Hom}_R(M,N)$ is considered.

3.1 Definition: Let $M \in \underline{\underline{M}}_R$, $N \in {}_R\underline{\underline{M}}$, and let G be an abelian group. A map $\varphi : M \times N \longrightarrow G$ is called an R-balanced map, if it is bilinear and satisfies $\varphi(mr,n) = \varphi(m,rn)$ for $m \in M$, $n \in N$, $r \in R$.

3.2 The universal mapping problem: Let $M \in \underline{\underline{M}}_R$, $N \in {}_R\underline{\underline{M}}$. Does there exist an abelian group G', and an R-balanced map $\psi : M \times N \longrightarrow G'$ such that for every R-balanced map $\varphi : M \times N \longrightarrow G$ there exists a unique Z-homomorphism $\sigma : G' \longrightarrow G$, which makes the diagram

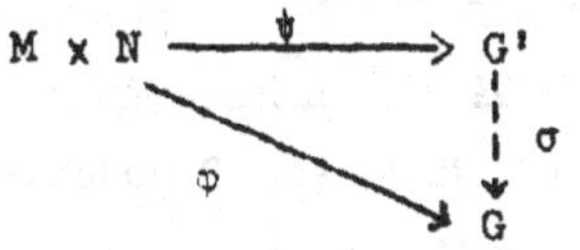

commute.

3.3 Definition: Let $M \in \underline{\underline{M}}_R$ and $N \in {}_R\underline{\underline{M}}$. Let C be the free abelian group generated by the symbols $\{(m,n) : m \in M, n \in N\}$, and let D be the $\underline{Z}$-submodule of C generated by all elements of the following form: $(m+m',n) - (m,n) - (m',n)$, $(m,n+n') - (m,n) - (m,n')$, $(mr,n) - (m,rn)$, $m,m' \in M$, $n,n' \in N$, $r \in R$. Then the tensor product of the right R-module M and the left R-module N, $M \otimes_R N$, is the $\underline{Z}$-module C/D. For $m \in M$, $n \in N$, the tensor product of m and n, $m \otimes n$, is the image of (m,n) in C/D.

3.4 Theorem: The abelian group $M \otimes_R N$ together with the map $\psi : M \times N \longrightarrow M \otimes_R N$; $\psi : (m,n) \longmapsto m \otimes n$, is a solution of the universal mapping problem (3.2). Moreover, it is, up to

$\underline{Z}$-isomorphism, the only solution.

Proof: An application of (Ex. 2,3) shows that $M \otimes_R N, \psi$ is a solution of (3.2), the uniqueness of the solution is easily seen from (3.2). #

3.5 Remark: The tensor product of two non-zero modules can be zero: e.g., if $M = \underline{Z}/2\underline{Z}$, $N = \underline{Z}/3\underline{Z}$, then $M \otimes_{\underline{Z}} N = 0$.

3.6 Lemma: Let $M \in \underline{\underline{M}}_R$, $N \in {}_R\underline{\underline{M}}$. By M^{op} (resp. N^{op}) we denote the module M (resp. N) if considered as left (resp. right) R^{op}-module (cf. (1.1)). Then there exists a unique natural $\underline{Z}$-isomorphism

$$\sigma : M \otimes_R N \longrightarrow N^{op} \otimes_{R^{op}} M^{op} \quad \text{such that}$$

$$\sigma : m \otimes n \longmapsto n \otimes m.$$

The proof is straightforward. #

3.7 Corollary (commutativity of the tensor product): If R is a commutative ring, and if M, N are R-modules, then there is a natural isomorphism $M \otimes_R N \cong N \otimes_R M$, as $\underline{Z}$-modules.

3.8 Lemma: Let M be a right R-module. Then

$$\varphi : M \otimes_R ({}_R R) \xrightarrow{\sim} M,$$
$$\varphi : m \otimes r \longmapsto mr$$

as right R-modules. This isomorphism is natural.

The proof is straightforward. #

3.9 Definition: Let $M, M' \in \underline{\underline{M}}_R$ and $N, N' \in {}_R\underline{\underline{M}}$. We define a map

$$\text{ten} : \text{Hom}_R(M,M') \times \text{Hom}_R(N,N') \longrightarrow \text{Hom}_{\underline{Z}}(M \otimes_R N, M' \otimes_R N')$$
$$\text{ten} : (\varphi, \psi) \longmapsto \varphi \otimes \psi,$$

where $\varphi \otimes \psi : M \otimes_R N \longrightarrow M' \otimes_R N'$

is induced from the R-balanced map

$$(\varphi,\psi) : M \times N \longrightarrow M' \otimes_R N'$$
$$(\varphi,\psi) : (m,n) \longmapsto \varphi m \otimes n\psi.$$

By (3.2) and (3.4) there exists a unique $\underline{\underline{Z}}$-homomorphism

$$\varphi \otimes \psi : M \otimes_R N \longrightarrow M' \otimes_R N'$$
$$\varphi \otimes \psi : m \otimes n \longmapsto \varphi m \otimes n\psi.$$

Hence the map ten is well defined; $\varphi \otimes \psi$ is called the <u>tensor product of the R-homomorphisms φ and ψ</u>. ten has the following properties:

(i) $(\varphi_1 + \varphi_2) \otimes \psi = \varphi_1 \otimes \psi + \varphi_2 \otimes \psi$, $\varphi_i \in \mathrm{Hom}_R(M,M')$, $i = 1,2$, $\psi \in \mathrm{Hom}_R(N,N')$,

(ii) $\varphi \otimes (\psi_1 + \psi_2) = \varphi \otimes \psi_1 + \varphi \otimes \psi_2$, $\varphi \in \mathrm{Hom}_R(M,M')$, $\psi_i \in \mathrm{Hom}_R(N,N')$, $i = 1,2$,

(iii) $\varphi'\varphi \otimes \psi\psi' = (\varphi' \otimes \psi')(\varphi \otimes \psi)$, $\varphi' \in \mathrm{Hom}_R(M',M'')$, $\varphi \in \mathrm{Hom}_R(M,M')$, $M'' \in \underline{\underline{M}}_R$, $\psi' \in \mathrm{Hom}_R(N',N'')$, $\psi \in \mathrm{Hom}_R(N,N')$, $N'' \in {}_R\underline{\underline{M}}$,
(Note: Homomorphisms of tensor products are written on the left.)

(iv) $1_M \otimes 1_N = 1_{M \otimes_R N}$

(v) $0 \otimes \psi = \varphi \otimes 0 = 0.$

3.10 <u>Remark</u>:

(i) The map ten of (3.9) is $\underline{\underline{Z}}$-balanced; thus it induces a $\underline{\underline{Z}}$-homomorphism

$$\mathrm{ten}' : \mathrm{Hom}_R(M,M') \otimes_{\underline{\underline{Z}}} \mathrm{Hom}_R(N,N') \longrightarrow \mathrm{Hom}_{\underline{\underline{Z}}}(M \otimes_R N, M' \otimes_R N').$$

(Generally, this is neither an epimorphism nor a monomorphism.)

(ii) Let $M,M',N,N' \in {}_R\underline{\underline{M}}$, then the map hom of (2.5) is $\underline{\underline{Z}}$-balanced; thus it induces a $\underline{\underline{Z}}$-homomorphism

$$\mathrm{hom}' : \mathrm{Hom}_R(M',M) \otimes_{\underline{\underline{Z}}} \mathrm{Hom}_R(N,N') \longrightarrow \mathrm{Hom}_{\underline{\underline{Z}}}(\mathrm{Hom}_R(M,N),\mathrm{Hom}_R(M',N'))$$

Similarly for right modules.

3.11 <u>Theorem</u> (<u>associativity of the tensor product</u>): Let R, S be two rings, M a right R-module, N an (R,S)-bimodule (cf. (1.4)) and L a left S-module. Then $M \otimes_R N \in \underline{\underline{M}}_S$ and $N \otimes_S L \in {}_R\underline{\underline{M}}$, and there exists a unique $\underline{\underline{Z}}$-homomorphism

$$\Phi : M \otimes_R (N \otimes_S L) \longrightarrow (M \otimes_R N) \otimes_S L;$$
$$\Phi : m \otimes (n \otimes \ell) \longmapsto (m \otimes n) \otimes \ell;$$

moreover, Φ is a natural isomorphism.

<u>Proof</u>: One checks easily that the definition

$$(m \otimes n)s = m \otimes ns, \; m \otimes n \in M \otimes_R N, \; s \in S$$

makes $M \otimes_R N$ into a right S-module. Similarly, $N \otimes_S L$ becomes a left R-module, so that the above expressions make sense. The uniqueness of the above map, if it exists, is clear, since $M \otimes_R (N \otimes_S L)$ is generated by the elements $m \otimes (n \otimes \ell)$, $m \in M$, $n \in N$, $\ell \in L$. For each $\ell \in L$, the map $\rho_\ell : N \to N \otimes_S L$; $n \longmapsto n \otimes \ell$ is an R-homomorphism, and the map $\sigma_\ell = 1_M \otimes \rho_\ell$ is a $\underline{\underline{Z}}$-homomorphism (cf. 3.9). The map $\Phi' : (M \otimes_R N) \times L \longrightarrow M \otimes_R (N \otimes_S L)$; $(x,\ell) \longmapsto \sigma_\ell(x)$ is R-balanced and thus induces the required $\underline{\underline{Z}}$-homomorphism Φ. Similarly a $\underline{\underline{Z}}$-homomorphism $\Psi : M \otimes_R (N \otimes_S L) \longrightarrow (M \otimes_R N) \otimes_S L$; $(m \otimes n) \otimes \ell \longmapsto m \otimes (n \otimes \ell)$, is obtained. Obviously Φ and Ψ are inverses of each other and are both natural. #

3.12 <u>Theorem</u>: For every exact sequence of left R-modules

$$M' \xrightarrow{\varphi} M \xrightarrow{\psi} M'' \longrightarrow 0$$

the sequence

$$M' \otimes_R N \xrightarrow{\varphi \otimes 1_N} M \otimes_R N \xrightarrow{\psi \otimes 1_N} M'' \otimes_R N \longrightarrow 0$$

is an exact sequence of $\underline{\underline{Z}}$-homomorphisms.

<u>Proof</u>: Since $(\psi \otimes 1_N)(\varphi \otimes 1_N) = 0$, we have $\mathrm{Im}(\varphi \otimes 1_N) \subset \mathrm{Ker}(\psi \otimes 1_N)$. Conversely, the R-balanced map $M'' \times N \longrightarrow (M \otimes_R N)/\mathrm{Im}(\varphi \otimes 1_N)$, $(m'',n) \longmapsto m \otimes n + \mathrm{Im}(\varphi \otimes 1_N)$, where m is such that $m\psi = m''$

factors through $M'' \otimes_R N$ (cf. (3.2)); i.e., we get a Z-homomorphism

$$\sigma : M'' \otimes_R N \longrightarrow (M \otimes_R N)/\mathrm{Im}(\varphi \otimes 1_N).$$

Since $\mathrm{Im}(\varphi \otimes 1_N) \subset \mathrm{Ker}(\psi \otimes 1_N)$, we can complete the following diagram commutatively (cf. Ex. 2,3):

$$\begin{array}{ccc} M \otimes_R N & \xrightarrow{\psi\otimes 1_N} & M'' \otimes_R N \\ & \searrow^{\iota} & \uparrow \rho \\ & & (M \otimes_R N)/\mathrm{Im}(\varphi \otimes 1_N), \end{array}$$

where ι is the canonical epimorphism. It is now easily seen that $\sigma\rho = 1_{(M\otimes_R N)/\mathrm{Im}(\varphi\otimes 1_N)}$ and $\rho\sigma = 1_{M''\otimes_R N}$. ♯

3.13 <u>Corollary</u>: For every exact sequence of left R-modules

$$E : N' \xrightarrow{\varphi} N \xrightarrow{\psi} N'' \longrightarrow 0$$

the sequence

$$E' : M \otimes_R N' \xrightarrow{1_M\otimes\varphi} M \otimes_R N \xrightarrow{1_M\otimes\psi} M \otimes_R N' \longrightarrow 0$$

is an exact sequence of $\underline{Z}$-modules.

The <u>proof</u> is done by considering N'^{op}, N^{op} and N''^{op} as R^{op} right R^{op} modules and applying (3.6) and (3.12). ♯

3.14 <u>Corollary</u>: Let $M = M_1 \oplus M_2$ be a right R-module and N a left R-module. Then $M \otimes_R N \cong (M_1 \otimes_R N) \oplus (M_2 \otimes_R N)$; this isomorphism is natural. (Similarly $M \otimes_R (N_1 \oplus N_2) \cong (M\otimes_R N_1) \oplus (M \otimes_R N_2)$.

<u>Proof</u>: The split exact sequence

$$0 \longrightarrow M_1 \xrightarrow{\iota_1} M \xrightarrow{\pi_2} M_2 \longrightarrow 0 \quad \text{(cf. (1.10))}$$

gives rise to the split exact sequence

$$0 \longrightarrow M_1 \otimes_R N \xrightarrow{\iota_1 \otimes 1_N} M \otimes_R N \xrightarrow{\pi_2 \otimes 1_N} M_2 \otimes_R N \longrightarrow 0$$

(cf. (3.12), (1.10) and (2.2)); i.e.,

$$(M_1 \oplus M_2) \otimes_R N \cong (M_1 \otimes_R N) \oplus (M_2 \otimes_R N).$$

Obviously, this is a natural isomorphism. #

3.15 Remark: If $0 \longrightarrow M' \xrightarrow{\varphi} M \xrightarrow{\psi} M'' \longrightarrow 0$ is an exact sequence of right R-modules, then the sequence

$$0 \longrightarrow M' \otimes_R N \xrightarrow{\varphi \otimes 1_N} M \otimes_R N \xrightarrow{\psi \otimes 1_N} M'' \otimes_R N \longrightarrow 0,$$

where N is a left R-module, is not necessarily exact. For example, $0 \longrightarrow 2\mathbb{Z} \xrightarrow{\varphi} \mathbb{Z} \xrightarrow{\psi} \mathbb{Z}/2\mathbb{Z} \longrightarrow 0$, with the canonical homomorphisms (cf. (2.2)), is exact; but $0 \longrightarrow \mathbb{Z}/2\mathbb{Z} \otimes_{\mathbb{Z}} 2\mathbb{Z} \xrightarrow{1 \otimes \varphi} \mathbb{Z}/2\mathbb{Z} \otimes_{\mathbb{Z}} \mathbb{Z} \xrightarrow{1 \otimes \psi} \mathbb{Z}/2\mathbb{Z} \otimes_{\mathbb{Z}} \mathbb{Z}/2\mathbb{Z} \longrightarrow 0$ is not exact, since $\mathbb{Z}/2\mathbb{Z} \otimes_{\mathbb{Z}} 2\mathbb{Z} \neq 0$, whereas $\mathrm{Im}(1 \otimes \varphi) = 0$.

3.16 Definition: A left R-module N is called flat, if for every exact sequence

$$0 \longrightarrow M' \xrightarrow{\varphi} M \xrightarrow{\psi} M'' \longrightarrow 0$$

of right R-modules, the sequence

$$0 \longrightarrow M' \otimes_R N \xrightarrow{\varphi \otimes 1_N} M \otimes_R N \xrightarrow{\psi \otimes 1_N} M'' \otimes_R N \longrightarrow 0$$

is an exact sequence of $\mathbb{Z}$-modules. Similarly for a right R-module.

3.17 Lemma: A finitely generated projective left R-module is flat.

The proof can be obtained by the technique used in proving (2.9), (iv) $\Longrightarrow$ (i), and it is left as an exercise. #

3.18 Lemma: Let $\underline{\underline{a}}$ be a right ideal of R and M a left R-module; let $\underline{\underline{a}}M$ be the $\underline{Z}$-submodule of M generated by the elements of the form αm, $\alpha \in \underline{\underline{a}}$, $m \in M$. Then there is a natural $\underline{Z}$-isomorphism

$$\varphi : R/\underline{\underline{a}} \otimes_R M \longrightarrow M/\underline{\underline{a}}M$$

$$\varphi : \alpha \otimes m \longmapsto \alpha m + \underline{\underline{a}}m.$$

Moreover, if $\underline{\underline{a}}$ is a two-sided R-ideal (i.e., if $\underline{\underline{a}}$ is an (R,R)-bimodule contained in R) then φ is an isomorphism of left R-modules.

Proof: The canonical epimorphism $R_R \xrightarrow{\psi} R/\underline{\underline{a}}$ induces the epimorphism

$$\psi \otimes 1_M : R_R \otimes_R M \longrightarrow R/\underline{\underline{a}} \otimes_R M.$$

Now,

$$\mathrm{Ker}(\psi \otimes 1_M) = \{(\Sigma r_i \otimes m_i : \Sigma\, r_i m_i \in \underline{\underline{a}}M)\}.$$

Under the isomorphism $R_R \otimes_R M \cong M$ (cf. 3.8) , $\mathrm{Ker}(\psi \otimes 1_M) \cong \underline{\underline{a}}\, M$; i.e., $M/\underline{\underline{a}}\, M \cong R/\underline{\underline{a}} \otimes_R M$. If, in addition, $\underline{\underline{a}}$ is a two-sided ideal, then $\underline{\underline{a}}M$ is a left R-module, and the above isomorphism is an isomorphism of left R-modules, as is easily seen. Trivially, φ is natural. $\#$

Exercises §3:

1.) Show that $\underline{Z}/2\underline{Z} \otimes_{\underline{Z}} \underline{Z}/3\underline{Z} = 0$.

2.) Show that the following isomorphisms are natural

(i) $\qquad M \otimes_R N \xrightarrow{\sim} N^{op} \otimes_{R^{op}} M^{op}$, $\qquad$ (3.6)

(ii) $\qquad M \otimes_R {}_R R \xrightarrow{\sim} M$, $\qquad$ (3.8)

(iii) $M \otimes_R (N \otimes_S L) \xrightarrow{\sim} (M \otimes_R N) \otimes_S L$, (3.11)

(iv) $M \otimes_R (N_1 \oplus N_2) \xrightarrow{\sim} M \otimes_R N_1 \oplus M \otimes_R N_2$, (3.14)

(v) $R/\underline{\underline{a}} \otimes_R M \xrightarrow{\sim} M/\underline{\underline{a}}\, M$ (3.18)

3.) Show that the sequence

$0 \longrightarrow \underline{Z}/2\underline{Z} \otimes_{\underline{Z}} 2\underline{Z} \xrightarrow{1\otimes\varphi} \underline{Z}/2\underline{Z} \otimes_{\underline{Z}} \underline{Z} \xrightarrow{1\otimes\psi} \underline{Z}/2\underline{Z} \otimes_{\underline{Z}} \underline{Z}/2\underline{Z} \longrightarrow 0$

is not exact, where $\varphi : 2\underline{Z} \longrightarrow \underline{Z}$ and $\psi : \underline{Z} \longrightarrow \underline{Z}/2\underline{Z}$ are the canonical homomorphisms.

4.) Verify the formulae (3.9,i,...,v).

5.) Let $M, N \in {}_R\underline{\underline{M}}^f$. Show:

(i) $M^* \otimes_R N$, $\mathrm{Hom}_R(M,N) \in {}_{\Omega(M)}\underline{\underline{M}}_{\Omega(N)}$, where $\Omega(X) = \mathrm{End}_R(X)$, $M^* = \mathrm{Hom}_R(M,R)$. If $M = N$, then $M^* \otimes_R M$ is a "ring"; but it does not necessarily have an identity!

(ii) The map

$$\mu : M^* \otimes_R N \longrightarrow \mathrm{Hom}_R(M,N);\ (\varphi\otimes n) \longmapsto (\varphi\otimes n)^\mu$$

where $m[(\varphi\otimes n)^\mu] = (m\varphi)n$ is a natural $(\Omega(M),\Omega(N))$-homomorphism. If $M = N$, then μ is a ring homomorphism, but not necessarily unitary. (Hint: To give $M^* \otimes_R M$ the structure of a ring observe that, for every $\varphi_o \otimes m_o \in M^* \otimes_R M$, the map $M^* \times M \longrightarrow M^* \otimes_R M$; $(\varphi,m) \longmapsto \varphi \otimes (m\varphi_o)m_o$ is R-balanced.)

6.) Let R, S be rings, $M \in \underline{\underline{M}}_R$, $N \in {}_R\underline{\underline{M}}_S$, $L \in \underline{\underline{M}}_S$. Show that there is a natural isomorphism of abelian groups:

$$\Phi : \mathrm{Hom}_S(M \otimes_R N, L) \longrightarrow \mathrm{Hom}_R(M, \mathrm{Hom}_S(N,L));\ \varphi \longmapsto \varphi^\Phi,$$

where $(n)(m\varphi^\Phi) = (n\otimes m)\varphi$ for $n \in N$, $m \in M$. This isomorphism preserves any structure that $\mathrm{Hom}_S(M \otimes_R N, L)$ has.

§4. Artinian and noetherian modules

The theorem of Jordan-Hölder for modules of finite length is stated, and the Krull-Schmidt theorem is proved for rings R, for which $\mathrm{End}_R(M)$ is completely primary if M is indecomposable. Nakayama's lemma is proved, and some properties of the Jacobson radical are derived.

4.1 Definition: A left R-module M is called artinian (resp. noetherian) if it satisfies one of the following equivalent conditions:

(i) Every non-empty set of submodules of M, partially ordered by inclusion, contains a minimal (resp. maximal) element.

(ii) Every descending (resp. ascending) chain of submodules of M becomes stationary; i.e., if

$$M_1 \supset M_2 \supset \ldots \supset M_i \supset \ldots$$

is a descending chain of submodules of M, then there exists a positive integer n, such that $M_k = M_\ell$ for all $k, \ell \geq n$.

4.2 Lemma: Let $M \in {}_R\underline{\underline{M}}$. Then M is noetherian if and only if every submodule of M is of finite type.

Proof: If M is a noetherian left R-module and N a submodule of M, let S be the set of submodules generated by finite subsets of N. By (4.1, (i)) S contains a maximal element N_o. For every element $n \in N$, one has $N_o + Rn = N_o$; hence $N = N_o$, and N is finitely generated. Conversely, let $\{M_i\}$ be an ascending chain of submodules of M. Then $M_o = \bigcup_{i=1,2\ldots} M_i$ is a submodule of M; hence of finite type by hypothesis; say, M_o is generated by $m_1,\ldots,m_n$. Then there exists n_o, such that $m_i \in M_{n_o}$, $1 \leq i \leq n$.

Hence the chain $\{M_i\}$ becomes stationary; i.e., M is noetherian. #

4.3 <u>Lemma</u>: Let $M \in {}_R\underline{\underline{M}}$, N a submodule of M. M is artinian (resp. noetherian) if and only if N and M/N are artinian (resp. noetherian).

<u>Proof</u>: <u>(i)</u> If M is artinian (resp. noetherian) so is N. Let $\varphi : M \longrightarrow M/N$ be the canonical homomorphism. If $\{\overline{M}_i\}$ is a descending (resp. ascending) chain of submodules of M/N, then the $M_i = \{m \in M : m\varphi \in \overline{M}_i\}$ form a descending (resp. ascending) chain of submodules of M, which becomes stationary by hypothesis; i.e., $M_k = M_\ell$ for $k, \ell \geq n_o$. But then also $\overline{M}_k = \overline{M}_\ell$ for $k, \ell \geq n_o$. Hence M/N is artinian (resp. noetherian).

<u>(ii)</u> Conversely, let $\{M_i\}$, $M_i \in M$, be a descending (ascending) chain. Set $\overline{M}_i = M_i N/N$; then $\{\overline{M}_i\}$ and $M_i \cap N$ are descending(ascending) chains of submodules of M/N and N resp., which become stationary by hypothesis. Hence there exists n_o such that $\overline{M}_k = \overline{M}_{n_o}$, and $M_k \cap N = M_{n_o} \cap N$ for every $k \geq n_o$. But then $M_k = M_{n_o}$ (cf. Ex. 4,2) for every $k \geq n_o$, and M is artinian (resp. noetherian). #

4.4 <u>Corollary</u>: A finite direct sum of left R-modules is artinian (resp. noetherian) if and only if each summand is artinian (resp. noetherian).

<u>Proof</u>: This follows immediately from (4.3). #

4.5 <u>Definitions</u>: A left R-module M is called a <u>simple</u> R-<u>module</u> if M contains no non-trivial submodule. Let M be a left R-module. A finite descending chain

$$M = M_1 \supsetneqq M_2 \supsetneqq \cdots \supsetneqq M_n \supsetneqq M_{n+1} = 0$$

is called a <u>composition series for M of length n</u>, if the factor modules M_i/M_{i+1}, $1 \leq i \leq n$, are simple R-modules. Two composition series are said to be <u>equivalent</u> if they have the same length n, and if the factors can be paired off in such a way that corresponding factors are isomorphic. M is said to have finite length (length n) if M has a composition series (of length n).

4.6 Theorem (Jordan-Hölder): If a left R-module M has finite length, then any two composition series of M are equivalent.

This is proved as for finite groups. #

4.7 Theorem: A left R-module M has finite length if and only if M is artinian and noetherian.

Proof: If M has finite length, then any two composition series of M have the same length, say n. Hence, every strictly decreasing (resp. increasing) chain of submodules has less than $n + 1$ terms; i.e., M is artinian (resp. noetherian).

Conversely: Since M is noetherian (cf. (4.3)), among its submodules, which are different from M, there exists a maximal submodule N. Obviously M/N is simple. Applying this procedure recurrently, (cf. (4.3)), we obtain a decreasing sequence $\{M_i\}$ of submodules of M such that M_i/M_{i+1} is a simple R-module. Since M is artinian, this sequence becomes stationary; i.e., M is of finite length. #

4.8 Definition: A ring S is called completely primary, if the non-units in S form a 2-sided ideal (a unit of S is an invertible element in $(R,\cdot)$). In this case the ideal of the non-units is the unique maximal right and left ideal of S. A commutative completely primary ring is called a local ring. A left R-module is decomposable if it is the direct sum of two non-trivial submodules.

4.9 Lemma: If M is an indecomposable left R-module of finite length, then $\mathrm{End}_R(M)$ is a completely primary ring.

Proof: We show first that a ring S in which every sum of non-units is a non-unit is completely primary. All we need show is that in such a ring t is a unit whenever st is one for some $s \in S$.

Thus let $r(st) = 1$. We show that $t(rs) = 1$: $e = trs = t(rst)rs$ is an idempotent - i.e., $e^2=e$-, and $1=e+(1-e)$. By hypothesis e or $(1-e)$ is a unit. If $(1-e)$ is a unit then $1-e=1$, since $1-e$ is also an idempotent and $e = 0$ contradicting the fact that $rset = (rst)(rst) = 1$. Thus e has to be a unit, i.e., $e = rst = 1$.

Now we come to the actual proof of (4.9). Let $\varphi, \psi \in \mathrm{End}_R(M)$ be such that $\varphi + \psi = 1_M$; with the above result it suffices to show that either φ of ψ is a unit. We put $I_n = \mathrm{Im}\,\varphi^n$, where $\varphi^n = \varphi\varphi\ldots\varphi$, n-times. Then we have the descending chain

$$M = I_o \supset I_1 \supset \ldots \supset I_n \supset I_{n+1} \supset \ldots,$$

which becomes stationary by hypothesis, say $I_n = I_{n_1}$ for $n \geq n_1$. On the other hand, $\{K_n = \mathrm{Ker}\,\varphi^n\}$ form an ascending chain

$$K_1 \subset K_2 \subset \ldots \subset K_n \subset K_{n+1} \subset \ldots$$

which becomes stationary by hypothesis, say $K_n = K_{n_2}$ for $n \geq n_2$. If we put $a = \max(n_1, n_2)$, then $\varphi^a \restriction_{I_a} : I_a \longrightarrow I_{2a} = I_a$ is epic and has zero kernel, i.e., $\varphi^a \restriction_{I_a}$ is an isomorphism. ($\beta \restriction_X$ denotes the restriction of β to X.) The existence of $(\varphi^a \restriction_{I_a})^{-1} \iota$, where ι is the canonical injection, $I_a \longrightarrow M$, shows that the exact sequence

$$0 \longrightarrow K_a \longrightarrow M \longrightarrow I_a \longrightarrow 0$$

is split. But M was assumed to be indecomposable, i.e., $I_a = 0$ or $I_a = M$. If $I_a = 0$, then $\varphi^a = 0$ and $\psi = (1-\varphi)$ has the inverse $1 + \varphi + \varphi^2 + \ldots + \varphi^{a-1}$. If $I_a = M$ then $\varphi^a : M \longrightarrow M$ is an isomorphism and thus φ is an isomorphism. ∮

4.10 <u>Theorem</u> (Krull-Schmidt, Azumaya [1]): Let R be a ring and $M \in {}_R\underline{\underline{M}}$ a noetherian module. If

$$M = \oplus_{i=1}^{m} M_i = \oplus_{j=1}^{n} N_j$$

are two decompositions of M into indecomposable summands, and if $\mathrm{End}_R(M_i)$, $1 \leq i \leq m$, are completely primary rings, then $m = n$, and - if necessary, after renumbering - $M_i \cong N_i$, $1 \leq i \leq m$.

<u>Proof</u>: Since M is noetherian, $m,n < \infty$. For the proof we shall use induction on m. Let

$$\pi_i : M \longrightarrow M_i, \ 1 \leq i \leq m,$$
$$\pi_j' : M \longrightarrow N_j, \ 1 \leq j \leq n,$$

be the projections associated with the above decompositions (cf. (1.10)). Then

$$1_M = \Sigma_{i=1}^{m} \pi_i \iota_i = \Sigma_{j=1}^{n} \pi_j' \iota_j', \quad (\text{cf. } (1.10))$$
$$0 = \pi_i \iota_j \pi_j \quad = \pi_k' \iota_\ell' \pi_\ell', \quad \text{for } i \neq j, \ k \neq \ell.$$

Thus

$$1_{M_1} = \iota_1 \pi_1 = \Sigma_{j=1}^{n} \iota_1 \pi_j' \iota_j' \pi_1 .$$

Since $\mathrm{End}_R(M_1)$ is completely primary, one of the $\iota_1 \pi_j' \iota_j' \pi_1$, $1 \leq j \leq n$, has to be a unit in $\mathrm{End}_R(M_1)$, say $\iota_1 \pi_1' \iota_1' \pi_1$.

<u>Claim</u>: $\iota_1 \pi_1' : M_1 \longrightarrow N_1$ and $\iota_1' \pi_1 : N_1 \longrightarrow M_1$ are isomorphisms.

To prove the claim we apply the X-Lemma (Ex. 4,9) to the diagram

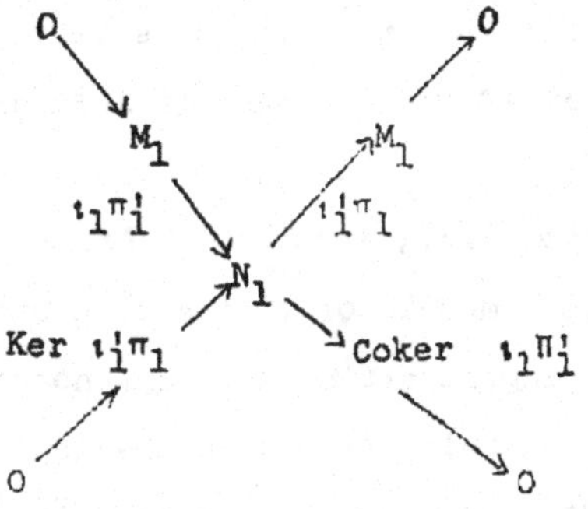

,

where $\iota_1\pi_1'\iota_1'\pi_1$ is an isomorphism, and thus, we conclude $N_1 = \operatorname{Ker} \iota_1'\pi_1 \oplus \operatorname{Im} \iota_1\pi_1'$. Since $\operatorname{Im} \iota_1\pi_1' \neq 0$, we must have $\operatorname{Ker} \iota_1'\pi_1 = 0$; i.e., $\iota_1'\pi_1$ and $\iota_1\pi_1'$ are isomorphisms, as was to be shown.

Now, to finish the proof of (4.10), we apply the X-Lemma to the diagram

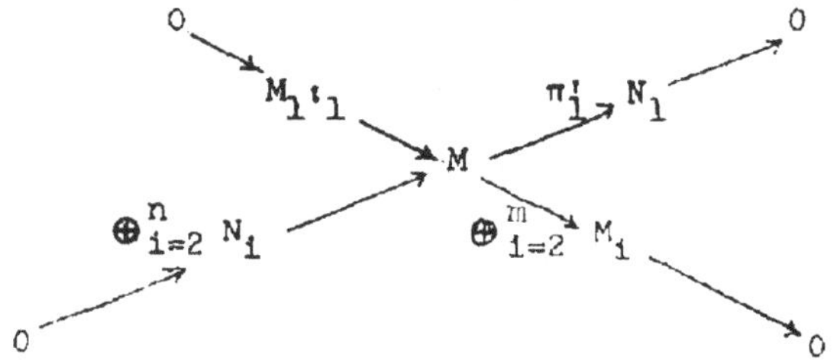

where $\iota_1\pi_1'$ is an isomorphism. Thus, $\bigoplus_{i=2}^{n} N_i \cong \bigoplus_{i=2}^{m} M_i$, and we may apply induction. #

4.11 <u>Definition</u>: A ring R is called <u>left artinian</u> (resp. <u>left noetherian</u>), if ${}_RR$ (cf. (1.3)) is artinian (resp. noetherian).

4.12 <u>Lemma</u>: Let R be a left artinian (resp. noetherian) ring and $\underline{\underline{a}}$ a two-sided ideal in R. Then $R/\underline{\underline{a}}$ is left artinian (resp. noetherian).

<u>Proof</u>: This is an immediate consequence of (4.3). #

4.13 <u>Lemma</u>: A finite direct sum of rings is left artinian (resp. noetherian) if and only if each summand is left artinian (resp. noetherian).

<u>Proof</u>: This follows from (4.4), (cf. Ex. 4,7).

4.14 <u>Theorem</u>: Every left module of finite type over a left artinian (resp. noetherian) ring is artinian (resp. noetherian).

<u>Proof</u>: Let R be left artinian (resp. noetherian) and $M \in {}_R\underline{\underline{M}}^f$. Then there exists a free left R-module F with a finite basis, such

that we have an epimorphism

$$\varphi : F \longrightarrow M.$$

Since $F \cong ({}_R R)^n$ (cf. (1.12)), F is artinian (resp. noetherian) by (4.4), and from (4.3) it follows that M is artinian (resp. noetherian). #

The statements in the rest of this section can be proved for arbitrary rings, using Zorn's lemma; however, for our purposes it suffices to prove them for left noetherian rings.

4.15 Definition: Let R be a left noetherian ring and $M \in {}_R\underline{\underline{M}}^f$. The radical of M, rad M, is defined as

$$\text{rad } M = \bigcap_{\{\varphi : M \to S,\ S \text{ simple}\}} \text{Ker } \varphi$$

where the intersection is taken over all homomorphisms φ from M into simple left R-modules S. Since R is noetherian, rad M is the intersection of all maximal left R-submodules of M. R is called semisimple if rad $R = 0$.

4.16 Remark: For $M \in {}_R\underline{\underline{M}}^f$, rad M is a characteristic submodule; i.e., if $\varphi \in \text{End}_R(M)$, then $\varphi\restriction_{\text{rad } M} : \text{rad } M \longrightarrow \text{rad } M$. Indeed, if $\sigma : M \longrightarrow S$, where S is a simple left R-module, then $\varphi\restriction_{\text{rad } M} : \text{rad } M \longrightarrow \text{Ker } \sigma$. In particular, rad R is a two-sided ideal, since $\text{End}_R(R) \cong R$ via right multiplication. Moreover, $(\text{rad } R)\cdot M \subset \text{rad } M$, since $(\text{rad } R)\cdot S = 0$ for every simple left R-module S; in fact, given $s \in S$ there is an R-homomorphism $\varphi_s : {}_R R \longrightarrow S$, $\varphi_s : r \longmapsto rs$; hence $\varphi_s : \text{rad } R \longrightarrow 0$, and since s was arbitrary, $(\text{rad } R)\cdot S = 0$.

4.17 Lemma: If $J \subset \text{rad } R$ is a two-sided ideal, then $\text{rad } (R/J) \cong (\text{rad } R)/J$; in particular, $R/\text{rad } R$ is semisimple.

Proof: This follows from the fact that J is contained in every maximal left ideal of R. #

4.18 Lemma (Nakayama's Lemma): The following conditions on a left ideal I in R are equivalent:

(i) $I \subset \text{rad } R$.

(ii) If $M \in {}_{\underline{\underline{R}}}M^f$, then $IM = M$ implies $M = 0$.

(iii) If $M \in {}_{\underline{\underline{R}}}M^f$ and if $N \subset M$, then $M = N + IM$ implies $N = M$.

(iv) $1 + I$ consists of left invertible elements.

Proof: (i)$\Longrightarrow$(ii). Since $I \subset \text{rad } R$, we have $IM \subset (\text{rad } R) \cdot M \subset \text{rad } M$; thus $\text{rad } M = M$, a contradiction unless $M = 0$, since $\text{rad } M$ is the intersection of the maximal submodules of M.

(ii)$\Longrightarrow$(iii). From $M = N + IM$ we conclude $M/N = (N+IM)/N = I(M/N)$, and the result follows from (ii).

(iii)$\Longrightarrow$(iv). Let $u = 1 + x \in 1 + I$, then $R = I + R \cdot u$, and with (iii) we obtain $R \cdot u = R$; i.e., there exists $v \in R$ such that $vu = 1$; but $1 = vu = v + vx$ implies $v = 1 - vx \in 1 + I$; thus v has a left inverse, and u is left invertible.

(iv)$\Longrightarrow$(i). If I were not contained in every maximal left ideal $\underline{\underline{m}}$ of R, then $I + \underline{\underline{m}} = R$ for some $\underline{\underline{m}}$, and $\underline{\underline{m}}$ would contain a left invertible element $1 + x$, with $x \in I$, a contradiction.

4.19 Lemma: Let R be a ring, $M,N \in {}_{\underline{\underline{R}}}M^f$. If $\varphi \in \text{Hom}_R(M,N)$ and if $\underline{\underline{a}}$ is a right R-ideal, $\underline{\underline{a}} \subset \text{rad } R$, and if

$$1_{R_R/\underline{\underline{a}}} \otimes \varphi : R_R/\underline{\underline{a}} \otimes_R M \longrightarrow R_R/\underline{\underline{a}} \otimes_R N$$

is an epimorphism, then φ is an epimorphism.

The proof is straightforward. ∤

Exercises §4:

1.) Let R be a ring and $M \in {}_{\underline{\underline{R}}}M$ an R-module of length n. If $0 \neq N \subsetneq M$, show that N and M/N are R-modules of length $< n$.

2.) Let $M, M', M'' \in {}_{R}\underline{\underline{M}}$, where R is a ring. If

$$0 \longrightarrow M' \xrightarrow{\varphi} M \xrightarrow{\psi} M'' \longrightarrow 0$$

is an exact sequence, show, that for $N \subset M$, the sequence

$$0 \longrightarrow N \cap M'\varphi \longrightarrow N \longrightarrow N\psi \longrightarrow 0$$

is exact. Use this to fill in the last step in the proof of (4.3).

3.) Let R be a ring and $M \in {}_{R}\underline{\underline{M}}$ a simple R-module. Show that $\mathrm{End}_R(M)$ is a skewfield (i.e., a ring, in which every non-zero element is invertible). This fact is known as Schur's lemma. Show that for a completely primary ring S, $S/\mathrm{rad}\, S$ is a skewfield.

4.) Let R be a ring. An ideal $\underline{\underline{a}}$ of R is called nil, if every $a \in \underline{\underline{a}}$ is nilpotent. Show that any nilpotent ideal is nil; but not conversely. When is nil = nilpotent?

5.) Let R be a left noetherian ring. Show that $\mathrm{rad}\, R = \{x \in R : 1 - r_1 x r_2$ is invertible in R, $\forall r_1, r_2 \in R\}$. Use this to show that for a unitary epic ring homomorphism $\varphi : R \longrightarrow R_1$, $(\mathrm{rad}\, R)\varphi \subset \mathrm{rad}(R_1)$.

6.) Let S be a noetherian and artinian ring. Show that

(i) $\mathrm{rad}\, S$ is nilpotent; i.e., $\exists\, n \in N$ such that

$$(\mathrm{rad}\, S)^n = 0 \quad ((\mathrm{rad}\, S)^2 = \{\sum_{\text{finite}} x_i y_i : x_i, y_i \in \mathrm{rad}\, S\})$$

(ii) $S/\mathrm{rad}\, S$ does not contain any nilpotent left ideals.

(Hint for (ii): Show first - using (4.15) - that every nilpotent left ideal N of S has to be contained in $\mathrm{rad}\, S$.

7.) Let R_i, $1 \leq i \leq n$, be rings and make $\bigoplus_{i=1}^{n} R_i = R$ into a ring by defining $(r_1, \ldots, r_n)(r_1', \ldots, r_n') = (r_1 r_1', \ldots, r_n r_n')$. Show that the projections π_i and the injections ι_i, $1 \leq i \leq n$ are ring homomorphisms. However, while the π_i are unitary ring homomorphisms; i.e., $1\pi_i = 1$, this is not the case with the ι_i.

8.) Under the hypotheses of (4.10), show that for every subset $M_1, \ldots, M_{i_0}$ of $\{M_i\}_{1 \leq i \leq m}$, there exists a subset $\{N_{j_\nu}\}_{1 \leq \nu \leq i_0}$ of $\{N_j\}_{1 \leq j \leq n}$ such that $M = N_{j_1} \oplus \ldots \oplus N_{j_{i_0}} \oplus M_{i_0+1} \oplus \ldots \oplus M_n$.

9.) X-Lemma: Let R be a ring and $M', M'', X, N', N'' \in {}_{\underline{\underline{R}}}M$. Assume that the following diagram of exact sequences is given:

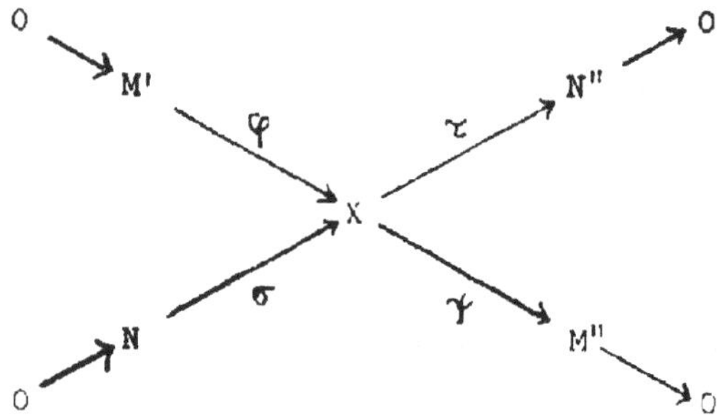

Show: If $\varphi\tau$ is an isomorphism, so is $\sigma\psi$.

10.) If R is a completely primary ring, then every $P \in {}_{\underline{\underline{R}}}P^{f}$ is free. (Hint: Show first that R is indecomposable as left R-module; then apply (4.10)).

§5 Integers

If A is an algebra over a commutative ring, conditions equivalent to "$a \in A$ is integral over R" are introduced; definition of the integral closure.

5.1 Definition: Let R be a commutative ring. An R-algebra A is a ring A together with a unitary ring homomorphism $\varphi : R \longrightarrow \text{center}(A)$, where $\text{center}(A) = \{x \in A : xa = ax \text{ for every } a \in A\}$. We may consider A as $(\text{Im } \varphi)$-algebra $(1 \in \text{Im } \varphi)$, and from now on we assume that $R \subset \text{center}(A)$.

5.2 Proposition: Let A be an R-algebra. For an element $a \in A$, the following conditions are equivalent:

(i) a satisfies a monic polynomial with coefficients in R; (i.e., a polynomial, whose leading coefficient is 1).

(ii) The subalgebra $R[a] = \{\sum_{\text{finite}} r_i a^i : r_i \in R\}$ of A is an R-module of finite type.

(iii) There exists a faithful R[a]-module, which is an R-module of finite type. (A left module M over a ring S is called faithful, if $\text{ann}_S(M) = (0)$.)

Proof: (i) obviously implies (ii).

(ii) $\Longrightarrow$ (iii): This is clear, since R[a] is a faithful R[a]-module in ${}_{R}\underline{\underline{M}}^f$.

(iii) $\Longrightarrow$ (i): Let $M \in {}_{R}\underline{\underline{M}}^f$ be a faithful R[a]-module, generated over R by $\{m_i\}_{1 \leq i \leq n}$. For $a \in A$ and for every $1 \leq i \leq n$, we may write

$$am_i = \sum_{j=1}^{n} r_{ij} m_j, \quad 1 \leq i \leq n, \; r_{ij} \in R.$$

Let $\underline{\underline{D}}$ be the following matrix with entries in R[a]:

$\underline{\underline{D}} = (r_{ij} - \delta_{ij} a)$, where δ_{ij} is the Kronecker symbol; i.e.,

$$\delta_{ij} = \begin{cases} 0 & \text{if } i \neq j \\ 1 & \text{if } i = j \end{cases}.$$

Since $R[a]$ is commutative, one has $\tilde{\underline{\underline{D}}}\underline{\underline{D}} = \det(\underline{\underline{D}}) \cdot \underline{\underline{E}}_n$, where $\underline{\underline{E}}_n$ is the n-dimensional identity matrix, and $\tilde{\underline{\underline{D}}} = (\tilde{d}_{ij})$ is the matrix of the cofactors; i.e., $\tilde{d}_{ij} = (-1)^{i+j}\det(\underline{\underline{D}}_{ji})$, and $\underline{\underline{D}}_{ji}$ is the matrix $\underline{\underline{D}}$ with the j-th row and the i-th column deleted. But

$$\underline{\underline{D}} \cdot \begin{pmatrix} m_1 \\ m_2 \\ \cdot \\ \cdot \\ \cdot \\ m_n \end{pmatrix} = 0 \text{ implies } \det(\underline{\underline{D}}) \cdot M = 0.$$

Since M is a faithful $R[a]$-module and since $\det(\underline{\underline{D}}) \in R[a]$, $\det(\underline{\underline{D}}) = 0$ and a is a root of the monic polynomial $\det(r_{ij} - \delta_{ij}X)$. #

5.3 Definition: Let R be a commutative ring and A an R-algebra. An element $a \in A$ is said to be integral over R, if a satisfies one of the equivalent conditions of (5.2). One says that A is integral over R if every element $a \in A$ is integral over R. A is called finite over R or a finite R-algebra if A is an R-module of finite type.

5.4 Remark: If the R-algebra A is finite over R, then A is integral over R by (5.2,iii). The converse of this statement is false; e.g., the ring of all algebraic integers is not a finite $\underline{\underline{Z}}$-algebra.

5.5 Lemma: Let R be a commutative ring and A a commutative R-algebra. The set of elements in A that are integral over R form an R-subalgebra of A.

Proof: Let $a,b \in A$ be integral over R. We have to show, that the same is true for $a \pm b$ and $a \cdot b$.

We have $R[a], R[b] \subset R[a,b] = R[a][b] = R[b][a] \subset A$.

Since $M = R[a]\, R[b] = \{ \sum_{\text{finite}} x_i y_i : x_i \in R[a], y_i \in R[b]\}$ contains the identity of A, M is a faithful $R[a,b]$-module. $M \in {}_R\underline{\underline{M}}^f$, since $R[a], R[b] \in {}_R\underline{\underline{M}}^f$ and since A is commutative. But then M is a faithful $R[a\pm b]$- as well as a faithful $R[ab]$-module, since $R[a\pm b]$ and

R[ab] are contained in R[a,b]. Now the statement follows from (5.2,iii). #

5.6 <u>Definition</u>: Let A be a commutative R-algebra. The ring of all elements in A which are integral over R is called <u>the integral closure of R in A</u>.

5.7 <u>Remark</u>: (i) If A is not commutative and if a,b ∈ A are integral over R, then a±b, a·b are integral over R, if ab = ba.

(ii) If A is a non-commutative R-algebra, then the product and the sum of integral elements need not be integral (cf. Ex. 5,2).

<u>Exercises §5</u>:

1.) Let R be a commutative ring and A, A' two R-algebras. Show

(i) $A \otimes_R A'$ is an R-algebra,

(ii) $A \otimes_R A'$ is integral over R if A and A' are commutative and integral over R.

2.) Let $\underline{\underline{Q}}_2$ be the ring of (2×2)matrices over $\underline{\underline{Q}}$. Give an example of two matrices in $\underline{\underline{Q}}_2$ which are integral over $\underline{\underline{Z}}$, such that neither their sum nor their product is integral over $\underline{\underline{Z}}$. (Hint: Consider matrices of the form $\underline{\underline{E}}_2 + \underline{\underline{N}}$, where $\underline{\underline{N}}$ is nilpotent and $\underline{\underline{E}}_2$ is the (2×2) identity matrix.)

3.) Prove the statement in (5.7)!

§6 Localization

If S is a multiplicative system in R, then, for $M \in {}_R\underline{\underline{M}}$, $M_S \cong R_S \otimes_S M$, where "$-_S$" denotes the localization at S. The quotient field of an integral domain is introduced. Localization preserves the properties "integral" and "noetherian".

6.1 Definition: For a ring R, a multiplicative system S is a multiplicatively closed subset of the center of R, containing $1 \in R$, but $0 \notin S$.

6.2 The universal problem of localization: Let R be a ring and S a multiplicative system in R. For $M \in {}_R\underline{\underline{M}}$ we are looking for a module $M_S \in {}_R\underline{\underline{M}}$ such that

(i) the elements of S act as automorphisms on M_S (via left multiplication),

(ii) there exists an R-homomorphism $\varphi_M : M \longrightarrow M_S$,

(iii) the pair (M_S, φ_M) is universal with respect to this property; i.e., given $N \in {}_R\underline{\underline{M}}$ such that the elements in S act as automorphisms on N, then the following diagram can be completed uniquely

$$\begin{array}{ccc} M & \xrightarrow{\varphi_M} & M_S \\ & \searrow^{\chi} & \downarrow{\scriptstyle\Psi} \\ & & N \end{array}$$

for every given R-homomorphism χ; i.e., χ factors through φ_M.

6.3 Theorem: (6.2) has an up to R-isomorphism unique solution M_S.

Proof: M_S consists of equivalence classes $(\frac{m}{s})$ of pairs (m,s), $m \in M$, $s \in S$, where $(m,s) \sim (m',s')$ if and only if there exists $t \in S$ such that $t\cdot(s'm-sm') = 0$. M_S is a left R-module, by

$$(\tfrac{m}{s}) + (\tfrac{m'}{s'}) = (\tfrac{s'm+sm'}{ss'}) \text{ and } r(\tfrac{m}{s}) = (\tfrac{rm}{s}).$$

In addition, if $M = R$, then R_S is a ring under $(\frac{r}{s})(\frac{r'}{s'}) = (\frac{rr'}{ss'})$. Thus, M_S becomes an R_S-module, and the map

$$\varphi_M : M \longrightarrow M_S,\ \varphi_M : m \longmapsto (\tfrac{m}{1})$$

is an R-homomorphism. The map ψ of (6.2) is defined by $\psi : \frac{m}{s} \longmapsto \frac{m\chi}{s}$. It is now easy to prove that (M_S, φ_M) satisfies the requirements of (6.2). The uniqueness follows from (6.2,iii)). #

6.4 Lemma: If S is a multiplicative system in R, and if $M \in {}_R\underline{\underline{M}}$, then there is a natural isomorphism $M_S \xrightarrow{\sim} R_S \otimes_R M$.

Proof: If we can show, that $R_S \otimes_R M$ together with $\varphi'_M : M \longrightarrow R_S \otimes_R M$, $\varphi'_M : m \longmapsto 1 \otimes m$ satisfies (6.2), then the result will follow from the universality. (6.2,i,ii) are obviously satisfied, and for (iii), we define $\psi : \frac{r}{s} \otimes m \longmapsto \frac{r}{s}(m\chi)$. The universality follows from the universal property of the tensor product (cf. (3.2)). #

6.5 Theorem: Let S be a multiplicative system in R. Then R_S is a flat right R-module.

Proof: Because of (3.12) it only remains to show that for a monomorphism $\alpha : M' \longrightarrow M$, the map $1_{R_S} \otimes \alpha : R_S \otimes_R M' \longrightarrow R_S \otimes_R M$ is monic. Because of (6.4), this amounts to showing that $\alpha_S : M'_S \longrightarrow M_S$. $(\frac{m'}{s}) \longmapsto (\frac{m'\alpha}{s})$ is monic. But $(\frac{m'}{s})\alpha_S = 0$ implies $t(\frac{m'\alpha}{s}) = 0$ for some $t \in S$; hence $t(m'\alpha) = 0$; i.e., $tm' = 0$, α being monic. Thus $(\frac{m'}{s}) = 0$ in M_S. #

6.6 Examples: Let R be an integral domain; i.e., a commutative ring without zero-divisors.

(i) If $S = R \setminus \{0\}$, then R_S is a field, since every non-zero element in R_S is invertible. R_S is called the quotient field K of R.

(ii) A prime ideal $\underline{\underline{p}}$ in R is an ideal such that $rr' \in \underline{\underline{p}}$ implies $r \in \underline{\underline{p}}$ or $r' \in \underline{\underline{p}}$. Every maximal ideal is prime, as is easily seen. Let now $\{\underline{\underline{p}}_i\}_{i \in I}$ be a set of prime ideals in R and $S = R \setminus \{\bigcup_{i \in I} \underline{\underline{p}}_i\}$; then S is a multiplicative system.

If I has only one element, we write $R_{\underline{\underline{p}}}$ for R_S and call $R_{\underline{\underline{p}}}$ the localization of R at the prime ideal $\underline{\underline{p}}$.

6.7 <u>Theorem</u>: Let R be an integral domain, S a multiplicative system in R and A a commutative R-algebra. If R' is the integral closure of R in A, then the integral closure of R_S in $R_S \otimes_R A$ is $R'_S = R_S \otimes_R R'$.
<u>Proof</u>: It should be observed, that $R_S \otimes_R A$ is naturally isomorphic to A_S, not only as R_S-module (cf. (6.4)) but also as ring. Thus we identify both structures. A_S is an R_S-algebra (cf. Ex. 5,1), and it suffices to show that R'_S is the integral closure of R_S in A_S.
Let $x/s \in R'_S$, $x \in R'$, then x is an integer in A; and hence $x/s = (1/s)x$ is an integer in A_S over R_S (cf. (5.5)). Conversely, let $a/s' \in A_S$, $a \in A$, be integral over R_S. Then $a = s'(a/s')$ is also integral over R_S; i.e., a satisfies a monic polynomial $X^n + b_{n-1}X^{n-1} +\dots+ b_o$, $b_i \in R_S$. Choose $0 \neq s \in S$ such that $sb_i \in R$, $1 \leq i \leq n$. Then a satisfies also $(sX)^n + b_{n-1}s(sX)^{n-1} +\dots+ b_o s^n$; and sa $\in$ A is integral over R; i.e., $a \in R'_S$ and so $a/s' \in R'_S$. #
6.8 <u>Remark</u>: If A is an R-algebra (not necessarily commutative), then A is integral over R if and only if A_S is integral over R_S (cf. (5.7) and (Ex. 5,3)).

<u>Exercises §6</u>:

1.) Let R be a ring and S a multiplicative system in R. Show that for $M \in {}_R\underline{\underline{M}}$, the relation

$(s,m) \sim (s',m')$ if there exists $t \in S$: $ts'm = tsm'$

is an equivalence relation on $S \times M$, and that M_S is an R_S-module. Give an example, where the map $M \longrightarrow M_S$, $m \longmapsto (m/1)$, is not a monomorphism.

2.) Prove the statement of (6.8).

§7 Dedekind domains

Every lattice over a Dedekind domain is projective; principal ideal domains are Dedekind domains, and semi-local Dedekind domains are principal ideal domains. The Chinese remainder theorem is proved.

7.1 Definitions: An integral domain R is called a Dedekind domain, if

(i) R is integrally closed in its quotient field K, i.e., R coincides with the integral closure of R in K (cf. (5.6)).

(ii) R is noetherian (cf. (4.11)).

(iii) Every prime ideal in R is maximal.

An R-module M is said to be R-torsion-free, if $rm = 0$, $r \in R$, $m \in M$, implies $r = 0$ or $m = 0$. M is called an R-torsion module, if $\forall m \in M, \exists\, 0 \neq r \in R$ such that $rm = 0$. (If $M \in {}_R\underline{\underline{M}}^f$, this is the same as saying $\mathrm{ann}_R(M) \neq 0$ (cf. (4.15)).) An R-lattice is a finitely generated torsion-free R-module. By ${}_R\underline{\underline{M}}^o$ we denote the class of R-lattices. The rank of an R-lattice M, rank(M), is defined to be $\dim_K(K \otimes_R M)$ $(<\infty)$, the K-dimension of the K-vector space $K \otimes_R M$. Since $K \otimes_R M$ is naturally isomorphic to

$$\{m/k : m \in M,\ k \in K,\ k \neq 0\} = KM \text{ (cf. (6.4))},$$

we identify KM and $K \otimes_R M$, and consider $M \subset KM = K \otimes_R M$, since $M \longrightarrow KM;\ m \longmapsto m/1$ is a monomorphism, M being a lattice. From (6.5) it follows that K is a flat R-module. For an R-lattice M, rank(M) = 0 if and only if M = 0. A fractional R-ideal in K is an R-lattice contained in K, and a fractional ideal is called an integral ideal if it is contained in R. For a fractional R-ideal $\underline{\underline{a}}$ we define $\underline{\underline{a}}^{-1} = \{x \in K : x\underline{\underline{a}} \in R\}$. Then $\underline{\underline{a}}^{-1}$ is a fractional ideal (cf. Ex. 7,1), since $\underline{\underline{a}}$ is of finite type. The fractional ideals are exactly the $M \in {}_R\underline{\underline{M}}^o$ with rank(M) = 1.

7.2 Theorem: Let R be a Dedekind domain. Then

(i) every fractional ideal $\underline{\underline{a}}$ has a unique prime decomposition $\underline{\underline{a}} = \prod_{i=1}^{n} \underline{\underline{p}}_i^{\alpha_i}$, where $\underline{\underline{p}}_i$, $1 \leq i \leq n$, are different prime ideals and the α_i, $1 \leq i \leq n$, are non-zero integers,

(ii) every R-lattice is a projective R-module.

Proof: This will follow from (IV, Ex. 4,1) and (IV, Ex. 4,2), where the above theorem is proved using techniques developed for maximal orders. However, there are direct proofs available (cf. e.g. Bourbaki [2], Ch. 7). #

7.3 Theorem: Let R be a Dedekind domain. Every R-lattice is isomorphic to a direct sum of a finite number of fractional ideals.

Proof: We need the concept of "pure" submodules:

7.4 Definition: Let M be an R-lattice and N a submodule of M (one should observe, that N is also an R-lattice (cf. (4.3) and (4.14)). N is called an R-pure submodule of M, if M/N is an R-lattice. Since every R-lattice is projective, this is equivalent to the splitting of the sequence $0 \longrightarrow N \longrightarrow M \longrightarrow M/N \longrightarrow 0$.

For the proof of (7.3) we use induction on rank(M): For rank(M) = 1 the statement is trivial (cf.(6.1)). Now, let rank(M) = n > 1. If $0 \neq m \in M$, then $N = M \cap Km$ (observe: $M \subset KM$) is a pure submodule of M. Thus $M = N \oplus (Km + M)/Km$ (cf. (2.9)), and rank(M) = rank(N) + rank((Km + M)/Km). Since N is isomorphic to a fractional ideal, the result follows now from the induction hypothesis. #

7.5 Lemma: Every principal ideal domain R is a Dedekind domain.

Proof: Let R be a principal ideal domain. Then R is noetherian, since every ideal is principal, and it is easily seen, that every prime ideal in R is maximal.

It remains to show that R is integrally closed. Let R' be the integral closure of R in the quotient field K of R. If $x \in R'$, then

$x = r/r'$, $r' \neq 0$, (cf. (6.3)) and x satisfies a monic polynomial

$$X^n + r_{n-1}X^{n-1} + \ldots + r_o, \; r_i \in R.$$

Since the cancellation law holds in R, we may assume, that r and r' do not have a common factor. Then, if r' is not a unit in R, there exists a maximal ideal Rm such that $r' \in Rm$ but $r \notin Rm$ (cf. (4.1)). Moreover

$$r^n + r_{n-1}r'r^{n-1} + \ldots + r'^n r_o = 0.$$

This implies $r^n \in Rm$. Since Rm is a prime ideal, $r \in Rm$, a contradiction. Thus there does not exist a maximal ideal containing r'; i.e., r' is a unit in R. Hence $x \in R$, and R is integrally closed in K. #

7.6 <u>Corollary</u>: Every lattice M over a principal ideal domain is free on rank(M) basis elements.

<u>Proof</u>: This follows immediately from (7.3), since every fractional ideal over a principal ideal domain is a free module with a basis consisting of one element (cf. Ex. 7.5). #

7.7 <u>Theorem</u> <u>(Chinese remainder theorem)</u>: Let S be a ring, $I_1, \ldots, I_n$ left ideals in S such that $I_i + I_j = S$ for all i, j, $i \neq j$, and $I_iI_j = I_jI_i$. Given n elements $s_1, \ldots, s_n$ in S. There exists an element $s \in S$ such that $s \equiv s_i \bmod(I_i)$; $1 \leq i \leq n$.

<u>Proof</u>: We use induction on n. For $n = 2$, we have $1 = x_1 + x_2$, $x_1 \in I_1$, $x_2 \in I_2$. If we put $s = s_2x_1 + s_1x_2$, then $s \equiv s_1 \bmod(I_1)$ and $s \equiv s_2 \bmod(I_2)$. Let us assume that the theorem is true for $n - 1$, $n > 2$. For every $i \geq 2$, we can find elements $x_i \in I_1$, $y_i \in I_i$ such that $x_i + y_i = 1$, $i \geq 2$. Now $\prod_{i=2}^n (x_i+y_i) = 1$, and $\prod_{i=2}^n (x_i+y_i) \in I_1 + \prod_{i=2}^n I_i$, since $I_iI_j = I_jI_i$. Thus $I_1 + \prod_{i=2}^n I_i = S$, and $I_1(\prod_{i=2}^n I_i) = (\prod_{i=2}^n I_i)I_1$. Now we apply the theorem for $n = 2$. There exists an element $y_1 \in S$ such that $y_1 \equiv 1 \bmod(I_1)$ and $y_1 \equiv 0 \bmod(\prod_{i=2}^n I_i)$. Similarly one can find

elements $y_2,\dots,y_n$, such that $y_i \equiv 1 \bmod(I_i), y_i \equiv 0 \bmod(\prod_{j\neq i} I_j)$. But then also $y_i \equiv 0 \bmod(I_j)$, $j\neq i$. Now, we put $s = \prod_{i=1}^n s_i y_i$, and it is easily checked, that s has the desired properties. #

7.8 Lemma: Let R be a Dedekind domain, which has only finitely many prime ideals. Then R is a principal ideal domain.

Proof: We remark that a commutative ring with only finitely many maximal ideals is called a semi-local ring. Because of (7.2), it suffices to show, that every prime ideal in R is principal. Let $\underline{\underline{p}}_1,\dots,\underline{\underline{p}}_n$ be the prime ideals in R. From (7.2) it follows that $\underline{\underline{p}}_1 \supsetneq \underline{\underline{p}}_1^2$. Let $p' \in \underline{\underline{p}}_1$, $p' \notin \underline{\underline{p}}_1^2$. Since the ideals $\underline{\underline{p}}_i$, $1\leq i\leq n$, are maximal, they satisfy the hypotheses of (7.7). Thus, (cf. Ex. 7,4) we can find an element $p \in R$ such that $p \equiv p' \bmod(\underline{\underline{p}}_1^2)$ $p \equiv 1 \bmod(\underline{\underline{p}}_i)$, $i=2,\dots,n$. Now, by (7.2) there exist $\alpha_i \geq 0$, $1\leq i\leq n$, such that $Rp = \prod_{i=1}^n \underline{\underline{p}}_i^{\alpha_i}$, where we set $\underline{\underline{p}}^0 = R$. This implies $p \equiv 0 \bmod(\underline{\underline{p}}_i)$ for all i for which $\alpha_i \geq 1$. Hence $\alpha_i = 0$, for $i \geq 2$, and for $i = 1$, we have $p \not\equiv 0 \bmod(\underline{\underline{p}}_1^2)$ $p \equiv 0 \bmod(\underline{\underline{p}}_1)$. Thus $\alpha_1 = 1$ and $\underline{\underline{p}}_1 = Rp$. #

Exercises §7:

In all these exercises, let R be a Dedekind domain with quotient field K.

1.) If $\underline{\underline{a}}$ is fractional ideal in K, show that $\underline{\underline{a}}^{-1} \in {}_R\underline{\underline{M}}^o$.

2.) Let $\underline{\underline{a}}$ be an integral ideal in R. Show that in the decomposition of $\underline{\underline{a}}$ into prime ideals:

$$\underline{\underline{a}} = \prod_{i=1}^n \underline{\underline{p}}_i^{\alpha_i}, \text{ (cf. (7.2))}, \alpha_i > 0, 1\leq i\leq n.$$

(Hint: Use - without proving it - the fact, that $\underline{\underline{a}}^{-1}\underline{\underline{a}} = R$ in a Dedekind domain.)

3.) Show that for any integral ideal $\underline{\underline{a}}$ in R, there are only finitely many prime ideals containing $\underline{\underline{a}}$.

4.) Let $\underline{\underline{a}},\underline{\underline{b}}$ be integral ideals in R. Give a necessary and sufficient condition for $\underline{\underline{a}} + \underline{\underline{b}} = R$, (this is expressed by saying, that $\underline{\underline{a}}$ and $\underline{\underline{b}}$ are <u>relatively prime</u>; notation $(\underline{\underline{a}},\underline{\underline{b}}) = 1$), in terms of the decomposition into prime ideals (7.2). Use this to show that $(\underline{\underline{a}},\underline{\underline{b}}) = 1$ implies $(\underline{\underline{a}}^n,\underline{\underline{b}}^m) = 1$, m,n positive integers. Let $\underline{\underline{a}}$ and $\underline{\underline{b}}$ be relatively prime ideals in R. Show that $\underline{\underline{a}} \cap \underline{\underline{b}} = \underline{\underline{ab}}$.

5.) Show that every fractional ideal in K is isomorphic to an integral ideal.

6.) Prove <u>Gauss' lemma</u>: Let R be a principal ideal domain and X an indeterminate over R. A polynomial $f(X) \in R[X]$ is called <u>primitive</u>, if the greatest common divisor of its coefficients is 1. The product of two primitive polynomials f(X) and g(X) is primitive. (Hint: If not, then $f(X)\cdot g(X)$ would be contained in $\underline{\underline{p}}R[X]$, where $\underline{\underline{p}}$ is a maximal ideal in R. Now consider congruences modulo $\underline{\underline{p}}$.)

§8 Localization of Dedekind domains

A Dedekind domain localizes to a Dedekind domain. A lattice is the intersection of all its localizations at prime ideals. A correspondence between the lattices over a Dedekind domain and over its localizations at the prime ideals is set up. The primary decomposition of a finite torsion module is derived.

8.1 **Lemma**: Let R be a commutative ring and S a multiplicative system in R; $\varphi : R \longrightarrow R_S$ is the canonical homomorphism of §6. Then

(i) $(\underline{a}(S) \cap (R)\varphi)_S = \underline{a}(S)$ for every ideal $\underline{a}(S)$ of R_S,

(ii) $(\underline{a})\varphi \subset \underline{a}_S \cap (R)\varphi$ for every ideal $\underline{a}$ of R,

(iii) if $\underline{a}$ is an ideal of R, then $\underline{a}_S = R_S \iff \underline{a} \cap S \neq \emptyset$.

The proofs are trivial. #

For the remainder of this section we shall assume that R is a Dedekind domain (cf. (7.1)).

8.2 **Corollary**: There is a one-to-one correspondence between the maximal ideals $\underline{m}$ of R such that $S \cap \underline{m} = \emptyset$ and the maximal ideals $\underline{m}'$ in R_S:

$$\Phi : \underline{m} \longrightarrow \underline{m}_S ; \ \Psi : \underline{m}' \longrightarrow \underline{m}' \cap R.$$

Proof: It is easily seen, that Φ and Ψ establish a one-to-one correspondence between prime ideals (cf. Ex. 8,4). But, over R the prime ideals are precisely the maximal ideals. #

8.3 **Lemma**: Let $\underline{p}_1, \dots, \underline{p}_n$ be a finite set of prime ideals in R. If $S = R \setminus \{\bigcup_{i=1}^{n} \underline{p}_i\}$ then R_S is a principal ideal domain.

Proof: By (8.2), the maximal ideals in R_S are $\underline{p}_{1_S}, \dots, \underline{p}_{n_S}$. Let $\underline{a}$ be any ideal in R_S. Then it follows from (7.2), that $\underline{a}_S = \prod_{i=1}^{n} \underline{p}_{i_S}^{\alpha_i}$, and $\underline{p}_{i_S} \underset{\neq}{\supset} \underline{p}_{i_S}^2$. Thus the proof of (7.8) is valid also in this situation and R_S is a principal ideal domain. In particular, R_S is a Dedekind domain (cf. (7.5)). #

8.4 Lemma:

(i) Let $\underline{\underline{p}}_1$, $\underline{\underline{p}}_2$ be prime ideals in R. If $\underline{\underline{p}}_1 = \underline{\underline{p}}_2$ then $(R_{\underline{\underline{p}}_1})_{\underline{\underline{p}}_2} = R_{\underline{\underline{p}}_1}$.

If $\underline{\underline{p}}_1 \neq \underline{\underline{p}}_2$, then $(R_{\underline{\underline{p}}_1})_{\underline{\underline{p}}_2} = (R_{\underline{\underline{p}}_2})_{\underline{\underline{p}}_1} = K$, the quotient field of R.

(ii) Let $S \subset S'$ be two multiplicative systems in R. Then $(R_S)_{S'} = R_{S'}$.

Proof: We point out that actually, we have only $(R_S)_{S'} \cong R_{S'}$. But since this isomorphism is natural, we identify both structures. The notation should be understood as $(R_{\underline{\underline{p}}_1})_{\underline{\underline{p}}_2} \cong R_{\underline{\underline{p}}_2} \otimes_R (R_{\underline{\underline{p}}_1} \otimes_R R)$, (cf. (6.4)), and $(R_S)_{S'} \cong R_{S'} \otimes_R (R_S \otimes_R R)$. This makes sense, since R_S is an R-module (cf. (6.3) and (1.3)). Moreover, $R_{\underline{\underline{p}}}$ is a subring of K in a natural way (cf. (5.2)). The verification of (8.4) is left as an exercise. #

8.5 Remark: Similar statements hold for modules over R.

8.6 Theorem: Let $\underline{\underline{S}}$ be the set of all prime ideals in R. If M is an R-lattice (cf. (7.1)), then $M = \bigcap_{\underline{\underline{p}} \in \underline{\underline{S}}} M_{\underline{\underline{p}}}$.

Remark: Let S, S' be two multiplicative systems in R, with $S \subset S'$. For $M \in {}_{\underline{\underline{R}}}\underline{\underline{M}}^o$, we can consider M_S in a natural way as a submodule of $M_{S'}$ (cf. the construction of $M_{S'}$); similarly $M_{\underline{\underline{p}}} \subset KM$.

Proof of (8.6): The map

$$\varphi : M \longrightarrow \bigcap_{\underline{\underline{p}} \in \underline{\underline{S}}} M_{\underline{\underline{p}}} \; ; \; m \longmapsto m/1$$

is a monomorphism, and thus, we can consider $M \subset \bigcap_{\underline{\underline{p}} \in \underline{\underline{S}}} M_{\underline{\underline{p}}} \subset M_{\underline{\underline{p}}} \subset KM$.

Now let $0 \neq x = \frac{m}{r} \in \bigcap_{\underline{\underline{p}} \in \underline{\underline{S}}} M_{\underline{\underline{p}}}$, and assume, that $0 \neq r \in R$ is not a unit in R. Then $rR \neq R$, and there are only finitely many prime ideals $\underline{\underline{p}}_1, \dots, \underline{\underline{p}}_n$ (cf. Ex. 7,3), containing rR. But also $x \in M_{\underline{\underline{p}}_i}$, $1 \leq i \leq n$;

i.e., $x = m_1/r_1$, $m_1 \in M$, $0 \neq r_1 \in R$, $r_1 \notin \underline{\underline{p}}_1$. We claim, that the ideal $\underline{\underline{a}}$, generated by $r, r_1, \dots, r_n$ is all of R. Assume $\underline{\underline{a}} \neq R$, then there is a maximal ideal $\underline{\underline{p}} \supset \underline{\underline{a}}$. Since $r \in \underline{\underline{a}} \subset \underline{\underline{p}}$, $\underline{\underline{p}} = \underline{\underline{p}}_i$ for some $1 \leq i \leq n$. But $r_i \in \underline{\underline{a}} \subset \underline{\underline{p}}$. Hence we have arrived at a contradiction, and $\underline{\underline{a}} = R$. Thus we have a relation $1 = \alpha r + \sum_{i=1}^{n} \alpha_i r_i$; $\alpha, \alpha_i \in R$. But then $m = \alpha r m + \sum_{i=1}^{n} \alpha_i r_i m = r(\alpha m + \sum_{i=1}^{n} \alpha_i m_i)$, since $r_i m = r m_i$; i.e., $m/r \in M$. #

8.7 __Lemma:__ Let M and N be R-lattices, such that $KM = KN$. Then $M_{\underline{\underline{p}}} = N_{\underline{\underline{p}}}$ for almost all $\underline{\underline{p}} \in \underline{\underline{S}}$.

__Proof:__ M and N are generated by, say, $\{m_i\}_{1 \leq i \leq s}$ and $\{n_j\}_{1 \leq j \leq t}$ resp. Since $KM = KN$, there exist $\{k_{ij} \in K : 1 \leq i \leq s, 1 \leq j \leq t\}$ such that $m_i = \sum_{j=1}^{n} k_{ij} n_j$, $1 \leq i \leq s$. However, for almost all $\underline{\underline{p}} \in \underline{\underline{S}}$, $k_{ij} \in R_{\underline{\underline{p}}}$, $1 \leq i \leq s$, $1 \leq j \leq t$, i.e., for almost all $\underline{\underline{p}} \in S$, $M_{\underline{\underline{p}}} \subset N_{\underline{\underline{p}}}$. Similarly, we have for almost all $\underline{\underline{p}} \in \underline{\underline{S}}$, $N_{\underline{\underline{p}}} \subset M_{\underline{\underline{p}}}$. #

8.8 __Lemma:__ Let $\{M(\underline{\underline{p}})\}_{\underline{\underline{p}} \in \underline{\underline{S}}}$ be a family of $R_{\underline{\underline{p}}}$-lattices, such that $KM(\underline{\underline{p}}) = V$ is the same for all $\underline{\underline{p}} \in \underline{\underline{S}}$. If there exists an R-lattice N in V, such that $N_{\underline{\underline{p}}} = M(\underline{\underline{p}})$ for almost all $\underline{\underline{p}} \in \underline{\underline{S}}$, then there exists an R-lattice M in V such that $M_{\underline{\underline{p}}} = M(\underline{\underline{p}})$ for all $\underline{\underline{p}} \in \underline{\underline{S}}$.

__Proof:__ Since $N_{\underline{\underline{p}}} = M(\underline{\underline{p}})$ for almost all $\underline{\underline{p}} \in \underline{\underline{S}}$, we can replace N by $r^{-1}N$, where $0 \neq r \in R$, is such that $r^{-1}N_{\underline{\underline{p}}} \supset M(\underline{\underline{p}})$ for every $\underline{\underline{p}} \in \underline{\underline{S}}$. Let $\underline{\underline{p}}_1, \dots, \underline{\underline{p}}_n$ be the prime ideals such that $N_{\underline{\underline{p}}_i} \neq M(\underline{\underline{p}}_i)$, $1 \leq i \leq n$, and put

$$M = N \cap M(\underline{\underline{p}}_1) \cap M(\underline{\underline{p}}_2) \cap \dots \cap M(\underline{\underline{p}}_n). \qquad \#$$

8.9 __Theorem:__ Let M and N be R-lattices such that $KN = KM$ and $N \subset M$. Then $M/N \cong \bigoplus_{\underline{\underline{p}} \in \underline{\underline{S}}} M_{\underline{\underline{p}}}/N_{\underline{\underline{p}}}$. This sum is finite and $M_{\underline{\underline{p}}}/N_{\underline{\underline{p}}}$ is called the

$\underline{p}$-primary component M/N.

Proof: Since KM = KN, M/N = X is an R-torsion module of finite type (cf. (7.1)), and hence, $\mathrm{ann}_R(X) = \prod_{i=1}^{n} \underline{p}_i^{\alpha_i}$, $\alpha_i > 0$, $\underline{p}_i \in \underline{S}$ (cf. (7.2)). We shall show by induction on n, that

$$M/N \cong \oplus_{i=1}^{n} M_{\underline{p}_i}/N_{\underline{p}_i}.$$

For n = 1, $(M/N)_{\underline{p}} = 0$ for every $\underline{p} \neq \underline{p}_1$, and $X_{\underline{p}_1} = X$. For this it suffices to show that for every $x \in X$, $s \in R\setminus\{\underline{p}_1\}$, there exists $x' \in X$ with $x = sx'$. But this is easily seen, since $(\underline{p}_1^{\alpha_1}, R\cdot s) = 1$. Moreover, $R_{\underline{p}_1}$ being a flat R-module (cf. (6.5)), $(M/N)_{\underline{p}_1} \cong M_{\underline{p}_1}/N_{\underline{p}_1}$. Thus, the statement is true for n = 1. Now, given X = M/N with n > 1, we set $X_1 = \{x \in X : \underline{p}_1^{\alpha_1}x = 0\} \subsetneq X$; then the canonical exact sequence $0 \longrightarrow X_1 \xrightarrow{\sigma} X \longrightarrow X/X_1 \longrightarrow 0$ splits; for, let $a \in \prod_{i=2}^{n} \underline{p}_i^{\alpha_i}$, then $(Ra, \underline{p}_1^{\alpha_1}) = 1$, and there exists $r \in R$ such that $rax_1 = x_1$ for every $x_1 \in X_1$. Now, we define $\tau : X \longrightarrow X_1$, $x \longmapsto rax$. Then $\sigma\tau = 1_{X_1}$ and $X \cong X_1 \oplus X/X_1$. Since $\mathrm{ann}_R(X/X_1) = \prod_{i=2}^{n} \underline{p}_i^{\alpha_i}$, (8.9) follows now by induction. #

Exercises §8:

In exercises 1, 2 and 3, R is a Dedekind domain with quotient field K.

1.) Let $M \in {}_R\underline{M}$ be an R-torsion module, and $N \in {}_R\underline{M}^o$. Show that $\mathrm{Hom}_R(M,N) = 0$.

2.) Let $X,Y \in {}_R\underline{M}^o$ such that KX = KY. Show that

(i) $R_{\underline{p}}(X+Y) = R_{\underline{p}}X + R_{\underline{p}}Y$ and, (ii) $R_{\underline{p}}(X\cap Y) = R_{\underline{p}}X \cap R_{\underline{p}}Y$.

3.) Let $M \in {}_R\underline{M}^f$ and set $tM = \{m \in M : \exists\, 0 \neq r \in R,\ rm = 0\}$; tM is called the torsion part of M. Show that

$tM = \oplus_{\underline{p}\in\underline{S}} R_{\underline{p}} \otimes_R tM$, and $M = M/tM \oplus tM$ with $M/tM \in {}_R\underline{P}^f$.

4.) Let R be a commutative ring and S a multiplicative system in R. Show that there is a one-to-one correspondence between the prime ideals of R that do not meet S and the prime ideals of R_S.

§ 9 Completions

Ideal-adic completions are introduced via projective limits. If the module is hausdorff, this completion is the topological ideal-adic completion. The completion functor is flat on hausdorff modules of finite type. If R is a Dedekind domain and $\underline{\underline{p}}$ a prime ideal in R, then the $\underline{\underline{p}}$-adic completion $\hat{R}_{\underline{\underline{p}}}$ is flat on R-modules of finite type, and $\hat{R}_{\underline{\underline{p}}}$ is also the completion of the localization. There is a one-to-one correspondence between the $R_{\underline{\underline{p}}}$-lattices and the $\hat{R}_{\underline{\underline{p}}}$-lattices. The results of §8 remain valid for completions.

9.1 <u>Definition</u>: A partially ordered set $(S,\prec)$ is called a <u>directed set</u>, if for every pair $\alpha, \beta \in S$, there exists $\gamma \in S$ with $\alpha \prec \gamma$ and $\beta \prec \gamma$. If R is a ring, and if $M_\alpha \in {}_{\underline{\underline{R}}}\underline{\underline{M}}^f$, $\alpha \in S$, and $\pi_{\beta\alpha} \in \mathrm{Hom}_R(M_\beta, M_\alpha)$ for $\alpha \prec \beta \in S$ are given, then $\{M_\alpha, \pi_{\beta\alpha}\}_{\alpha \prec \beta \in S}$ is called a <u>projective system of left R-modules</u>, if for $\alpha \prec \beta \prec \gamma \in S$

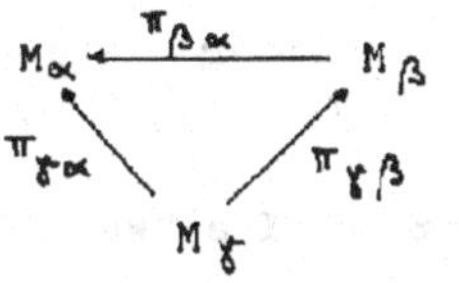

commutes,

and if $\pi_{\alpha\alpha} = 1_{M_\alpha}$, $\alpha \in S$.

9.2 <u>Universal problem of the projective limit</u>: Given a projective system $\{M_\alpha, \pi_{\beta\alpha}\}_{\alpha\prec\beta \in S}$; does there exist $M \in {}_{\underline{\underline{R}}}\underline{\underline{M}}$ and $\pi_\alpha \in \mathrm{Hom}_R(M, M_\alpha)$, $\alpha \in S$, with the following universal property: Whenever $N \in {}_{\underline{\underline{R}}}\underline{\underline{M}}$ and $\chi_\alpha \in \mathrm{Hom}_R(N, M_\alpha)$, $\alpha \in S$, are such that

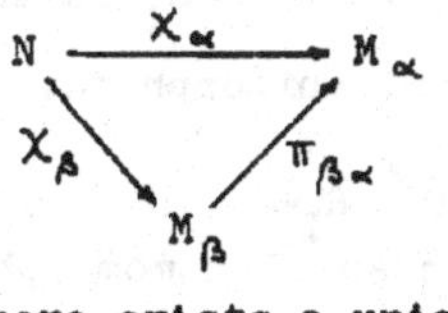

commutes for $\alpha \prec \beta \in S$, there exists a unique $\chi \in \mathrm{Hom}_R(N,M)$ completing the diagram

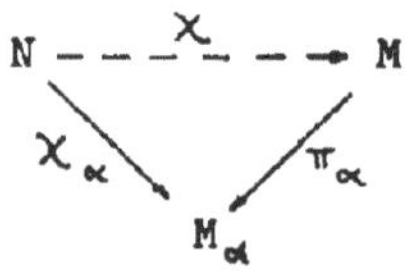

for every $\alpha \in S$.

9.3 <u>Lemma</u>: The universal problem (9.2) has an - up to natural isomorphism - unique solution M, called the <u>projective limit</u> of $\{M_\alpha, \pi_{\beta\alpha}\}_{\alpha<\beta\in S}$; notation: $M = \varprojlim M_\alpha$, $\chi = \varprojlim \chi_\alpha$.

<u>Proof</u>: In the product of the $\{M_\alpha\}_{\alpha\in S}$, $P = \prod_{\alpha\in S} M_\alpha$, with projections $\pi'_\alpha : P \longrightarrow M_\alpha$ (cf. Ex. 1,2) we consider the submodule $M = \{(m_\alpha)_{\alpha\in S} \in P : [(m_\alpha)_{\alpha\in S}]\,\pi'_\beta = [(m_\alpha)_{\alpha\in S}]\,\pi'_\gamma \pi_{\gamma\beta}, \beta<\gamma\in S\}$ $= \{(m_\alpha)_{\alpha\in S} \in P : m_\beta = m_\gamma \pi_{\gamma\beta}\ , \beta<\gamma\in S\}$. If we put $\pi_\alpha = \pi'_\alpha|_M \in \mathrm{Hom}_R(M, M_\alpha)$, and define $\chi : N \longrightarrow M$ by $\chi : n \longmapsto (n\chi_\alpha)_{\alpha\in S} \in M$, then it is easily checked, that $(M, \pi_\alpha)_{\alpha\in S}$ is the unique solution of (9.2). #

For our purposes it suffices to consider a special type of projective limit.

9.4 <u>Definition</u>: Let R be a ring and I a two-sided ideal in R. For $M \in {}_R\underline{\underline{M}}$, we define

$$\pi_{n,m} : M/I^nM \longrightarrow M/I^mM \text{ for } m \leq n \in \underline{\underline{N}}$$
$$\pi_{n,m} : m+I^nM \longmapsto m+I^mM.$$

Then $\{(M/I^mM), \pi_{n,m}\}_{m\in\underline{\underline{N}}}$ is a projective system of left R-modules. We call $\hat{M}_I = \varprojlim(M/I^mM)$ the <u>I-adic completion of M</u>, and M is said to be <u>I-complete</u>, if $\pi = \varprojlim \pi_m : M \longrightarrow \hat{M}_I$ is an isomorphism, where $\pi_m : M \longrightarrow M/I^mM$ is the canonical epimorphism.

9.5 <u>Lemma</u>: $\hat{R}_I$ is a ring and $\hat{M}_I \in {}_{\hat{R}_I}\underline{\underline{M}}$.

<u>Proof</u>: For $\hat{r} \in \hat{R}_I$, $m \in \underline{\underline{N}}$, we have an R-homomorphism

$$\chi_m(\hat{r}) : {}_R\hat{R}_I \longrightarrow R/I^m, \quad \hat{x} \longmapsto \hat{x}\pi_m \cdot \hat{r}\pi_m \ , \text{ i.e.,}$$

$\chi_m(\hat{r}) = \pi_m(\hat{r}\,\pi_m)$. Since $\pi_{n,m} : R/I^n \longrightarrow R/I^m$ is a ring homomorphism, the following diagram is commutative

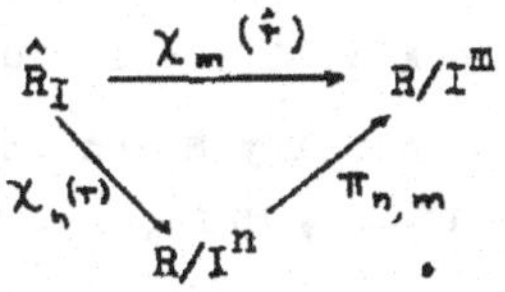

From (9.2) we obtain an R-homomorphism $\chi(\hat{r}) : \hat{R}_I \longrightarrow \hat{R}_I$ which acts as right multiplication by $\hat{r}$ in $\hat{R}_I$. Similarly one makes $\hat{M}_I$ into a left $\hat{R}_I$-module. #

9.6 Definitions: For $M \in {}_{\underline{\underline{R}}}\underline{\underline{M}}$ and a two-sided ideal I we make M into a topological space under the I-adic topology. A base of neighborhoods of $0 \in M$ is given by the sets $I^n M$, $n \in \underline{\underline{N}}$. The neighborhoods of $m \in M$ are obtained by translations. M is hausdorff under this topology if and only if $\bigcap_{n \in \underline{\underline{N}}} I^n M = 0$; we then say that M is I-hausdorff, and I is called hausdorff if $\bigcap_{n \in \underline{\underline{N}}} I^n = 0$. If M is I-hausdorff, then the following distance function

$$d : M \times M \longrightarrow \underline{\underline{R}},\ d(m,m) = 0 \text{ and}$$
$$d(m,m') = 2^{-n} \text{ if and only if } m-m' \in I^n M \setminus I^{n+1} M$$

makes M into a metric R-module. The completion (via Cauchy sequences) of M with respect to d is called the topological I-adic completion. It is easily seen (cf. Ex. 9.2), that for an I-hausdorff module, the I-adic completion is naturally isomorphic to the topological I-adic completion.

9.7 Lemma: If $M \in {}_{\underline{\underline{R}}}\underline{\underline{M}}$ is I-hausdorff, we have a canonical monomorphism $\chi : M \longrightarrow \hat{M}_I$. If, in addition, I is hausdorff and $M \in {}_{\underline{\underline{R}}}\underline{\underline{M}}^f$, we have $\hat{R}_I M = \hat{M}_I$, provided we identify M and $\operatorname{Im}\chi$.

Proof: Let $\chi_n : M \longrightarrow M/I^n M$ be the canonical epimorphism, and put $\chi = \varprojlim \chi_n$. If $m \in \operatorname{Ker}\chi$, then the commutativity of

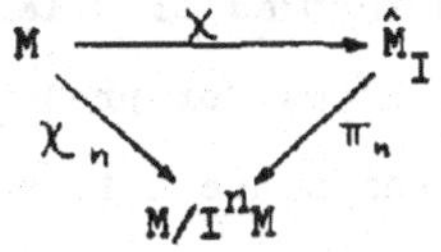

shows that $m \in I^nM$ for every $n \in \underline{\underline{N}}$; i.e., $m = 0$, since M is I-hausdorff; and χ is monic. Let us identify M and $\mathrm{Im}\,\chi$; then $\hat{R}_I M \subset \hat{M}_I$. Let $M, M' \in {}_{R}\underline{\underline{M}}$ and let $\pi_{mn}: M/I^m \longrightarrow M/I^n$, $\pi_n: \hat{M}_I \longrightarrow M/I^nM$ be the homomorphisms defined in (9.4), similarly for π'_{mn} and π'_n. For $n \in \underline{\underline{N}}$, $\sigma \in \mathrm{Hom}_R(M,M')$, we denote by σ_n the homomorphism induced from $1_{R/I^n} \otimes \sigma$. Since $\sigma_m \pi'_{mn} = \pi_{mn}\sigma_n$, $n<m \in \underline{\underline{N}}$, the maps $\pi_n \sigma_n: \hat{M}_I \longrightarrow M'/I^nM'$ satisfy the conditions of (9.3) and we may define $\hat{\sigma} = \varprojlim \pi_n \sigma_n$. $\hat{\sigma}$ is then the unique $x \in \mathrm{Hom}_{\hat{R}_I}(\hat{M}_I, \hat{M}'_I)$ such that $\pi_n \sigma_n = x \pi'_n$. It is easy to show, using the universality of the projective limit, that $\hat{\sigma}\hat{\tau} = \widehat{\sigma\tau}$ for $\tau \in \mathrm{Hom}_R(M',M'')$. Moreover, if σ is an epimorphism then so is $\hat{\sigma}$. For, in general $\{\mathrm{Coker}\,\sigma_n, \bar{\pi}'_{mn}\}_{n<m\in \underline{\underline{N}}}$ *) is a projective system and we have $\varprojlim \mathrm{Coker}\,\sigma_n = \mathrm{Coker}\,\hat{\sigma}$, $\varprojlim \pi'_n \mathrm{coker}\,\sigma_n = \mathrm{coker}\,\hat{\sigma}$ (cf. Ex. 9,4). But, if σ is epic then so is every σ_n, $n \in \underline{\underline{N}}$, because the tensor product is right exact, and it follows that $\mathrm{Coker}\,\hat{\sigma} = 0$, i.e., $\hat{\sigma}$ is an epimorphism. We assume now, that $M \in {}_{R}\underline{\underline{M}}^f$ is I-hausdorff and that I is hausdorff, and we take a free module $R^{(t)}$ which maps onto M, and we obtain the commutative diagram

$$\begin{array}{ccccc}
\hat{R}_I \otimes_R R^{(t)} & \longrightarrow & \hat{R}_I \otimes_R M & \longrightarrow & 0 \\
\downarrow & & \downarrow & & \\
\hat{R}_I^{(t)} & \longrightarrow & \hat{R}_I M & & \\
\downarrow & & \downarrow & & \\
\widehat{R^{(t)}}_I & \longrightarrow & \hat{M}_I & \longleftarrow & 0 .
\end{array}$$

But is is easily seen, that $\hat{R}_I^{(t)} \overset{nat}{\cong} \widehat{R^{(t)}}_I$, and diagram chasing shows that $\hat{R}_I M = \hat{M}_I$. #

9.8 <u>Lemma</u>: If R is left noetherian and if $M \in {}_{R}\underline{\underline{M}}^f$ is I-hausdorff, then $\hat{R}_I \otimes_R M \overset{nat}{\cong} \hat{M}_I$. (It should be remarked, that it suffices to assume M of finite presentation instead of R left noetherian.)

<u>Proof</u>: From (9.7), the result follows for projective left R-modules of finite type. Since $M \in {}_{R}\underline{\underline{M}}^f$, and since R is noetherian, we can find $P_1, P_2 \in {}_{R}\underline{\underline{P}}^f$ such that

---

*) We define $\bar{\pi}'_{mn}: \mathrm{Coker}\,\sigma_m \longrightarrow \mathrm{Coker}\,\sigma_n$ by $x + \mathrm{Im}\,\sigma_m \longmapsto x\pi'_{mn} + \mathrm{Im}\,\sigma_n$.

$$P_2 \xrightarrow{\sigma} P_1 \xrightarrow{\tau} M \longrightarrow 0$$

is an exact sequence of left R-modules. From the commutativity of the diagram with exact top row

$$\begin{array}{ccccccc}
\hat{R}_I \otimes_R P_2 & \longrightarrow & \hat{R}_I \otimes_R P_1 & \longrightarrow & \hat{R}_I \otimes_R M & \longrightarrow & 0 \\
\downarrow \wr & & \downarrow \wr & & \downarrow \varphi & & \\
\hat{P}_2 & \xrightarrow{\hat{\sigma}} & \hat{P}_1 & \xrightarrow{\hat{\tau}} & \hat{M}_I & \longrightarrow & 0 \\
 & & & & \downarrow & & \\
 & & & & 0 & &
\end{array}$$

we conclude, that φ is an isomorphism. From the universal property it follows that φ is natural. #

9.9 <u>Lemma</u>: If R is noetherian, then $\hat{R}_I \otimes_R -$ is flat on the class of I-hausdorff modules of finite type.

<u>Proof</u>: It suffices to show that for a monic map $\sigma : M' \longrightarrow M$ of two I-hausdorff modules of finite type, $1_{\hat{R}_I} \otimes \sigma$ is monic. But this is an immediate consequence of (9.8) and the commutative diagram

$$\begin{array}{ccccc}
 & & \hat{R}_I \otimes_R M' & \longrightarrow & \hat{R}_I \otimes_R M \\
 & & \downarrow \wr & & \downarrow \wr \\
 & & \hat{M}'_I & \longrightarrow & \hat{M}_I \\
 & & \downarrow \wr & & \downarrow \wr \\
0 & \longrightarrow & \hat{R}_I M' & \longrightarrow & \hat{R}_I M
\end{array} \quad . \quad \#$$

9.10 <u>Lemma</u>: Let R be left noetherian, I-hausdorff and $M \in {}_R\underline{\underline{M}}$ I-hausdorff of finite type. Then $\hat{M}_I$ is $\hat{I}_I$-complete and $\hat{M}_I$ is topologically $\hat{I}_I$-complete.

<u>Proof</u>: We shall first show that

$$\varprojlim_m [(\varprojlim_n M/I^nM)/(\varprojlim_n I/I^nI)^m \varprojlim_n M/I^nM] \cong \varprojlim M/I^nM.$$

Because of (9.8) and (9.9), if suffices to show $\varprojlim M/I^nM \cong$ $\cong \varprojlim(\hat{R}_I \otimes_R M/I^nM)$. But $\hat{R}_I \otimes_R M/I^nM \cong \hat{R}_I \otimes_R R/I^n \otimes_R M \cong \widehat{R/I^n}_I \otimes_R M$. since R/I^n is I-hausdorff. But it is easily seen that $\widehat{R/I^n}_I = R/I^n$. Thus, $\hat{R}_I \otimes_R M/I^nM = M/I^nM$. This shows that $\hat{M}_I$ is I-complete. Let M_I^* be the topological completion of M in the I-adic topology (cf. (9.6)). Then the following map is easily checked to be an isomorphism

$$\varphi: M_I^{\#} \longrightarrow \varprojlim M/I^n M$$

$$\varphi: \langle m_n \rangle \longmapsto (m_n + I^n M)_{n \in \mathbb{N}},$$

where $\langle m_n \rangle$ denotes the equivalence class of the Cauchy sequence $(m_n)_{n \in \mathbb{N}}$. #

9.11 <u>Lemma</u>: Let I_1 and I_2 be two-sided ideals in R such that $I_1^{n_1} \subset I_2$ and $I_2^{n_2} \subset I_1$. If $M \in {}_R\mathfrak{M}$ is I_1-hausdorff, then M is I_1-complete if and only if M is I_2-complete.

<u>Proof</u>: Let $I_1^{n_1} \subset I_2$ and assume M is I_2-complete. Then the commutative diagram

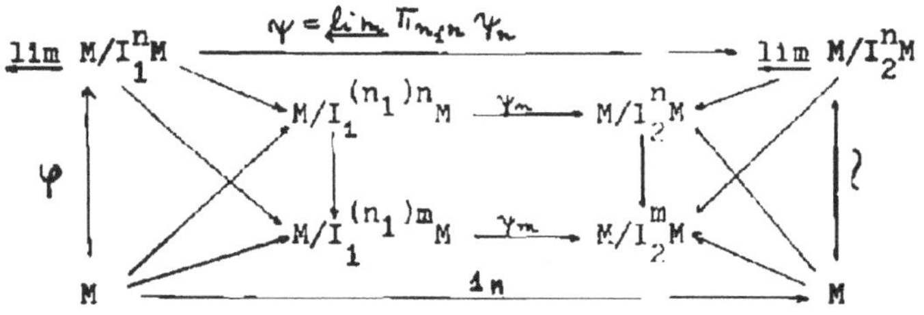

shows that φ is an isomorphism by the universality of $\varprojlim$.. #

9.12 <u>Lemma</u>: Let R be a noetherian integral domain, and $\underline{\underline{m}}$ a maximal ideal in R. Then $\underline{\underline{m}}$ is hausdorff, and $\hat{R}_{\underline{\underline{m}}}$ is the $\underline{\underline{m}}R_{\underline{\underline{m}}}$-completion of $R_{\underline{\underline{m}}}$.

<u>Proof</u>: Since R is noetherian, $\underline{\underline{m}} \cdot (\bigcap_{n \in \mathbb{N}} \underline{\underline{m}}^n) = \bigcap_{n \in \mathbb{N}} \underline{\underline{m}}^n$, as follows from Herstein's lemma (cf. Ex. 9,1). If we localize at $R \setminus \underline{\underline{m}}$ and apply Nakayama's lemma, we get $\bigcap_{n \in \mathbb{N}} \underline{\underline{m}}^n = 0$; i.e., $\underline{\underline{m}}$ is hausdorff. Thus, $R \subset \hat{R}_{\underline{\underline{m}}}$, and one finds, that the elements of $R \setminus \underline{\underline{m}}$ are invertible in $R_{\underline{\underline{m}}}$. From the universal property of the localization (6,2) we get the commutative diagram

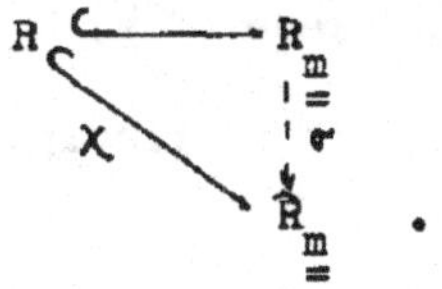

Since χ is a monomorphism, so is σ. Moreover, from the proof of (8.9) it follows, that $R/\mathfrak{m} \cong R_{\mathfrak{m}}/\mathfrak{m}R_{\mathfrak{m}}$. Thus, $\hat{R}_{\mathfrak{m}}$ is also the $\mathfrak{m}R_{\mathfrak{m}}$-completion of $R_{\mathfrak{m}}$. #

9.13 <u>Lemma</u>: Let R be a Dedekind domain and $\mathfrak{p}$ a prime ideal in R, then

(i) every R-lattice is $\mathfrak{p}$-hausdorff,

(ii) $\hat{R}_{\mathfrak{p}}$ is a local principal ideal domain with quotient field $\hat{K}_{\mathfrak{p}} =: \hat{R}_{\mathfrak{p}} \otimes_R K$.

<u>Proof</u>: (i) is proved like (9.12). $\hat{R}_{\mathfrak{p}} \hookrightarrow \hat{R}_{\mathfrak{p}} \otimes_R K = (\hat{R}_{\mathfrak{p}})_S$ where $S = R \setminus \{0\}$ is a multiplicative system in $\hat{R}_{\mathfrak{p}}$. Since $\mathfrak{p}$ is hausdorff, $K \hookrightarrow (\hat{R}_{\mathfrak{p}})_S$. We can extend the $\mathfrak{p}$-adic topology on $\hat{R}_{\mathfrak{p}}$ to all of $(\hat{R}_{\mathfrak{p}})_S$ in a natural way.

Let $\hat{I}$ be a maximal ideal in $(\hat{R}_{\mathfrak{p}})_S$. Then $\hat{I} \cap \hat{R}_{\mathfrak{p}} = \hat{I}_{\mathfrak{p}}$ is a prime ideal in $\hat{R}_{\mathfrak{p}}$ (cf. Ex. 8,4). Then $I_{\mathfrak{p}} = R_{\mathfrak{p}} \cap \hat{I}_{\mathfrak{p}}$ is a dense subspace of $\hat{I}_{\mathfrak{p}}$, and thus, $\widehat{(I_{\mathfrak{p}})} = \hat{I}_{\mathfrak{p}}$, since $\hat{I}_{\mathfrak{p}} = \hat{R}_{\mathfrak{p}} \otimes_{R_{\mathfrak{p}}} I_{\mathfrak{p}}$ (cf. (9.8)). Hence $I_{\mathfrak{p}} = \mathfrak{p} \cdot R_{\mathfrak{p}}$ since in $R_{\mathfrak{p}}$ every ideal is a power of $\mathfrak{p}R_{\mathfrak{p}}$. But then $\hat{I} \cap (R \setminus \{0\}) \neq \emptyset$ since R is dense in $\hat{R}_{\mathfrak{p}}$; i.e., $\hat{I} = (\hat{R}_{\mathfrak{p}})_S$, a contradiction. Thus, $(\hat{R}_{\mathfrak{p}})_S$ is a field and $\hat{R}_{\mathfrak{p}}$ is an integral domain. Consequently, $(\hat{R}_{\mathfrak{p}})_S$ contains the quotient field of $\hat{R}_{\mathfrak{p}}$; but it is clear, that the quotient field of $\hat{R}_{\mathfrak{p}}$ has to contain $(\hat{R}_{\mathfrak{p}})_S$. If now $\hat{\mathfrak{a}}$ is an ideal in $\hat{R}_{\mathfrak{p}}$, then $\mathfrak{a} = R_{\mathfrak{p}} \cap \hat{\mathfrak{a}}$ is such that $\hat{R}_{\mathfrak{p}}\mathfrak{a} \subset \hat{\mathfrak{a}}$; but $\hat{R}_{\mathfrak{p}}\mathfrak{a} = \hat{R}_{\mathfrak{p}} \otimes_R \mathfrak{a}$ is the completion of $\underline{\underline{a}}$, and since $\underline{\underline{a}}$ is a dense subspace of $\hat{\underline{\underline{a}}}$, $\hat{R}_{\underline{\underline{p}}}\underline{\underline{a}} = \hat{\underline{\underline{a}}}$. But $R_{\underline{\underline{p}}}$ is a principal ideal domain, and hence $\hat{R}_{\underline{\underline{p}}}$ is a principal ideal domain. In particular, $\hat{R}_{\underline{\underline{p}}}$ is local. #

9.14 <u>Theorem</u>: Let R be a Dedekind domain with quotient field K and $\underline{\underline{p}}$ a prime ideal in R. If V is a finite dimensional K-vectorspace then there is a one-to-one, inclusion preserving correspondence between the $R_{\underline{\underline{p}}}$-lattices in V and the $\hat{R}_{\underline{\underline{p}}}$-lattices in $\hat{V}_{\underline{\underline{p}}} = \hat{R}_{\underline{\underline{p}}} \otimes_R V$. The correspondence is given by: $M_{\underline{\underline{p}}} \longmapsto \hat{M}_{\underline{\underline{p}}}$, $\hat{N}_{\underline{\underline{p}}} \longmapsto V \cap N_{\underline{\underline{p}}}$.

<u>Proof</u>: For an $R_{\underline{\underline{p}}}$-lattice $M_{\underline{\underline{p}}}$ in V we have the $\underline{\underline{p}}R_{\underline{\underline{p}}}$-adic topology on $M_{\underline{\underline{p}}}$, which can be extended to a topology on V. Similarly for an $\hat{R}_{\underline{\underline{p}}}$-lattice $\hat{N}_{\underline{\underline{p}}}$ in $\hat{V}_{\underline{\underline{p}}}$. If $M_{\underline{\underline{p}}}$ is an $R_{\underline{\underline{p}}}$-lattice in V, then $\hat{M}_{\underline{\underline{p}}}$ is an $\hat{R}_{\underline{\underline{p}}}$-lattice in $\hat{V}_{\underline{\underline{p}}}$. Moreover, $M'_{\underline{\underline{p}}} = V \cap \hat{M}_{\underline{\underline{p}}} \supset M_{\underline{\underline{p}}}$ and since $M_{\underline{\underline{p}}}$ is closed in V, $M'_{\underline{\underline{p}}} = M_{\underline{\underline{p}}}$. If now $\hat{N}_{\underline{\underline{p}}}$ is an $\hat{R}_{\underline{\underline{p}}}$-lattice in $\hat{V}_{\underline{\underline{p}}}$, then $M_{\underline{\underline{p}}} = V \cap \hat{N}_{\underline{\underline{p}}}$ is an $R_{\underline{\underline{p}}}$-modul in V (i.e., $KM_{\underline{\underline{p}}} = V$); moreover, since $M_{\underline{\underline{p}}}$ is dense in $\hat{N}_{\underline{\underline{p}}}$, $\hat{R}_{\underline{\underline{p}}} M_{\underline{\underline{p}}} = \hat{M}_{\underline{\underline{p}}} = \hat{N}_{\underline{\underline{p}}}$. It remains to show, that $M_{\underline{\underline{p}}} \in {}_{R_{\underline{\underline{p}}}}\underline{\underline{M}}^f$. But $\hat{N}_{\underline{\underline{p}}} = \bigoplus_{i=1}^n \hat{R}_{\underline{\underline{p}}} \hat{n}_i$, $\hat{n}_i \in \hat{N}_{\underline{\underline{p}}}$, $n = \dim_K(V)$; then $\hat{n}_i = \sum_{j=1}^n \hat{r}_{ij} \otimes v_j$, where the $\{v_j\}_{1 \le i \le n}$ form a K-basis of V. Then

$$\hat{N}_{\underline{\underline{p}}} \subset \bigoplus_{j=1}^n \hat{R}_{\underline{\underline{p}}} v_j,$$

and hence

$$M_{\underline{\underline{p}}} \subset \bigoplus_{j=1}^n \hat{R}_{\underline{\underline{p}}} v_j \cap V = \bigoplus_{j=1}^n R_{\underline{\underline{p}}} v_j.$$

Since $R_{\underline{\underline{p}}}$ is noetherian, $M_{\underline{\underline{p}}} \in {}_{R}\underline{\underline{M}}^f$. #

9.15 <u>Remark</u>: Let R be a Dedekind domain and $\underline{\underline{p}}$ a prime ideal in R, then $\hat{R}_{\underline{\underline{p}}}$ is flat with respect to $\underline{\underline{p}}$-hausdorff modules of finite type, in particular R-lattices (cf. (9.9),(9.13)). But, if M is an R-torsion-module of finite type, such that $\mathrm{ann}_R(M) + \underline{\underline{p}} = R$, then $\underline{\underline{p}}^n M = M$ for every $n \in \underline{\underline{N}}$. Thus, M is not hausdorff, and we can not apply (9.9). Still it is true, that $\hat{R}_{\underline{\underline{p}}}$ is flat with respect to R-modules of finite type.

9.16 <u>Lemma</u>: Let R be a Dedekind domain and $\underline{\underline{p}}$ a prime ideal in R. Then every $R_{\underline{\underline{p}}}$-module $M_{\underline{\underline{p}}}$ of finite type is $\underline{\underline{p}}R_{\underline{\underline{p}}}$-hausdorff.

<u>Proof</u>: From (4.16) and (8.2) it follows, that $\underline{\underline{p}}R_{\underline{\underline{p}}} = \mathrm{rad}\, R_{\underline{\underline{p}}}$. Now, $X = \bigcap_{n \in \underline{\underline{N}}} \underline{\underline{p}}^n M_{\underline{\underline{p}}}$ has the property, that $\underline{\underline{p}}R_{\underline{\underline{p}}} X = X$ (cf. Ex. 9,1). Thus by Nakayama's lemma (4.18), X = 0. #

9.17 <u>Lemma</u>: Let R be a Dedekind domain, and $\underline{\underline{p}}$ a prime ideal in R. If

$$0 \longrightarrow M' \xrightarrow{\varphi} M \xrightarrow{\psi} M'' \longrightarrow 0$$

is an exact sequence of R-modules of finite type, then

$$0 \longrightarrow \hat{R}_{\underline{\underline{p}}} \otimes_R M' \xrightarrow{1 \otimes \varphi} \hat{R}_{\underline{\underline{p}}} \otimes_R M \xrightarrow{1 \otimes \psi} \hat{R}_{\underline{\underline{p}}} \otimes_R M'' \longrightarrow 0$$

is an exact sequence of $\hat{R}_{\underline{\underline{p}}}$-modules.

The <u>proof</u> is left as an exercise. #

9.18 <u>Remark</u>: From the above theorems it follows, that the results of §8 remain valid, if $R_{\underline{\underline{p}}}$ is replaced by $\hat{R}_{\underline{\underline{p}}}$.

<u>Exercises §9</u>:

1.) Prove <u>Herstein's Lemma</u>: Let R be a commutative noetherian ring, $M \in {}_R\underline{\underline{M}}^f$ and $\underline{\underline{a}}$ an ideal in R. If $X = \bigcap_{n \in \underline{\underline{N}}} \underline{\underline{a}}^n M$, then $\underline{\underline{a}}X = X$. (Hint: The set $\{N \underset{\neq}{\subset} M : N \cap X = \underline{\underline{a}}X\}$ is not empty and thus contains a maximal element U. Define $\forall\, a \in \underline{\underline{a}}, M_k(a) = \{m \in M : a^k m \in U\}, \forall\, k \in \underline{\underline{N}}$. Then for some r, $M_r = \bigcup_k M_k$ and $(a^r M + U) \cap X = \underline{\underline{a}}X$, i.e., $a^r M \subset U$ and $a^r M \subset X$. Now, let $\underline{\underline{a}} = \sum_{i=1}^n R\, a_i$, and pick t large enough so that, $a_i^t M \subset U, \forall i$. Then $\underline{\underline{a}}^{tn} \subset \sum_{i=1}^n R\, a_i^t = \underline{\underline{a}}'$, and $X \subset \underline{\underline{a}}^{tn} M \subset \underline{\underline{a}}' M \subset U$, hence $X = \underline{\underline{a}}X$.)

2.) Let $M \in {}_{\underline{\underline{R}}}\underline{\underline{M}}^f$ be I-hausdorff. Show that the I-adic completion and the topological I-adic completion are naturally isomorphic. Here I is a two-sided ideal in the ring R.

3.) Dualize (9.1) and (9.2) to define the injective limit. Show its existence.

4.) Let the notation be as in (9.7). Prove that:

(i) $\hat{\sigma} \in \mathrm{Hom}_{\hat{R}_I}(\hat{M}_I, \hat{M}'_I)$,

(ii) $\hat{\sigma}\hat{\tau} = \widehat{\sigma\tau}$

(iii) $\hat{\sigma}$ is an epimorphism whenever σ is one. (Hint: use the remarks of the proof of (9.7), the universality of $\varprojlim \mathrm{Coker}\, \sigma_n$, (ii), and the universality of Coker $\hat{\sigma}$.)

(iv) discuss the kernel of $\hat{\sigma}$.

Chapter II

HOMOLOGICAL ALGEBRA

§1. Categories and functors

Elementary definitions and examples for categories and functors are given; additive functors preserve finite products; examples of functors that preserve additional structure. Kernels, cokernels, etc. and abelian categories are considered in the exercises. Fiber products and fiber coproducts are introduced.

1.1 Definition: Let $\underline{\underline{C}}$ be a class of "objects" $A, B, C, \ldots$ together with a function and a family of set functions defined as follows:

(i) objects $(\underline{\underline{C}}) \times$ objects $(\underline{\underline{C}}) \longrightarrow$ sets

$$(A,B) \longmapsto \mathrm{morph}_{\underline{\underline{C}}}(A,B).$$

The elements $\varphi \in \mathrm{morph}_{\underline{\underline{C}}}(A,B)$ are called the $\underline{\underline{C}}$-morphisms from A to B; $\varphi : A \longrightarrow B$. We shall write morphisms on the right.

(ii) For each triple $A,B,C \in \mathrm{ob}(\underline{\underline{C}})$ $(\mathrm{ob}(\underline{\underline{C}}) = \text{objects } (\underline{\underline{C}}))$ a function:

$$\mathrm{morph}_{\underline{\underline{C}}}(A,B) \times \mathrm{morph}_{\underline{\underline{C}}}(B,C) \longrightarrow \mathrm{morph}_{\underline{\underline{C}}}(A,C)$$

$$(\varphi, \psi) \longmapsto \varphi\psi.$$

$\varphi\psi$ is called the composite of φ and ψ.

$\underline{\underline{C}}$ is called a category, if the following two axioms hold:

Associativity: If $\varphi : A \longrightarrow B$, $\psi : B \longrightarrow C$, $\chi : C \longrightarrow D$ are morphisms, then

$$(\varphi\psi)\chi = \varphi(\psi\chi).$$

Identity: For every $A \in ob(\underline{\underline{C}})$, there exists $1_A \in morph\ (A,A)$ such that for each $\varphi \in morph_{\underline{\underline{C}}}(A,B)$ and $\psi \in morph_{\underline{\underline{C}}}(C,A)$

$\varphi = 1_A\varphi$ and $\psi = \psi 1_A$.

1.2 Examples: I. Let R be a ring

(i) ${}_R\underline{\underline{M}}$ = category of left R-modules where

$ob({}_R\underline{\underline{M}}) = \{M : M = \text{left R-module}\}$, $morph_{{}_R\underline{\underline{M}}}(M',M) = Hom_R(M',M)$.

(ii) ${}_R\underline{\underline{M}}^f$ = category of finitely generated left R-modules.

(iii) ${}_R\underline{\underline{P}}^f$ = category of finitely generated projective left R-modules.

(iv) If S is also a ring, then ${}_R\underline{\underline{M}}_S$ = category of (R,S)-bimodules: $ob({}_R\underline{\underline{M}}_S) = \{M : M = (R,S)\text{-bimodule}\}$,
$morph_{{}_R\underline{\underline{M}}_S}(M',M) = Hom_{R,S}(M',M) = \{\varphi \in Hom_{\underline{\underline{Z}}}(M',M) : r(m'^{\varphi})s =$
$(rm's)^{\varphi};\quad m' \in M',\ r \in R,\ s \in S\}$.

(For bimodules we write the homomorphisms generally as exponents.)

II. $\underline{\underline{A}}$ = category of abelian groups where the morphisms are group-homomorphisms, ($\underline{\underline{A}} = {}_{\underline{\underline{Z}}}\underline{\underline{M}}$).

1.3 Definition: Let $\underline{\underline{C}}$ and $\underline{\underline{D}}$ be categories. A covariant functor (contravariant functor) $\underline{\underline{F}} : \underline{\underline{C}} \longrightarrow \underline{\underline{D}}$ is a pair consisting of an object function and a family of morphism functions

$$\underline{\underline{F}} : ob(\underline{\underline{C}}) \longrightarrow ob(\underline{\underline{D}}),\ F : A \longmapsto \underline{\underline{F}}(A),$$
$$\underline{\underline{F}} : morph_{\underline{\underline{C}}}(A,B) \longrightarrow morph_{\underline{\underline{D}}}(\underline{\underline{F}}(A),\underline{\underline{F}}(B)),\ \underline{\underline{F}} : \varphi \longmapsto \underline{\underline{F}}(\varphi),$$
$$[(\underline{\underline{F}} : morph_{\underline{\underline{C}}}(A,B) \longrightarrow morph_{\underline{\underline{D}}}(\underline{\underline{F}}(B),\underline{\underline{F}}(A))]$$

satisfying the following two requirements: (i) $\underline{\underline{F}}(1_A) = 1_{\underline{\underline{F}}(A)}$,

(ii) $\underline{\underline{F}}(\varphi\psi) = \underline{\underline{F}}(\varphi)\underline{\underline{F}}(\psi)$ $[\underline{\underline{F}}(\varphi\psi) = \underline{\underline{F}}(\psi)\underline{\underline{F}}(\varphi)]$.

Condition (ii) can be expressed as follows: Given a commutative diagram in $\underline{\underline{C}}$

$$\begin{array}{ccc} A & \xrightarrow{\varphi} & B \\ & \overset{\chi}{\searrow} & \Big\downarrow{\scriptstyle \psi} \\ & & C \end{array},$$

then the diagram

$$\begin{array}{ccc} \underline{\underline{F}}(A) & \xrightarrow{\underline{\underline{F}}(\varphi)} & \underline{\underline{F}}(B) \\ & \overset{\underline{\underline{F}}(\chi)}{\searrow} & \Big\downarrow{\scriptstyle \underline{\underline{F}}(\psi)} \\ & & \underline{\underline{F}}(C) \end{array}$$

$$\begin{array}{ccc} \underline{\underline{F}}(A) & \xleftarrow{\underline{\underline{F}}(\varphi)} & \underline{\underline{F}}(B) \\ & \overset{\underline{\underline{F}}(\chi)}{\nwarrow} & \Big\uparrow{\scriptstyle \underline{\underline{F}}(\psi)} \\ & & \underline{\underline{F}}(C) \end{array}$$

is commutative.

1.4 Remark: In all cases that we consider, $morph(A,B)$ as well as $morph(\underline{\underline{F}}(A),\underline{\underline{F}}(B))$ is an abelian group. Then we can also require from a functor

(iii) $\underline{\underline{F}}(\varphi+\varphi') = \underline{\underline{F}}(\varphi) + \underline{\underline{F}}(\varphi')$, and call $\underline{\underline{F}}$ an additive functor.

More precisely, since all our categories are categories of modules, we are only interested in additive categories with finite direct sums (finite biproducts); i.e., categories $\underline{\underline{C}}$ in which $morph_{\underline{\underline{C}}}(A,B)$ is an abelian group, such that

α.) the composition of morphisms is distributive, when defined; i.e.,

$\varphi(\psi_1+\psi_2) = \varphi\psi_1 + \varphi\psi_2$, $(\varphi_1+\varphi_2)\psi = \varphi_1\psi + \varphi_2\psi$.

β.) there exists a unique zero object 0 such that $morph_{\underline{\underline{C}}}(0,A)$ and $morph_{\underline{\underline{C}}}(B,0)$ have exactly one element each, denoted by 0, for every $A, B \in ob(\underline{\underline{C}})$.

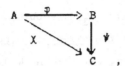

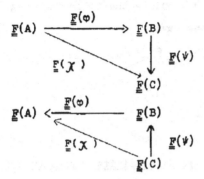

γ.) To each pair of objects $A_1, A_2 \in ob(\underline{\underline{C}})$ there exists an object $A_1 \oplus A_2 \in ob(\underline{\underline{C}})$ called the *direct sum* (*coproduct*) of A_1 and A_2, with a pair of maps $\iota_1 : A_1 \longrightarrow A_1 \oplus A_2$ and $\iota_2 : A_2 \longrightarrow A_1 \oplus A_2$, such that, given $\varphi \in morph_{\underline{\underline{C}}}(A_1, B)$, $\psi \in morph_{\underline{\underline{C}}}(A_2, B)$, one can complete the following diagram commutatively in one and only one way:

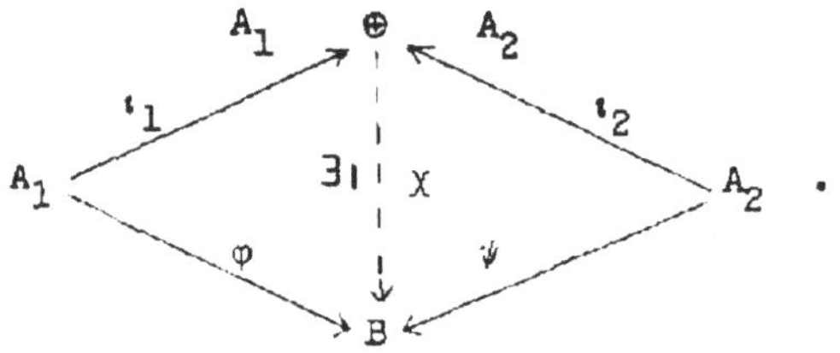

This means that *in* $\underline{\underline{C}}$ *there exist finite direct sums* (cf. Ex. I, 1, 2).

For additive categories one obviously requires that a functor be additive. *From now on, all categories and functors under consideration are additive*.

1.5 *Examples*:

I. Let R be a ring, and let $M \in {}_R\underline{\underline{M}}$ be fixed. (For categories of modules we write $M \in {}_R\underline{\underline{M}}$ instead of $M \in ob({}_R\underline{\underline{M}})$.) Then

(i) $\underline{hom_R(M,-)} : {}_R\underline{\underline{M}} \longrightarrow \underline{\underline{A}}$,

$hom_R(M,-) : N \longmapsto Hom_R(M,N)$,

$hom_R(M,-) : Hom_R(N',N) \longrightarrow Hom_{\underline{\underline{Z}}}(Hom_R(M,N'), Hom_R(M,N))$,

$hom_R(M,-) : \varphi \longmapsto hom(1_M, \varphi)$,

is an additive *covariant functor*.

(ii) $\underline{hom_R(-,M)} : {}_R\underline{\underline{M}} \longrightarrow \underline{\underline{A}}$,

$hom_R(-,M) : N \longmapsto Hom_R(N,M)$,

$hom_R(-,M) : Hom_R(N',N) \longrightarrow Hom_{\underline{\underline{Z}}}(Hom_R(N,M), Hom_R(N',M))$,

$hom_R(-,M) : \varphi \longmapsto hom(\varphi, 1_M)$,

is an additive *contravariant functor*.

II. Let $M \in \mathbf{M}_R$ be fixed.

(iii) $\underline{M \otimes_R -} : {}_R\mathbf{M} \longrightarrow \mathbf{A}$

$M \otimes_R - : N \longmapsto M \otimes_R N$

$M \otimes_R - : \mathrm{Hom}_R(N',N) \longrightarrow \mathrm{Hom}_{\mathbf{Z}}(M \otimes_R N, M \otimes_R N')$

$M \otimes_R - : \varphi \longmapsto 1_M \otimes \varphi$

is an additive *covariant functor*.

(iv) Similarly, for $M \in {}_R\mathbf{M}$, $- \otimes_R M : \mathbf{M}_R \longrightarrow \mathbf{A}$ is an additive *covariant functor*.

1.6 *Lemma*: An additive functor preserves (finite) direct sums.

Proof: Let $\mathbf{F} : \mathbf{C} \longrightarrow \mathbf{D}$ be a contravariant functor. For $A_1, A_2 \in \mathrm{ob}(\mathbf{C})$, we know $\mathbf{F}(A_1) \oplus \mathbf{F}(A_2)$ together with $\tilde{\pi}_i \in \mathrm{morph}_{\mathbf{D}}(\mathbf{F}(A_1) \oplus \mathbf{F}(A_2), \mathbf{F}(A_i))$, $i = 1,2$ is a product (cf. Ex. 1,5). Thus, we can complete the following diagram commutatively

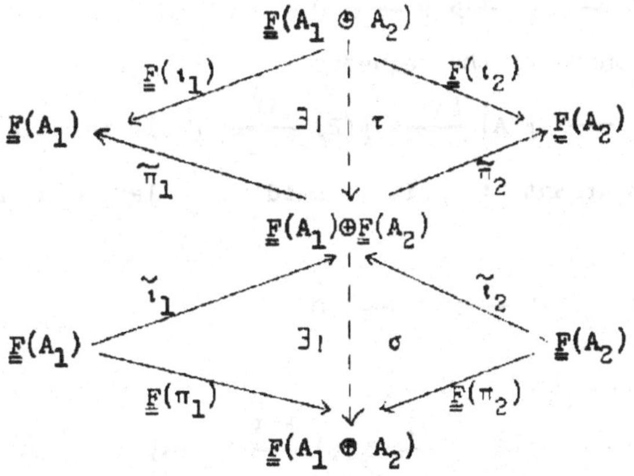

and consequently,

$$1_{\mathbf{F}(A_1 \oplus A_2)} = \mathbf{F}(\iota_1)\mathbf{F}(\pi_1) + \mathbf{F}(\iota_2)\mathbf{F}(\pi_2) = \tau\tilde{\pi}_1\tilde{\iota}_1\sigma + \tau\tilde{\pi}_2\tilde{\iota}_2\sigma$$

$$= \tau(\tilde{\pi}_1\tilde{\iota}_1 + \tilde{\pi}_2\tilde{\iota}_2)\sigma = \tau\sigma.$$

Similarly one shows that $\sigma\tau = 1_{\underline{F}(A_1) \oplus \underline{F}(A_2)}$; thus $\underline{F}(A_1) \oplus \underline{F}(A_2)$ with $\tilde{\iota}_1$ and $\tilde{\iota}_2$, and $\underline{F}(A_1 \oplus A_2)$ with $\underline{F}(\pi_1)$ and $\underline{F}(\pi_2)$ are both coproducts in $\underline{D}$; whence, by the universal property of coproducts, they are naturally isomorphic. Similarly for a covariant $\underline{F}$. #

1.7 Corollary: Let R and S be rings. If

$$0 \longrightarrow A \xrightarrow{\varphi} B \xrightarrow{\psi} C \longrightarrow 0$$

is a split exact sequence in ${}_R\underline{M}$ and if $\underline{F} : {}_R\underline{M} \longrightarrow {}_S\underline{M}$ is a covariant (contravariant) functor, then

$$0 \longrightarrow \underline{F}(A) \xrightarrow{\underline{F}(\varphi)} \underline{F}(B) \xrightarrow{\underline{F}(\psi)} \underline{F}(C) \longrightarrow 0$$
$$(0 \longrightarrow \underline{F}(C) \xrightarrow{\underline{F}(\psi)} \underline{F}(B) \xrightarrow{\underline{F}(\varphi)} \underline{F}(A) \longrightarrow 0)$$

is a split exact sequence.

1.8 Definitions: (i) Let R and S be rings; a covariant functor $\underline{F} : {}_R\underline{M} \longrightarrow {}_S\underline{M}$ is called left exact if the exactness of the sequence

$$0 \longrightarrow A \xrightarrow{\varphi} B \xrightarrow{\psi} C \quad \text{in } {}_R\underline{M}$$

implies the exactness of the sequence

$$0 \longrightarrow \underline{F}(A) \xrightarrow{\underline{F}(\varphi)} \underline{F}(B) \xrightarrow{\underline{F}(\psi)} \underline{F}(C) \quad \text{in } {}_S\underline{M}.$$

If $\underline{F}$ is contravariant, then it is said to be left exact, if the exactness of

$$A \longrightarrow B \longrightarrow C \longrightarrow 0$$

implies that

$$0 \longrightarrow \underline{F}(C) \xrightarrow{\underline{F}(\psi)} \underline{F}(B) \xrightarrow{\underline{F}(\varphi)} \underline{F}(A)$$

is an exact sequence. Right exactness is defined similarly. $\underline{F}$ is called exact, if it is left exact as well as right exact.

(ii) A functor $\underline{\underline{F}} : {}_{\underline{\underline{R}}}\underline{\underline{M}} \longrightarrow {}_{\underline{\underline{S}}}\underline{\underline{M}}$ is called <u>faithful</u> if

$$\underline{\underline{F}}(\varphi) = 0 \Longrightarrow \varphi = 0, \ \varphi \in \mathrm{Hom}_R(M,M'), \ M,M' \in {}_{\underline{\underline{R}}}\underline{\underline{M}}.$$

This automatically implies $M = 0$ if $\underline{\underline{F}}(M) = 0$, for $M \in {}_{\underline{\underline{R}}}\underline{\underline{M}}$.

1.9 <u>Theorem</u>: Let R be a ring.

(i) For $M \in {}_{\underline{\underline{R}}}\underline{\underline{M}}$, both $\hom_R(M,-)$ and $\hom_R(-,M)$ are left exact.

(ii) For $M \in \underline{\underline{M}}_{\underline{\underline{R}}}$, $N \in {}_{\underline{\underline{R}}}\underline{\underline{M}}$, $M \otimes_R -$ and $- \otimes_R N$ are right exact.

(iii) $M \in {}_{\underline{\underline{R}}}\underline{\underline{M}}^f$ is projective $\Longleftrightarrow$ $\hom_R(M,-)$ is exact.

(iv) $M \in \underline{\underline{M}}_{\underline{\underline{R}}}$ is flat $\Longleftrightarrow$ $M \otimes_R -$ is exact.

<u>Proof</u>: (i) follows from (I, 2.6) and (I, 2.7), (ii) follows from (I, 3.12) and (I, 3.13), (iii) is the translation of (I, 2.9), and (iv) is the definition of flat (cf. I, (3.16)). #

1.10 <u>Definition</u>: Let $\underline{\underline{C}}$ and $\underline{\underline{D}}$ be categories and $\underline{\underline{F}}_1, \underline{\underline{F}}_2 : \underline{\underline{C}} \longrightarrow \underline{\underline{D}}$ functors (either both covariant or both contravariant). A family

$$\mu = \{\mu_A\}_{A \in \mathrm{ob}(\underline{\underline{C}})}, \ \mu_A \in \mathrm{morph}_{\underline{\underline{D}}}(\underline{\underline{F}}_1(A), \underline{\underline{F}}_2(A))$$

is called a <u>natural transformation</u> of the functors $\underline{\underline{F}}_1$ and $\underline{\underline{F}}_2$: $\mu : \underline{\underline{F}}_1 \longrightarrow \underline{\underline{F}}_2$, if for every $\alpha \in \mathrm{morph}_{\underline{\underline{C}}}(A,B)$, $A,B \in \mathrm{ob}(\underline{\underline{C}})$, the following diagram is commutative:

$$\begin{array}{ccc} \underline{\underline{F}}_1(A) & \xrightarrow{\mu_A} & \underline{\underline{F}}_2(A) \\ {\scriptstyle \underline{\underline{F}}_1(\alpha)} \downarrow & & \downarrow {\scriptstyle \underline{\underline{F}}_2(\alpha)} \\ \underline{\underline{F}}_1(B) & \xrightarrow{\mu_B} & \underline{\underline{F}}_2(B) \end{array}$$

(This is the diagram for a covariant $\underline{\underline{F}}_1$, $\underline{\underline{F}}_2$; similarly for contravariant $\underline{\underline{F}}_1$, $\underline{\underline{F}}_2$.)

If in μ, each μ_A is an isomorphism, i.e., $\forall \ \mu_A$,

$\exists\ \nu_A \in \mathrm{morph}_{\mathbf{D}}(\mathbf{F}_2(A),\mathbf{F}_1(A))$ such that $\nu_A\mu_A = 1_{\mathbf{F}_2(A)}$ and $\mu_A\nu_A = 1_{\mathbf{F}_1(A)}$ (cf. Ex. 1,1)), then μ is called a <u>natural equivalence</u>; <u>notation</u> for natural equivalence: $\mathbf{F}_1 \sim \mathbf{F}_2$. In that case, one can identify $\mathbf{F}_1(A)$ and $\mathbf{F}_2(A)$, $A \in \mathrm{ob}(\mathbf{C})$ and $\mathrm{morph}_{\mathbf{D}}(\mathbf{F}_1(A),\mathbf{F}\ (B))$ and $\mathrm{morph}_{\mathbf{D}}(\mathbf{F}_2(A),\mathbf{F}_2(B))$.

1.11 <u>Remark</u>: (1.10) justifies, that in Ch. I we have identified some modules; e.g., $A \otimes_R (B \otimes_S C)$ with $(A \otimes_R B) \otimes_S C$ and $M_{\mathbf{D}}$ with $R_{\mathbf{D}} \otimes_R M$. From now on, we shall in general identify naturally equivalent functors.

1.12 <u>Lemma</u>: Let $\mathbf{F} : {}_S\mathbf{M} \longrightarrow \mathbf{A}$ be a covariant [contravariant] functor ($\mathbf{A}$ = category of abelian groups). If $M \in {}_S\mathbf{M}_R$, where S and R are rings, then $\mathbf{F}(M) \in \mathbf{M}_R$ [$\mathbf{F}(M) \in {}_R\mathbf{M}$].

<u>Proof</u>: We give only a proof for contravariant $\mathbf{F}$, the proof for covariant $\mathbf{F}$ being similar. For every $r \in R$, we define $\varphi_r : M \longrightarrow M$, $\varphi_r : m \longmapsto mr$; then $\varphi_r \in \mathrm{morph}_{{}_S\mathbf{M}}(M,M)$, and thus $\mathbf{F}(\varphi_r) : \mathbf{F}(M) \longrightarrow \mathbf{F}(M)$ is a morphism in $\mathbf{A}$. We now put, for $x \in \mathbf{F}(M)$, $rx = x\,\mathbf{F}(\varphi_r)$. This gives $\mathbf{F}(M)$ the structure of a left R-module. We only have to check the associativity:
$(r_1r_2)x = x\mathbf{F}(\varphi_{r_1r_2}) = x\mathbf{F}(\varphi_{r_1}\varphi_{r_2}) = x\mathbf{F}(\varphi_{r_2})\mathbf{F}(\varphi_{r_1}) = r_1(r_2x)$. #

1.13 <u>Definition of the fiber product</u> (<u>pullback</u>): Let $\mathbf{C}$ be a category and $C_1 \xrightarrow{\alpha} C \xleftarrow{\beta} C_2$ a diagram in $\mathbf{C}$. $P \in \mathrm{ob}(\mathbf{C})$ together with $\varphi_1 \in \mathrm{morph}_{\mathbf{C}}(P,C_1)$ and $\varphi_2 \in \mathrm{morph}_{\mathbf{C}}(P,C_2)$ is called a <u>fiber product</u> (<u>pullback</u>) of the diagram $C_1 \xrightarrow{\alpha} C \xleftarrow{\beta} C_2$ if:

(i)
$$\begin{array}{ccc} P & \xrightarrow{\varphi_1} & C_1 \\ {\scriptstyle\varphi_2}\downarrow & & \downarrow{\scriptstyle\alpha} \\ C_2 & \xrightarrow{\beta} & C \end{array}$$
is commutative,

(ii) Given a commutative square

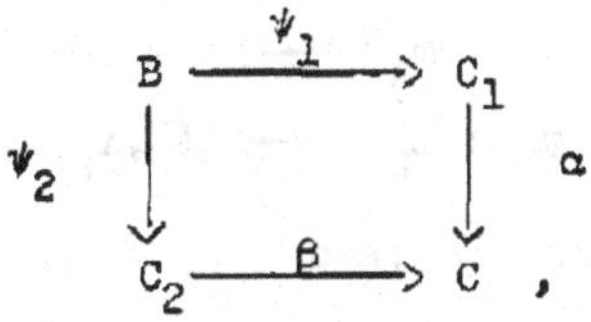

there exists a unique $\sigma \in \mathrm{morph}_{\underline{\underline{C}}}(B,P)$ such that the following diagram is commutative

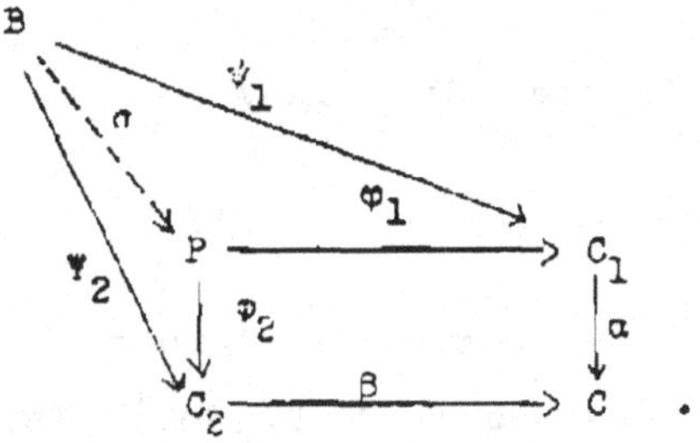

1.14 <u>Definition</u>: The dual concept, which is obtained from that of the fiber product by reversing the arrows, is called the <u>fiber coproduct</u> (<u>pushout</u>). It should be observed that the fiber product and the fiber coproduct are unique up to isomorphism if they exist.

1.15 <u>Theorem</u>: Let R be a ring and $\underline{\underline{C}} = {}_R\underline{\underline{M}}$.

(i) For every diagram $M_1 \xrightarrow{\alpha} M \xleftarrow{\beta} M_2$ in ${}_R\underline{\underline{M}}$ there exists a fiber product in ${}_R\underline{\underline{M}}$; in short, fiber products exist in ${}_R\underline{\underline{M}}$, namely:

$$P = \{(m_1,m_2) : m_1\alpha = m_2\beta\} \text{ with}$$

$$\varphi_i : P \longrightarrow M_i;\ \varphi_i : (m_1,m_2) \longmapsto m_i,\ i = 1,2.$$

(ii) For every diagram $M_1 \xleftarrow{\alpha} M \xrightarrow{\beta} M_2$ in ${}_R\underline{\underline{M}}$ there exists a fiber coproduct in ${}_R\underline{\underline{M}}$,

$$Q = (M_1 \oplus M_2)/M_0,$$

where M_0 is the left R-submodule of $M_1 \oplus M_2$ generated by the elements of the form $(m\alpha,-m\beta)$. The maps associated with Q are

$$\varphi_1 : M_1 \longrightarrow Q;\ \varphi_1 : m_1 \longmapsto (m_1, 0) + M_o$$
$$\varphi_2 : M_2 \longrightarrow Q;\ \varphi_2 : m_2 \longmapsto (0, m_2) + M_o\ .$$

__Proof__: Trivially, the diagrams

$$\begin{array}{ccc} P & \xrightarrow{\varphi_1} & M_1 \\ {\scriptstyle\varphi_2}\downarrow & & \downarrow{\scriptstyle\alpha} \\ M_2 & \xrightarrow{\beta} & M \end{array} \qquad \text{and} \qquad \begin{array}{ccc} Q & \xleftarrow{\varphi_1} & M_1 \\ {\scriptstyle\varphi_2}\uparrow & & \uparrow{\scriptstyle\alpha} \\ M_2 & \xleftarrow{\beta} & M \end{array}$$

are commutative. As for the universality, let

$$\begin{array}{ccc} B & \xrightarrow{\psi_1} & M_1 \\ {\scriptstyle\psi_2}\downarrow & & \downarrow{\scriptstyle\alpha} \\ M_2 & \xrightarrow{\beta} & M \end{array}$$ be a commutative diagram.

We define $\sigma : B \longrightarrow P;\ \sigma : b \longmapsto (b\psi_1, b\psi_2)$. Then $\sigma\varphi_1 = \psi_1$ and $\sigma\varphi_2 = \psi_2$. The uniqueness of σ is clear, since the φ_i are "projections" $P \longrightarrow M_i$. Observe: P is a subdirect sum*) of M_1 and M_2. For the fiber coproduct, let

$$\begin{array}{ccc} B & \xleftarrow{\psi_1} & M_1 \\ {\scriptstyle\psi_2}\uparrow & & \uparrow{\scriptstyle\alpha} \\ M_2 & \xleftarrow{\beta} & M \end{array}$$ be a commutative diagram.

We define $\sigma : Q \longrightarrow B;\ \sigma : (m_1, m_2) + M_o \longmapsto m_1\psi_1 + m_2\psi_2$. Then σ is well defined, and its uniqueness follows easily from the commutative diagrams:

$$\begin{array}{ccc} Q & \xrightarrow{\sigma'} & B \\ {\scriptstyle\varphi_1}\nwarrow & & \nearrow{\scriptstyle\psi_1} \\ & M_1 & \end{array} \qquad \text{and} \qquad \begin{array}{ccc} Q & \xrightarrow{\sigma'} & B \\ {\scriptstyle\varphi_2}\nwarrow & & \nearrow{\scriptstyle\psi_2} \\ & M_2 & \end{array},$$

and from the fact that Q is generated by $\operatorname{Im}\varphi_1 \cup \operatorname{Im}\varphi_2$.

1.16 __Lemma__: Let R be a ring, and consider ${}_R\underline{\underline{M}}$.

(i) In the fiber product, if β is an epimorphism, so is φ_1.

---

*) A submodule M of $A \oplus B$ is called a __subdirect sum__ of A and B if $M\pi_A = A$ and $M\pi_B = B$, where π_A and π_B are the projections.

(ii) In the fiber coproduct, if β is a monomorphism, so is φ_1.

Proof: This follows readily from (1.15).

Remark: For (1.15) and (1.16) it suffices that the morphisms in $\underline{\underline{C}}$ are set maps, and $\underline{\underline{C}}$ is an additive category in which kernels and cokernels exist.

Exercises §1:

1.) Prove the statements of (1.5).

2.) Let $\underline{\underline{C}}$ be any category and $\alpha \in \text{morph}_{\underline{\underline{C}}}(A,B)$, then α is called a monomorphism if $\varphi\alpha = \psi\alpha \Longrightarrow \varphi = \psi$, $\forall\, \varphi,\psi \in \text{morph}_{\underline{\underline{C}}}(D,A)$, $D \in \text{ob}(\underline{\underline{C}})$, epimorphism if $\alpha\varphi = \alpha\psi \Longrightarrow \varphi = \psi$, $\forall\, \varphi,\psi \in \text{morph}_{\underline{\underline{C}}}(B,D)$, $D \in \text{ob}(\underline{\underline{C}})$, isomorphism if $\exists\, \beta \in \text{morph}_{\underline{\underline{C}}}(B,A)$ such that $\alpha\beta = 1_A$ and $\beta\alpha = 1_B$.

Show:

(i) $\varphi\psi$ monic $\Longrightarrow \varphi$ monic, $\varphi\psi$ epic $\Longrightarrow \psi$ epic, and every isomorphism is both monic and epic.

(ii) In any category whose morphisms are set maps, (e.g., any category of algebraic structures with structure preserving maps), every injection is monic, every surjection is epic, and every map that is both monic and epic is an isomorphism. Note that the last property is to be taken with a grain of salt in the case of structures with partially defined operations or relations.

(iii) Not in all categories of algebraic structures does monic $\Longrightarrow$ injective, epic $\Longrightarrow$ surjective, monic and epic $\Longrightarrow$ isomorphic. (Hint: In the category $\underline{\underline{D}}$ of divisible*) abelian groups and group homomorphisms the canonical map $\underline{\underline{Q}} \longrightarrow \underline{\underline{Q}}/\underline{\underline{Z}}$ is monic. In the category $\underline{\underline{R}}$ of rings which do not necessarily have an identity and ring homomorphisms which are not necessarily

---

*) We recall that an additively written abelian group G is called divisible if, for every $a \in G$ and $n \in \underline{\underline{Z}}$, there exists $b \in G$ such that $a = n\,b$.

unitary, the canonical injection $\underline{\underline{Z}} \longrightarrow \underline{\underline{Q}}$ is both epic and monic (cf. Ex. 1,3d)).

3.) Let $\underline{\underline{C}}$ be a category with 0's (cf. (1.4), axiom β)), and let $\varphi \in \mathrm{morph}_{\underline{\underline{C}}}(A,B)$. An object $K \in \mathrm{ob}(\underline{\underline{C}})$ together with a morphism $\kappa \in \mathrm{morph}_{\underline{\underline{C}}}(K,A)$ is called a <u>kernel for φ</u> if $\kappa\varphi = 0$, and every commutative diagram

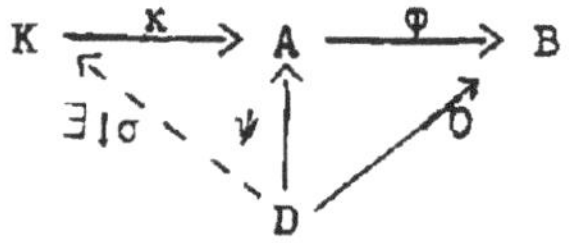

can be completed uniquely by $\sigma \in \mathrm{morph}_{\underline{\underline{C}}}(D,K)$. We sometimes write $K = \mathrm{Ker}\ \varphi$ and $\kappa = \ker \varphi$. Dually, an object $C \in \mathrm{ob}(\underline{\underline{C}})$ together with a morphism $\gamma \in \mathrm{morph}_{\underline{\underline{C}}}(B,C)$ is said to be a <u>cokernel for φ</u> if $\varphi\gamma = 0$ and every commutative diagram

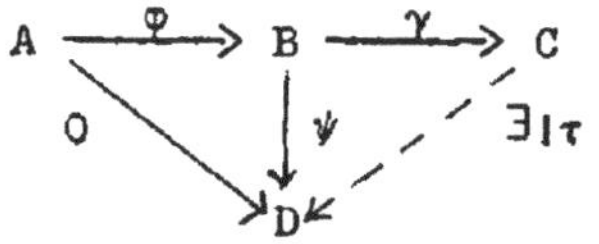

can be completed uniquely by $\tau \in \mathrm{morph}_{\underline{\underline{C}}}(C,D)$. We sometimes write $C = \mathrm{Coker}\ \varphi$ and $\gamma = \mathrm{coker}\ \varphi$. Show that: (i) Kernels and cokernels, if they exist, are unique up to "natural" isomorphisms, where a natural isomorphism between kernels $(K,\kappa) \longrightarrow (K',\kappa')$ is given by an isomorphism $\sigma \in \mathrm{morph}_{\underline{\underline{C}}}(K,K')$, for which the diagram

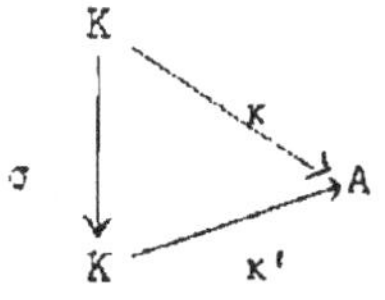

commutes. Similarly for isomorphisms of cokernels.

(ii) Kernels are monic and cokernels are epic if they exist. (As with the 0's we shall indulge in some abuse of language by calling K as well as κ a kernel of φ, or even the kernel of φ, whenever the meaning is clear from the context. Similarly for cokernels.)

(iii) Every monomorphism has a kernel, namely 0, and every epimorphism has a cokernel, namely 0.

(iv) For additive categories the converse of (iii) holds too: $\ker \varphi = 0 \Longrightarrow \varphi$ is monic, and $\operatorname{coker} \varphi = 0 \Longrightarrow \varphi$ is epic; i.e., φ is monic if and only if $\sigma = 0$ whenever $\sigma\varphi = 0$, and similarly for epimorphisms. (Note that the categories of the examples (2.(iii)) are additive.)

(v) Not every monomorphism is a kernel and not every epimorphism is a cokernel. (Hint: Use the examples of (2.(iii)) once more: If $\underline{\underline{Q}} \longrightarrow \underline{\underline{Q}}/\underline{\underline{Z}}$ were the kernel of φ, then φ would have to be 0, but $1_{\underline{\underline{Q}}/\underline{\underline{Z}}}$ does not factor through $\underline{\underline{Q}}$, and if $\underline{\underline{Z}} \longrightarrow \underline{\underline{Q}}$ were the cokernel of ψ, then ψ would have to be 0, but $1_{\underline{\underline{Z}}}$ does not factor through $\underline{\underline{Q}}$.)

4.) Whenever they exist, the cokernel of the kernel of a morphism φ is called the <u>coimage</u> of φ, $\operatorname{coker}(\ker \varphi) = \operatorname{coim} \varphi$, and the kernel of the cokernel of φ is called its <u>image</u>, $\ker(\operatorname{coker} \varphi) = \operatorname{im} \varphi$.

A category is called <u>abelian</u> if it is additive, has finite direct sums, kernels and cokernels, and if every monomorphism is a kernel and every epimorphism is a cokernel.

Show that in an abelian category:

(i) There exists to every morphism a unique natural isomorphism $\sigma : \operatorname{Coim} \varphi \longrightarrow \operatorname{Im} \varphi$, so that the following diagram commutes

$$\begin{array}{ccc} A & \xrightarrow{\varphi} & B \\ {\scriptstyle \text{coim}\,\varphi}\downarrow & & \uparrow{\scriptstyle \text{im}\,\varphi} \\ X & \xrightarrow{\exists!\sigma} & Y \end{array}$$

i.e., the homomorphism theorem holds. (This is often used as an axiom, AB_5, for abelian categories. To every morphism φ there exists then a monomorphism α and an epimorphism β such that $\varphi = \beta\alpha$. We shall call a category <u>semi-exact</u> if it has cokernels and kernels and if it has this property.)

(ii) Every monomorphism α is a kernel, namely $\alpha = \ker(\operatorname{coker} \alpha)$, every epimorphism β is a cokernel, namely $\beta = \operatorname{coker}(\ker \beta)$, and every morphism that is a monomorphism as well as an epimorphism is an isomorphism.

5.) (i) Show, that in an additive category $\underline{\underline{C}}$ there exists to every direct sum $(A_1 \oplus A_2; \iota_1, \iota_2)$ a pair of morphisms $\pi_i \in \operatorname{morph}_{\underline{\underline{C}}}(A_1 \oplus A_2, A_i)$, $i = 1,2$, such that

$$\iota_j \pi_i = \begin{cases} 0 & \text{if } j \neq i \\ 1_{A_j} & \text{if } j = i, \text{ and} \end{cases}$$

$$\pi_1\iota_1 + \pi_2\iota_2 = 1_{A_1 \oplus A_2} \quad .$$

(ii) Conversely, show that these conditions characterize $A_1 \oplus A_2$.

(iii) Define direct sums via π_1 and π_2; i.e., in categorical language, define finite <u>products</u>, (Products that are coproducts (sums) are called <u>biproducts</u>. Thus, in an additive category, all finite sums and all finite products are biproducts.)

(iv) Show that the ι_j's are monic and that the π_j's are epic.

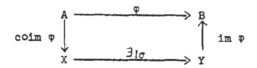

6.) (i) Define the concept of a <u>bifunctor</u>: $\underline{\underline{C}} \times \underline{\underline{D}} \longrightarrow \underline{\underline{E}}$.

(ii) Show that, for an additive category $\underline{\underline{C}}$ with direct sums

$$- \oplus - : \underline{\underline{C}} \times \underline{\underline{C}} \longrightarrow \underline{\underline{C}},$$

$$- \oplus - : (A_1, A_2) \longmapsto A_1 \oplus A_2$$

is a bifunctor with

$$-\oplus-: (\varphi, \psi) \longmapsto \varphi \oplus \psi = \pi_1 \varphi \iota_1' + \pi_2 \psi \iota_2',$$

for $\varphi \in \mathrm{morph}_{\underline{\underline{C}}}(A_1, A_1')$, $\psi \in \mathrm{morph}_{\underline{\underline{C}}}(A_2, A_2')$ and the appropriate morphisms π_j, ι_j'.

Note that alternately, $\varphi \oplus \psi$, can be defined as the unique morphism in $\mathrm{morph}_{\underline{\underline{C}}}(A_1 \oplus A_2, A_1' \oplus A_2')$ that makes the following diagram commute:

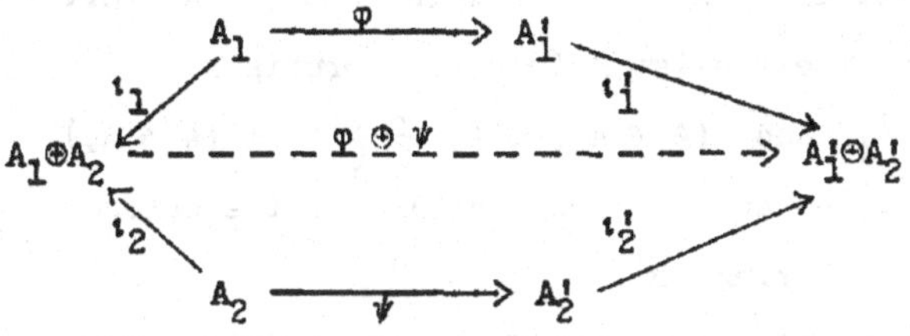

(Similarly for products.)

(iii) Show that the morphisms σ and τ of the proof of (1.6) do indeed define a natural equivalence of bifunctors.

7.) In a category $\underline{\underline{C}}$ with finite products the <u>diagonal</u> $\Delta_A : A \longrightarrow A \oplus A$ is defined as the unique morphism that completes the diagram commutatively:

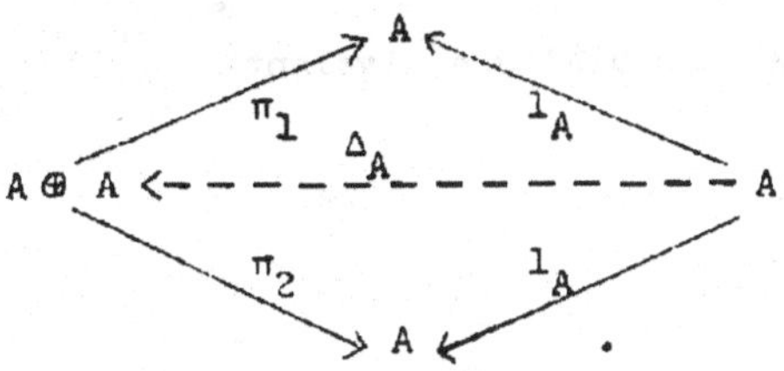

Observe that $\Delta_A = \iota_1 + \iota_2$ if the product is a biproduct, and if

$\underline{\underline{C}}$ is additive.

(i) Dualize this concept to define the codiagonal ∇_A.

(ii) Using (i) and the diagram of (Ex. 1,6 (ii)), show that in a category $\underline{\underline{C}}$ with biproducts a unique "addition of morphisms" can be defined by $\varphi + \psi = \Delta_A(\varphi \oplus \psi)\nabla_{A'}$, so that the sets $\mathrm{morph}_{\underline{\underline{C}}}(A,A')$ become semigroups. If moreover, every monomorphism of $\underline{\underline{C}}$ that is an epimorphism is an isomorphism, then these semigroups are groups.

(iii) Show that in a category $\underline{\underline{C}}$ with biproducts, for $\Delta = \Delta_A$, $\nabla = \nabla_A$ and $\varphi \in \mathrm{morph}_{\underline{\underline{C}}}(A,A)$ $\varphi\Delta = \Delta(\varphi \oplus \varphi)$, $\nabla\varphi = (\varphi \oplus \varphi)\nabla$, $\Delta(1_A \oplus \Delta) = \Delta(\Delta \oplus 1_A)$ and $(\nabla + 1_A)\nabla = (1_A + \nabla)\nabla$ where we have identified $(A \oplus A) \oplus A$ with $A \oplus (A \oplus A)$ (cf. (iv)).

(iv) Show that there exist natural isomorphisms $A_1 \oplus A_2 \xrightarrow{\sim} A_2 \oplus A_1$ and $(A_1 \oplus A_2) \oplus A_3 \xrightarrow{\sim} A_1 \oplus (A_2 \oplus A_3)$.

(v) Define $\oplus$, Δ and ∇ explicitly for the category ${}_{\underline{\underline{R}}}\underline{\underline{M}}$ of left modules over a ring R.

8.) Show that, for a ring R, the following functors are naturally equivalent to the identity functor:

(i) $\mathrm{Hom}_R({}_RR,-) : {}_{\underline{\underline{R}}}\underline{\underline{M}} \longrightarrow {}_{\underline{\underline{R}}}\underline{\underline{M}}$ (cf. I, (1.7)),

(ii) $\mathrm{Hom}_R(\mathrm{Hom}_R(-,{}_RR),{}_RR) : {}_{\underline{\underline{R}}}\underline{\underline{P}}^f \longrightarrow {}_{\underline{\underline{R}}}\underline{\underline{P}}^f$, (cf. I, (2.12)),

(iii) $- \otimes_R {}_RR : \underline{\underline{M}}_R \longrightarrow \underline{\underline{M}}_R$ (cf. I, (3.8)).

(iv) If R is an integral domain and S a multiplicative system: Show that the functors $R_S \otimes_R - : {}_{\underline{\underline{R}}}\underline{\underline{M}} \longrightarrow {}_{R_S}\underline{\underline{M}}$ and $-_S : {}_{\underline{\underline{R}}}\underline{\underline{M}} \longrightarrow {}_{R_S}\underline{\underline{M}}$ are naturally equivalent.

§2. Homology.

Homology is defined and the exact triangle theorem and the exact prism theorem are proved for complexes and graded complexes of modules. In the exercises, a categorical approach to homology in additive categories with kernels, cokernels and O's is outlined.

In this section R is a fixed ring.

2.1 Definition: A complex (M,δ) consists of $M \in {}_R\underline{\underline{M}}$ and $\delta \in \mathrm{End}_R(M)$ such that $\delta^2 = 0$; δ is called a differentiation. We define the cycles of (M,δ): $Z \overset{\text{def}}{=} \mathrm{Ker}\ \delta$, boundaries of (M,δ) : $B \overset{\text{def}}{=} \mathrm{Im}\ \delta$. Since $\delta^2 = 0$, $B \subset Z$, and $H(M,\delta) \overset{\text{def}}{=} Z/B \in {}_R\underline{\underline{M}}$ is called the homology group of the complex (M,δ). Given two complexes (M,δ), (M',δ'); a chain map $\varphi : (M,\delta) \longrightarrow (M',\delta')$ is a map $\varphi' \in \mathrm{Hom}_R(M,M')$ such that the following diagram is commutative:

$$D:\quad \begin{array}{ccc} M & \xrightarrow{\varphi'} & M' \\ \delta \downarrow & & \downarrow \delta' \\ M & \xrightarrow{\varphi'} & M' \end{array}$$

Since, in general, there is no ambiguity, we shall identify φ and φ'. The complexes and chain maps form a category.

2.2 Lemma: A chain map $\varphi : (M,\delta) \longrightarrow (M',\delta')$ induces a homomorphism $\hat{\varphi} : H(M,\delta) \longrightarrow H(M',\delta')$, $\varphi : z + B \longmapsto z\varphi + B'$, of left R-modules.

The proof is straightforward. #

2.3 Lemma: Let (M,δ), (M',δ') and (M'',δ'') be complexes and let $\varphi : (M,\delta) \longrightarrow (M',\delta')$ and $\Psi : (M',\delta') \longrightarrow (M'',\delta'')$ be chain maps. Then

$$\widehat{(\varphi\Psi)} = \hat{\varphi}\hat{\Psi}$$
$$(\widehat{\varphi+\Psi}) = \hat{\varphi} + \hat{\Psi}$$
$$\hat{1}_M = 1_{H(M,\delta)} .$$

The proof follows by applying the definition of $\hat{}$. #

This shows that $(M,\delta) \longrightarrow H(M,\delta)$ together with the operation "$\wedge$" is a covariant functor from the category of complexes and chain maps into ${}_R\underline{M}$, the category of left R-modules.

2.4 Definition: Let (M,δ) and (M',δ') be complexes and let $\varphi,\Psi : (M,\delta) \longrightarrow (M',\delta')$ be chain maps. Then φ is said to be homotopic to Ψ (notation, $\varphi \simeq \Psi$), if there exists $\rho \in \mathrm{Hom}_R(M,M')$ such that $\varphi - \Psi = \rho\delta' + \delta\rho$. "Being homotopic" is an equivalence relation.

2.5 Lemma: Let $\varphi,\Psi : (M,\delta) \longrightarrow (M',\delta')$ be two homotopic chain maps of complexes. Then $\hat{\varphi} = \hat{\Psi}$.

The proof follows from an easy computation. #

2.6 Definition: A sequence

$$0 \longrightarrow (M',\delta') \xrightarrow{\varphi} (M,\delta) \xrightarrow{\Psi} (M'',\delta'') \longrightarrow 0$$

of complexes and chain maps is said to be exact, if

$$\begin{array}{ccccccccc}
0 & \longrightarrow & M' & \xrightarrow{\varphi'} & M & \xrightarrow{\Psi'} & M'' & \longrightarrow & 0 \\
 & & \delta' \downarrow & & \delta \downarrow & & \delta'' \downarrow & & \\
0 & \longrightarrow & M' & \xrightarrow{\varphi} & M & \xrightarrow{\Psi} & M'' & \longrightarrow & 0
\end{array}$$

is a commutative diagram with exact rows.

2.7 Theorem (Exact triangle theorem): Given an exact sequence

$$E : 0 \longrightarrow (M',\delta') \xrightarrow{\varphi} (M,\delta) \xrightarrow{\Psi} (M'',\delta'') \longrightarrow 0$$

of complexes and chain maps.

Then there exists an exact triangle (cf. I, (2.1))

$$\begin{array}{ccc} H(M',\delta') & \xrightarrow{\hat{\varphi}} & H(M,\delta) \\ {\scriptstyle \Delta_E}\nwarrow & & \swarrow{\scriptstyle \hat{\psi}} \\ & H(M'',\delta'') & \end{array}$$

(this means the triangle is exact at every corner); Δ_E is called the <u>connecting homomorphism</u>.

<u>Proof</u>: <u>Definition of</u> Δ_E. We have the commutative diagram with exact rows

$$\begin{array}{ccccccccc} 0 & \longrightarrow & M' & \xrightarrow{\varphi} & M & \xrightarrow{\psi} & M'' & \longrightarrow & 0 \\ & & \downarrow{\scriptstyle \delta'} & & \downarrow{\scriptstyle \delta} & & \downarrow{\scriptstyle \delta''} & & \\ 0 & \longrightarrow & M' & \xrightarrow{\varphi} & M & \xrightarrow{\psi} & M'' & \longrightarrow & 0 \end{array} .$$

By (Ex. 2,1), there exists an R-homomorphism

$\Delta_E' : \operatorname{Ker} \delta'' \longrightarrow \operatorname{Coker} \delta'$, i.e.,

$$\Delta_E' : Z'' \longrightarrow M'/B'$$
$$\Delta_E' : z'' \longmapsto m' + B',$$

where the construction of m' is indicated by the diagram

$$\begin{array}{ccccccc} & & & & M & \xrightarrow{\psi} & M'' \longrightarrow 0 \\ & & & & \downarrow{\scriptstyle \delta} & & \\ 0 & \longrightarrow & M' & \xrightarrow{\varphi} & M & & \end{array} \qquad \begin{array}{ccc} & m & \overset{\psi}{\longmapsto} z'' \\ & \downarrow{\scriptstyle \delta} & \\ m' \overset{\varphi}{\longmapsto} & m\delta & \end{array} .$$

Moreover,

$$m'\delta'\varphi = m'\varphi\delta$$
$$= m\delta\delta = 0.$$

Since φ is monic,

$$m'\delta' = 0;\ \text{i.e.,}\quad m' \in Z'.$$

Then Δ_E' induces a map

$$\Delta_E'' : Z'' \longrightarrow H(M',\delta'),$$
$$\Delta_E'' : z'' \longmapsto m' + B'.$$

Moreover, if $z'' \in Z''$ is in B'', then $z''\Delta_E'' = 0$, and Δ_E'' induces an R-homomorphism $\Delta_E : H(M'',\delta'') \longrightarrow H(M',\delta')$, the <u>connecting homomorphism</u>.

The <u>proof of the exactness of the triangle in</u> (2.7) is left as an exercise. ∎

2.8 <u>Theorem</u>: Let

$$\begin{array}{lccccccccc} E : 0 & \longrightarrow & (M',\delta') & \xrightarrow{\varphi} & (M,\delta) & \xrightarrow{\psi} & (M'',\delta'') & \longrightarrow & 0 \\ & & \downarrow{\scriptstyle\rho} & & \downarrow{\scriptstyle\sigma} & & \downarrow{\scriptstyle\tau} & & \\ \tilde{E} : 0 & \longrightarrow & (N',\tilde{\delta}') & \xrightarrow{\varphi'} & (N,\tilde{\delta}) & \xrightarrow{\psi'} & (N'',\tilde{\delta}'') & \longrightarrow & 0 \end{array}$$

be a commutative diagram, with chain maps ρ, σ, τ and exact sequences of complexes and chain maps. Then the following prism has commutative sides and exact triangles:

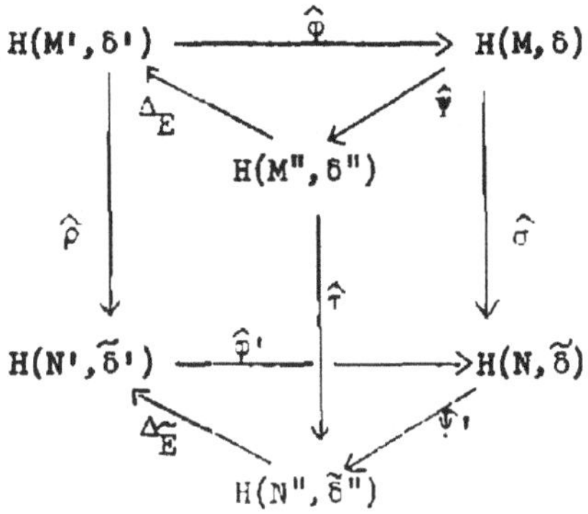

.

In other words, the functor H induces a functor from exact sequences of complexes to exact triangles.

__Proof__: (i) From the commutative diagram below $\Delta_E\hat{\rho} = \hat{\tau}\Delta_{\tilde{E}}$ follows immediately (cf. construction of $\Delta!$).

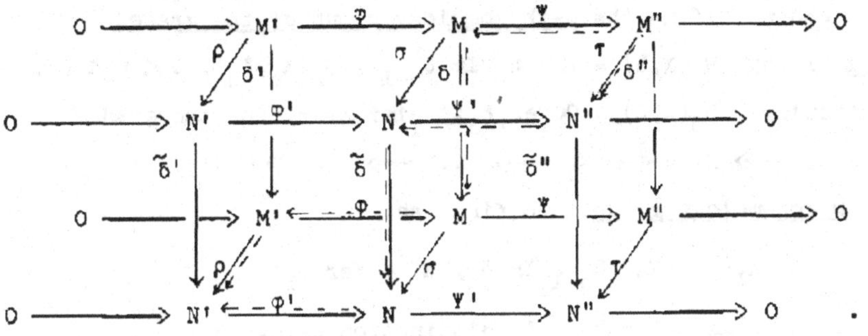

(ii) $\hat{\rho}\varphi' = (\widehat{\rho\varphi'}) = (\widehat{\varphi\sigma}) = \hat{\varphi}\hat{\sigma}$.

(iii) $\hat{\psi}\hat{\tau} = (\widehat{\psi\tau}) = (\sigma\psi') = \hat{\sigma}\hat{\psi}'$ (cf. (2.3)). #

2.9 __Definitions__: Let $X_i \in {}_R\underline{\underline{M}}$, $0 \leq i < \infty$, $\delta_i \in \mathrm{Hom}_R(X_i, X_{i-1})$, $0 < i < \infty$, such that $\delta_i\delta_{i-1} = 0$, $0 < i < \infty$. The sequence

$$X : \ldots \longrightarrow X_{i+1} \xrightarrow{\delta_{i+1}} X_i \xrightarrow{\delta_i} X_{i-1} \rightarrow \ldots \xrightarrow{\delta_1} X_0 \xrightarrow{\delta_0} 0$$

is called a __graded complex of R-modules__ and the δ_i, $0 < i < \infty$, are called __differentiations__.

With each graded complex X, we may associate a __complex__ (X,δ), (cf. (2.1)), in the following way: Let $X = \bigoplus_{i=0}^{\infty} X_i$ be the coproduct of the family $\{X_i : 0 \leq i < \infty\}$, (cf. Ex. I, 1,2). Then $X \in {}_R\underline{\underline{M}}$ is called a __graded left R-module__, and $\delta : X \longrightarrow X$,

$$\delta : \sum_{i=0}^{\infty} x_i \longmapsto \sum_{i=0}^{\infty} x_i\delta_i \quad \text{is a differentiation on } X,$$

(cf. (2.1)). It should be observed that in $\sum_{i=0}^{\infty} x_i \in X$ only finitely many entries are different from zero. Since δ maps

X_i into X_{i-1}, δ is said to be <u>homogeneous of degree -1</u> on the graded module X. The homology group of (X,δ),

$H(X,\delta) = \operatorname{Ker} \delta/\operatorname{Im} \delta = \bigoplus_{i=0}^{\infty} \operatorname{Ker} \delta_i/\operatorname{Im} \delta_{i+1}$, is also a graded module, and we define the <u>n-th homology group of the graded complex</u> X by $H_n(X,\delta) = \operatorname{Ker} \delta_n/\operatorname{Im} \delta_{n+1} = H_n(X_n,\delta_n)$, $0 < n < \infty$, in particular, $H_0(X,\delta) = \operatorname{Coker} \delta_1$. Similarly, for the graded complex $0 \longrightarrow X_0 \xrightarrow{\delta_1} X_1 \longrightarrow \cdots \longrightarrow X_n \xrightarrow{\delta_n} X_{n+1} \longrightarrow \cdots$ the <u>n-th cohomology group</u> is defined as

$$H^n(X,\delta) = \operatorname{Ker} \delta_{n+1}/\operatorname{Im} \delta_n,\ H^0 = \operatorname{Ker} \delta_1;$$

however, we shall not make this distinction here.

Let (X,δ), (X',δ'') be two graded complexes of R-modules. A <u>chain map</u> of these graded complexes, $\varphi : (X,\delta) \longrightarrow (X',\delta')$ is a family $\varphi_i \in \operatorname{Hom}_R(X_i,X_i')$, $0 < i < \infty$, such that for every $0 < i < \infty$, the following diagram is commutative:

$$\begin{array}{ccc} X_i & \xrightarrow{\delta_i} & X_{i-1} \\ \downarrow{\scriptstyle \varphi_i} & & \downarrow{\scriptstyle \varphi_{i-1}} \\ X_i' & \xrightarrow{\delta_i'} & X_{i-1}' \end{array} ;$$

i.e., $\varphi\delta' = \delta\varphi$ in symbolic notation. Again, the graded complexes and chain maps form a category.

From (2.2) it follows that a chain map $\varphi : (X,\delta) \longrightarrow (X',\delta')$ of graded complexes induces an R-homomorphism of graded left R-modules:

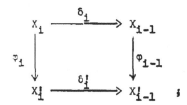

$$\begin{array}{rl} \hat{\varphi} : & H(X,\delta) \longrightarrow H(X',\delta') \\ \hat{\varphi}_i : & H(X_i,\delta_i) \longrightarrow H(X_i',\delta_i') \\ & \quad \| \qquad\qquad\qquad \| \\ & H_i(X,\delta) \longrightarrow H_i(X',\delta') \quad ; \end{array}$$

one says that φ is <u>homogeneous of degree zero</u>. If (X',δ'), (X,δ), (X'',δ'') are graded complexes of R-modules, and $\varphi : (X',\delta') \longrightarrow (X,\delta)$, $\Psi : (X,\delta) \longrightarrow (X'',\delta'')$ are chain maps, then

$$0 \longrightarrow (X',\delta') \xrightarrow{\varphi} (X,\delta) \xrightarrow{\Psi} (X'',\delta'') \longrightarrow 0$$

is an exact sequence of graded complexes, if, for every i, the diagram

$$\begin{array}{ccc} 0 & & 0 \\ \downarrow & & \downarrow \\ X_i' & \xrightarrow{\delta_i'} & X_{i-1} \\ \varphi_i \downarrow & & \downarrow \varphi_{i-1} \\ X_i & \xrightarrow{\delta_i} & X_{i-1} \\ \Psi_i \downarrow & & \downarrow \Psi_{i-1} \\ X_i'' & \xrightarrow{\delta_i''} & X_{i-1}'' \\ \downarrow & & \downarrow \\ 0 & & 0 \end{array}$$

is commutative, and the columns are short exact sequences of R-modules.

2.10 <u>Theorem (Exact homology sequence)</u>: Let

$$0 \longrightarrow (X',\delta') \xrightarrow{\varphi} (X,\delta) \xrightarrow{\Psi} (X'',\delta'') \longrightarrow 0$$

be an exact sequence of graded complexes. Then

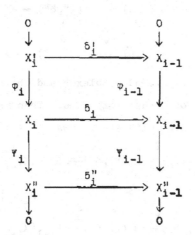

$$\begin{array}{ccc} H(X',\delta') & \xrightarrow{\hat{\varphi}} & H(X,\delta) \\ & {\scriptstyle\Delta}\nwarrow \quad \swarrow{\scriptstyle\hat{\psi}} & \\ & H(X'',\delta'') & \end{array}$$

is an exact triangle, where Δ is the <u>connecting homomorphism</u> (cf. (2.7)); i.e., in terms of the modules, we have the long exact sequence $(1 < i < \infty)$

$$\ldots \xrightarrow{\Delta_{i+1}} H_i(X_i',\delta_i') \xrightarrow{\hat{\varphi}_i} H_i(X_i,\delta_i) \xrightarrow{\hat{\psi}_i} H_i(X_i'',\delta_i'') \longrightarrow$$
$$\xrightarrow{\Delta_i} H_{i-1}(X_{i-1}',\delta_{i-1}) \xrightarrow{\hat{\varphi}_{i-1}} \ldots \quad .$$

<u>Proof</u>: This is an immediate consequence of (2.7) (cf. Ex. 2,2).#

2.11 <u>Theorem</u>: Let

$$\begin{array}{ccccccccc} 0 & \longrightarrow & (X',\delta') & \xrightarrow{\varphi} & (X,\delta) & \xrightarrow{\psi} & (X'',\delta'') & \longrightarrow & 0 \\ & & {\scriptstyle\rho}\downarrow & & {\scriptstyle\sigma}\downarrow & & {\scriptstyle\tau}\downarrow & & \\ 0 & \longrightarrow & (Y',\tilde{\delta}') & \xrightarrow{\varphi'} & (Y,\tilde{\delta}) & \xrightarrow{\psi'} & (Y'',\tilde{\delta}'') & \longrightarrow & 0 \end{array}$$

be a commutative diagram of graded complexes and chain maps, where the rows are exact sequences of graded complexes. Then we have the following commutative diagram with exact rows $(1 < i < \infty)$:

$$\begin{array}{ccccccccccc} \ldots\longrightarrow & H_i(X_i',\delta_i') & \xrightarrow{\hat{\varphi}_i} & H_i(X_i,\delta_i) & \xrightarrow{\hat{\psi}_i} & H_i(X_i'',\delta_i'') & \xrightarrow{\Delta_i} & H_{i-1}(X_{i-1}',\delta_{i-1}') & \longrightarrow \ldots \\ & {\scriptstyle\hat{\rho}_i}\downarrow & & {\scriptstyle\hat{\sigma}_i}\downarrow & & {\scriptstyle\hat{\tau}_i}\downarrow & & {\scriptstyle\hat{\rho}_{i-1}}\downarrow & \\ \ldots\longrightarrow & H_i(X_i',\tilde{\delta}_i') & \xrightarrow{\hat{\varphi}_i'} & H_i(X_i,\tilde{\delta}_i) & \xrightarrow{\hat{\psi}_i'} & H_i(X_i'',\tilde{\delta}'') & \xrightarrow{\tilde{\Delta}_i} & H_{i-1}(X_{i-1}',\tilde{\delta}_{i-1}') & \longrightarrow \ldots \end{array}$$

<u>Proof</u>: This is an immediate consequence of (2.8) (cf. Ex. 2,3).#

2.12 <u>Lemma</u>: Let $\underline{\underline{F}} : {}_{\underline{\underline{R}}}\underline{\underline{M}} \longrightarrow {}_{\underline{\underline{S}}}\underline{\underline{M}}$ be an exact covariant functor (i.e., $\underline{\underline{F}}$ is left exact and right exact, cf. (1.8)). Let

$$(X,\delta) : \ldots\longrightarrow X_n \xrightarrow{\delta_n} X_{n-1} \longrightarrow\ldots\longrightarrow X_1 \xrightarrow{\delta_1} X_0 \xrightarrow{\delta_0} M \longrightarrow 0$$

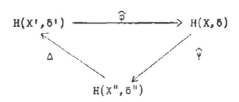

be a complex. (The slight change in the indices (cf. (2.9)) is self-explanatory.) Then

$$(\underline{\underline{F}}(X),\underline{\underline{F}}(\delta)) : \ldots \longrightarrow \underline{\underline{F}}(X_n) \xrightarrow{\underline{\underline{F}}(\delta_n)} \underline{\underline{F}}(X_{n-1}) \longrightarrow \ldots \longrightarrow \underline{\underline{F}}(X_o) \xrightarrow{\underline{\underline{F}}(\delta_o)} \underline{\underline{F}}(M) \longrightarrow 0$$

is a complex, and we have $\underline{\underline{F}}(H(X,\delta)) \overset{\text{nat.}}{\cong} H(\underline{\underline{F}}(X),\underline{\underline{F}}(\delta))$.

<u>Proof</u>: Because of the connection between a complex and a graded complex (cf. (2.8)), it suffices to show: If (M,δ) is a complex, then

$$\underline{\underline{F}}(H(M,\delta)) \overset{\text{nat.}}{\cong} H(\underline{\underline{F}}(M),\underline{\underline{F}}(\delta));$$

but this follows from (Ex. 2,4). #

<u>Exercises §2:</u>

1.) Let R be a ring and let

$$\begin{array}{ccccccc}
 & & M' & \xrightarrow{\varphi} & M & \xrightarrow{\psi} & M'' & \longrightarrow & 0 \\
 & & \alpha \downarrow & & \beta \downarrow & & \gamma \downarrow & & \\
0 & \longrightarrow & N' & \xrightarrow{\tilde{\varphi}} & N & \xrightarrow{\tilde{\psi}} & N'' & &
\end{array}$$

be a commutative diagram of left R-modules with exact rows. Show that there exists an R-homomorphism

$$\Delta' : \operatorname{Ker} \gamma \longrightarrow \operatorname{Coker} \alpha,$$

defined schematically by

$$\begin{array}{ccccc}
 & & \underline{m} & \xrightarrow{\psi} & m'' \\
 & & \beta \downarrow & & \\
n' & \xrightarrow{\tilde{\varphi}} & m\beta & &
\end{array}$$

$\Delta' : m'' \longmapsto n' + \operatorname{im} \alpha.$

(This exercise is known as the "<u>serpent lemma</u>".)

2.) Prove (2.10). (Hint: The diagram for the construction of Δ_i now has the following form:

$$\begin{array}{ccccccccc}
0 & \longrightarrow & X_i' & \xrightarrow{\varphi_i} & X_i & \xrightarrow{\psi_i} & X_{i-1} & \longrightarrow & 0 \\
 & & \downarrow{\scriptstyle \delta_i'} & & \downarrow{\scriptstyle \delta_i} & & \downarrow{\scriptstyle \delta_i''} & & \\
0 & \longrightarrow & X_{i-1}' & \xrightarrow{\varphi_{i-1}} & X_{i-1} & \xrightarrow{\psi_{i-1}} & X_{i-1}'' & \longrightarrow & 0 .
\end{array}$$

3.) Prove (2.11).

4.) Let $\underline{\underline{F}} : {}_{\underline{\underline{R}}}M \longrightarrow {}_{\underline{\underline{S}}}M$ be an exact covariant functor between two categories of modules. Show, that for a complex (M,δ), $M \in {}_{\underline{\underline{R}}}M$, we have

$$H(\underline{\underline{F}}(M),\underline{\underline{F}}(\delta)) \cong \underline{\underline{F}}(H(M,\delta)),$$

where this is a natural isomorphism.

We shall give now a more <u>categorical approach to homology</u>, i.e., to (2.1) - (2.8): We assume $\underline{\underline{C}}$ to be any (additive) category with kernels, cokernels and 0's.

5.) We call a sequence (φ,ψ), $\xrightarrow{\varphi}\ \xrightarrow{\psi}$, <u>exact</u> if and only if

$$\varphi\psi = \ker\varphi \cdot \operatorname{coker}\psi = 0.$$

(a) Show that the following are equivalent.

(i) $\xrightarrow{\varphi}\ \xrightarrow{\psi}$ is exact.

(ii) there exists a pair of morphisms μ, ν so that diagram (1) commutes.

(1) 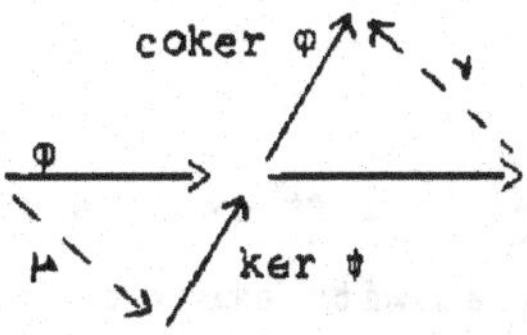

(2) 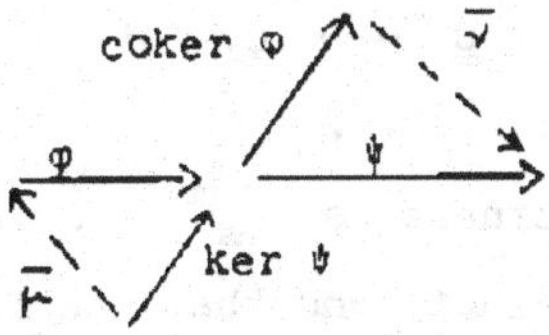

(iii) there exists a pair of morphisms $\bar{\mu}$, $\bar{\nu}$ so that diagram (2) commutes.

(iv) Ker Ψ = Im φ (where we write = for the natural isomorphisms discussed in Ex. 1.3 (a)).

We call a sequence exact if all its consecutive pairs of morphisms are exact.

(b) Show that a pair of morphisms $\xrightarrow{\varphi}$, $\xrightarrow{\Psi}$ can be connected to an exact sequence $\xrightarrow{\varphi} \xrightarrow{\chi} \xrightarrow{\Psi}$ if and only if there exists an isomorphism σ : Coker $\varphi \xrightarrow{\sim}$ Ker Ψ.

(c) Show that an exact functor between semiexact categories (cf. Ex. 1,4) preserves all exact sequences, and that a covariant left (right) exact functor between such categories preserves kernels (cokernels), i.e., it preserves exact sequences $0 \longrightarrow A \longrightarrow B \longrightarrow C$; dually for contravariant.

6.) Let $\underline{\underline{C}}$ be a category as in 5.). Define the category $\underline{\underline{C}}^m$ of morphisms and diagrams as follows. The objects are the morphisms of $\underline{\underline{C}}$, or, to be more explicit, the triples (A,A',α) with $A,A' \in ob(\underline{\underline{C}})$, $\alpha \in morph_{\underline{\underline{C}}}(A,A')$. The morphisms in $morph_{\underline{\underline{C}}^m}(\alpha,\beta)$ are the commutative diagrams

$$\begin{array}{ccc} A & \xrightarrow{\varphi} & B, \\ \alpha \downarrow & & \downarrow \beta \\ A' & \xrightarrow{\Psi} & B' \end{array}$$

i.e., they are induced by pairs of morphisms φ, Ψ of $\underline{\underline{C}}$ for which the above diagrams commute.

(a) Show that $\underline{\underline{Ker}} : \underline{\underline{C}}^m \longrightarrow \underline{\underline{C}}$, with $\alpha \longmapsto$ Ker α

and $\qquad \underline{\underline{\text{Coker}}} : \underline{\underline{C}}^m \longrightarrow \underline{\underline{C}}$, with $\alpha \longrightarrow \text{Coker}\ \alpha$

are (additive) functors.

(b) Define exactness for $\underline{\underline{C}}^m$ and show that $\underline{\underline{\text{Ker}}}$ is left exact and $\underline{\underline{\text{Coker}}}$ is right exact, and that the following diagram with exact rows (and obviously exact columns) can uniquely be completed to a commutative diagram, with all rows exact.

$$\begin{array}{ccccccccc}
 & & \text{Ker}\ \alpha & \overset{\hat{\Phi}}{\dashrightarrow} & \text{Ker}\ \beta & \overset{\hat{\Psi}}{\dashrightarrow} & \text{Ker}\ \gamma & \dashrightarrow & \text{Coker}\ \hat{\Psi} \\
 & & \downarrow & & \downarrow & & \downarrow & & \\
 & & A & \xrightarrow{\varphi} & B & \xrightarrow{\psi} & C & \longrightarrow & 0 \\
 & & \alpha \downarrow & & \beta \downarrow & & \gamma \downarrow & & \\
0 & \longrightarrow & A' & \xrightarrow{\mu} & B' & \xrightarrow{\nu} & C' & & \\
 & & \downarrow & & \downarrow & & \downarrow & & \\
\text{Ker}\ \hat{u} & \dashrightarrow & \text{Coker}\ \alpha & \overset{\hat{u}}{\dashrightarrow} & \text{Coker}\ \beta & \overset{\hat{\nu}}{\dashrightarrow} & \text{Coker}\ \gamma & &
\end{array}$$

(c) Assume now that the morphisms of $\underline{\underline{C}}$ are set maps such that all monomorphisms are injective and all epimorphisms are surjective. Show that there exists an isomorphism $\sigma : \text{Coker}\ \hat{\Psi} \longrightarrow \text{Ker}\ \hat{u}$, defined by $\sigma : x \longmapsto y$, along the schema of the serpent lemma, i.e., $x = \bar{c}$, $c = b\Psi$, $b\beta = a\mu$, $\bar{a} = y$, where we use "$-$" to indicate the cokermorphisms. It has to be shown that σ is well defined and maps into $\text{Ker}\ \hat{u}$. Then define $\tau : \text{Ker}\ \hat{\mu} \longrightarrow \text{Coker}\ \hat{\Psi}$, by $\tau : u \longmapsto v$, via $u = \bar{a}$, $a\mu = b\beta$, $\overline{b\Psi} = v$, and show that $\sigma\tau = 1_{\text{Coker}\ \hat{\Psi}}$ and $\tau\sigma = 1_{\text{Ker}\ \hat{u}}$. (Note that this can be done abstractly in any semiexact category (cf. Ex. 1,4(a)), but it is extremely tedious and quite unrewarding.)

7.) Let (M, δ) be a complex in $\underline{\underline{C}}$, i.e., $M \in \text{ob}\ \underline{\underline{C}}$ and $\delta \in \text{morph}_{\underline{\underline{C}}}(M, M)$ with $\delta^2 = 0$.

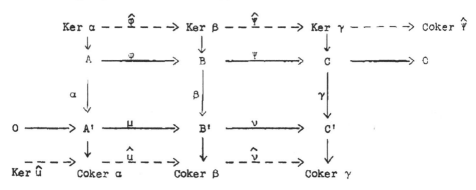

(a) Show that there exist two unique morphisms δ_0 and ∂ such that the following diagrams commute:

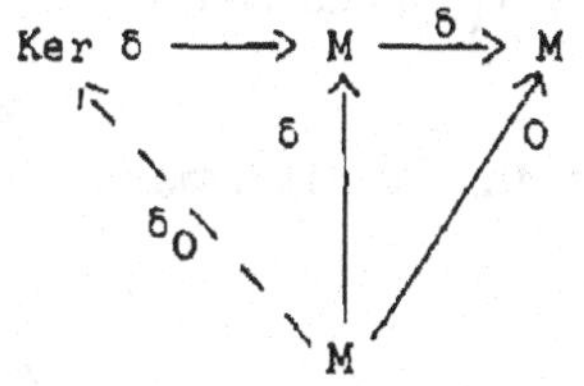

and

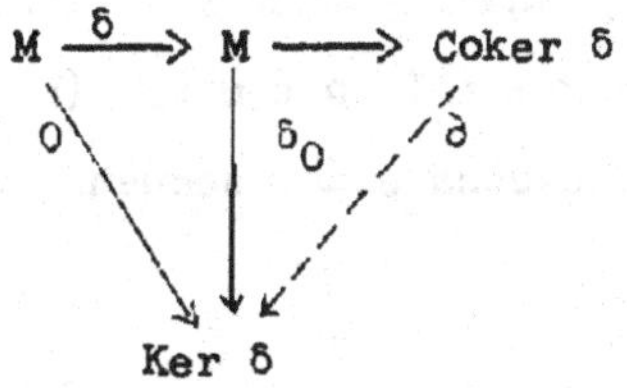

and that $\operatorname{Ker}\delta_0 = \operatorname{Ker}\delta$, $\operatorname{Coker}\partial = \operatorname{Coker}\delta_0 = \operatorname{Ker}\partial = H(M,\delta)$. Thus we have the two exact sequences

$$0 \longrightarrow \operatorname{Ker}\delta \xrightarrow{\ker\delta_0} M \xrightarrow{\delta_0} \operatorname{Ker}\delta \xrightarrow{\operatorname{coker}\delta_0} H(M,\delta) \longrightarrow 0,$$

and

$$0 \longrightarrow H(M,\delta) \xrightarrow{\ker\partial} \operatorname{Coker}\delta \xrightarrow{\partial} \operatorname{Ker}\delta \xrightarrow{\operatorname{coker}\partial} H(M,\delta) \longrightarrow 0.$$

(b) Conversely to every exact sequence

$$0 \longrightarrow Z \longrightarrow M \xrightarrow{d} Z \longrightarrow H \longrightarrow 0$$

there exists a unique complex, namely (M,δ) with $\delta = d\cdot\ker d$, so that $H = H(M,\delta)$. Use this to show that an exact functor preserves homology; i.e., do Ex. 4 formally.

(c) Use 6(a) and 7(a) to show that

if $\varphi : (M,\delta) \longrightarrow (M',\delta')$ is a chain map, then there exist unique maps φ_0 and φ completing the following diagram

$$\begin{array}{ccccccccc}
0 \longrightarrow & \operatorname{Ker}\delta & \xrightarrow{\ker\delta_0} & M & \xrightarrow{\delta_0} & \operatorname{Ker}\delta & \xrightarrow{\operatorname{coker}\delta_0} & H(M,\delta) & \longrightarrow 0 \\
& \downarrow{\scriptstyle\varphi_0} & & \downarrow{\scriptstyle\varphi} & & \downarrow{\scriptstyle\varphi_0} & & \downarrow{\scriptstyle\hat{\varphi}} & \\
0 \longrightarrow & \operatorname{Ker}\delta' & \xrightarrow{\ker\delta_0'} & M' & \xrightarrow{\delta_0'} & \operatorname{Ker}\delta' & \xrightarrow{\operatorname{coker}\delta_0'} & H(M',\delta') & \longrightarrow 0
\end{array}$$

Use this to show that $(M,\delta) \longmapsto H(M,\delta)$ with $\varphi \longmapsto \hat{\varphi}$ is an additive functor.

(d) Show that, in case $\underline{\underline{C}}$ is additive, $\hat{\varphi} = \hat{\Psi}$ for homotopic morphisms φ, Ψ, by showing that $(\rho\delta'+\delta\rho)_0 = \ker \delta \cdot \rho \cdot \delta'_0$, and hence $\widehat{(\rho\delta'+\delta\rho)} = 0$, for all $\rho \in \operatorname{morph}_{\underline{\underline{C}}}(M,M')$.

(e) Using the second exact sequence of 7(a) obtain the diagram, (cf. 6(b)),

$$\begin{array}{ccccccccc}
 & & H(M',\delta') & \xrightarrow{\hat{\varphi}} & H(M,\delta) & \xrightarrow{\hat{\Psi}} & H(M'',\delta'') & \longrightarrow & \operatorname{Coker} \hat{\Psi} \\
 & & \downarrow{\scriptstyle \ker \partial'} & & \downarrow{\scriptstyle \ker \partial} & & \downarrow{\scriptstyle \ker \partial''} & & \\
 & & \operatorname{Coker} \delta' & \longrightarrow & \operatorname{Coker} \delta & \longrightarrow & \operatorname{Coker} \delta'' & \longrightarrow & 0 \\
 & & \downarrow{\scriptstyle \partial'} & & \downarrow{\scriptstyle \partial} & & \downarrow{\scriptstyle \partial''} & & \\
0 & \longrightarrow & \operatorname{Ker} \delta' & \longrightarrow & \operatorname{Ker} \delta & \longrightarrow & \operatorname{Ker} \delta'' & & \\
 & & \downarrow{\scriptstyle \operatorname{coker} \partial'} & & \downarrow{\scriptstyle \operatorname{coker} \partial} & & \downarrow{\scriptstyle \operatorname{coker} \partial''} & & \\
\operatorname{Ker} \hat{\varphi} & \longrightarrow & H(M',\delta') & \longrightarrow & H(M,\delta) & \longrightarrow & H(M'',\delta'') & &
\end{array}$$

and use 6(c) to prove the exact triangle theorem and the prism theorem. (Note that it is only at this point that we are relying on our concrete assumptions about $\underline{\underline{C}}$.)

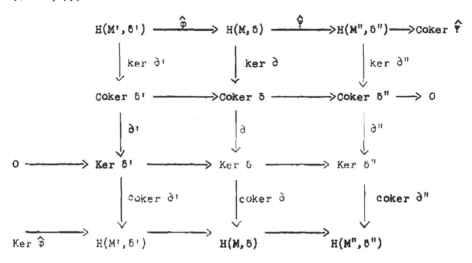

§3. <u>Derived functors.</u>

It is proved that $Ext^n_R(-,N)$ and $Tor^n_R(-,N)$ are functors ${}_R\underline{\underline{M}} \longrightarrow \underline{\underline{A}}$, resp. $\underline{\underline{M}}_R \longrightarrow \underline{\underline{A}}$. The long exact sequences and the exact prism theorem are derived for $Ext^n_R(-,N)$ and $Tor^R_n(-,N)$.

<u>Remark</u>: Since this chapter deals with homological algebra only as far as it is used later for applications to orders, where only finitely generated modules are considered, we define projective resolutions only for finitely generated modules.

<u>In this section, R is a noetherian ring.</u>

3.1 <u>Definition</u>: Let $M \in {}_R\underline{\underline{M}}^f$. A <u>projective resolution</u> for M is an exact sequence

$$P : \ldots \longrightarrow P_2 \xrightarrow{\delta_2} P_1 \xrightarrow{\delta_1} P_o \xrightarrow{\delta_o} M \longrightarrow 0$$

where $P_i \in {}_R\underline{\underline{P}}^f$, (i.e., the P_i are projective left R-modules of finite type).

3.2 <u>Lemma</u>: For $M \in {}_R\underline{\underline{M}}^f$ there exist projective resolutions.

<u>Proof</u>: Pick $P_o \in {}_R\underline{\underline{P}}^f$; e.g., a free module of finite type, such that M is the homomorphic image of P_o; say $M = P_o\delta_o$. Then we obtain the exact sequence

$$0 \longrightarrow \text{Ker } \delta_o \xrightarrow{\iota_o} P_o \xrightarrow{\delta_o} M \longrightarrow 0.$$

Since $P_o \in {}_R\underline{\underline{M}}^f$ and since P_o is noetherian, Ker $\delta_o \in {}_R\underline{\underline{M}}^f$, (cf. I, (4.2)); and we can find $P_1 \in {}_R\underline{\underline{P}}^f$ and δ_1 such that $P_1\delta_1 = \text{Ker } \iota_o$. Now we proceed this way and define inductively a chain of short exact sequences

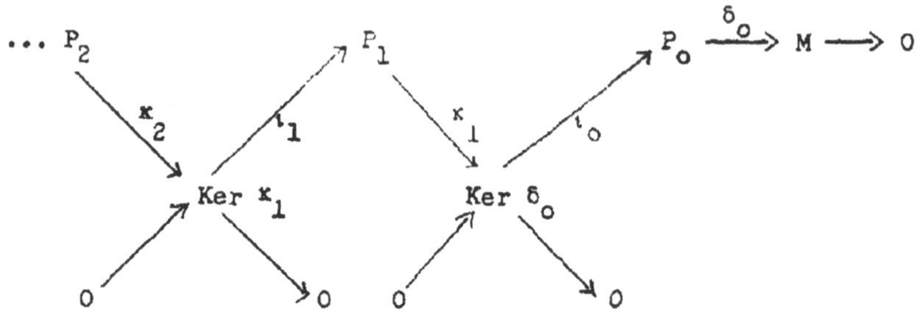

where $P_i \in {}_{\underline{\underline{R}}}\underline{\underline{P}}^f$. If we put, for $i \geq 1$, $\delta_i = \kappa_i \iota_{i-1}$, then

$$\cdots P_i \xrightarrow{\delta_i} P_{i-1} \longrightarrow \cdots \longrightarrow P_1 \xrightarrow{\delta_1} P_o \xrightarrow{\delta_o} M \longrightarrow 0$$

is a projective resolution for M. #

Remark: We point out that there is no uniqueness in the choice of a projective resolution, and that in general a projective resolution has infinite length. It has finite length, if for some i, $\operatorname{Ker} \delta_i \in {}_{\underline{\underline{R}}}\underline{\underline{P}}^f$. Clearly, we can consider a projective resolution

$$P : \cdots \longrightarrow P_2 \xrightarrow{\delta_2} P_1 \xrightarrow{\delta_1} P_o \xrightarrow{\delta_o} M \longrightarrow 0$$

as a graded complex (P, δ) (cf. (2.9)). From the exactness of P it follows that $H(P, \delta) = 0$. A graded complex (X, δ) with $H(X, \delta) = 0$ is called an acyclic graded complex. $H(X, \delta) = 0$ if and only if X is exact.

3.3 Definition: Let S_1, S_2 be rings and

$$\underline{\underline{F}} : {}_{S_1}\underline{\underline{M}} \longrightarrow {}_{S_2}\underline{\underline{M}}$$

a covariant (contravariant) additive functor (cf. (1.3)). If

$$X : \cdots \longrightarrow X_i \xrightarrow{\delta_i} X_{i-1} \longrightarrow \cdots \longrightarrow X_1 \xrightarrow{\delta_1} X_0 \xrightarrow{\delta_0} M \longrightarrow 0$$

is an acyclic graded complex of left R-modules, then

$$\underline{\underline{F}}X : \dots \longrightarrow \underline{\underline{F}}X_i \xrightarrow{\underline{\underline{F}}\delta_i} \underline{\underline{F}}X_{i-1} \longrightarrow \dots \longrightarrow \underline{\underline{F}}X_1 \xrightarrow{\underline{\underline{F}}\delta_1} \underline{\underline{F}}X_o \xrightarrow{\underline{\underline{F}}\delta_o} \underline{\underline{F}}M \longrightarrow 0$$

is a graded complex (cf. (2.9)), because $\underline{\underline{F}}$ is an additive functor (cf. (1.3) and (1.4, iii)). Thus we may form the homology groups $H(\underline{\underline{F}}X, \underline{\underline{F}}\delta)$ of $\underline{\underline{F}}(X)$.

As will be shown in (3.5), the correspondence

$$S_1\underline{\underline{M}} \longrightarrow S_2\underline{\underline{M}}$$
$$M \longmapsto H(\underline{\underline{F}}(X), \underline{\underline{F}}(\delta)) ,$$

which turns out to be independent of the choice of the projective resolution, gives rise to the so-called <u>derived functor of</u> $\underline{\underline{F}}$.

3.4 <u>Examples</u>: (i) Let for some fixed $N \in {}_R\underline{\underline{M}}$ $hom_R(-,N) : {}_R\underline{\underline{M}}^f \longrightarrow \underline{\underline{A}}$, $M \mapsto Hom_R(M,N)$ (cf. (1.5,i)). If

$$P : \dots \longrightarrow P_2 \xrightarrow{\delta_2} P_1 \xrightarrow{\delta_1} P_o \xrightarrow{\delta_o} M \longrightarrow 0$$

is a projective resolution for M (cf. (3.1), (3.2)), we apply $hom_R(-,N)$ to it and obtain the complex

$$hom(P,N) : 0 \longrightarrow Hom_R(M,N) \xrightarrow{hom(\delta_o, 1_N)} Hom_R(P_o,N) \xrightarrow{hom(\delta_1, 1_N)}$$
$$Hom_R(P_1,N) \xrightarrow{hom(\delta_2, 1_N)} Hom_R(P_2,N) \longrightarrow \dots .$$

Since $hom(-,N)$ is left exact (cf. (1.9)), $Ker(hom(\delta_1, 1_N)) = Im(hom(\delta_o, 1_N))$; thus $H_i(hom(P,N), hom(\delta, 1_N)) = 0$, $i = 0, 1$, and we are only interested in the homology groups of

$$hom(P,N)' : 0 \longrightarrow Hom_R(P_o,N) \xrightarrow{\delta_1^*} Hom_R(P_1,N) \xrightarrow{\delta_2^*} Hom_R(P_2,N) \longrightarrow \dots ,$$

where $\delta_i^* = hom(\delta_i, 1_N)$.

The homology groups of $hom(P,N)'$ are denoted by

$$H_i(\hom(P,N)',\delta^*) = \operatorname{Ext}^i_R(M,N)_P = \operatorname{Ker}\delta^*_{i+1}/\operatorname{Im}\delta^*_i,$$

$i = 0, 1, 2, \dots;\ \delta^*_o = 0,$ and the map

$$\operatorname{Ext}^i_R(-,N)_P : {}_R\underline{\underline{M}}^f \longrightarrow \underline{\underline{A}},\ N \longmapsto \operatorname{Ext}^i_R(-,N)_P$$

is called the __i-th right derived functor of__ $\hom(-,N)$, $i = 0, 1, 2, \dots$, induced by the resolution P; however, this will turn out to be independent of P.

(ii) For some fixed $N \in {}_R\underline{\underline{M}}$ let $-\otimes_R N : \underline{\underline{M}}^f_R \longrightarrow \underline{\underline{A}}$, $M \longmapsto M \otimes_R N$ (cf. (1.5,iii)). If

$$P : \dots \longrightarrow P_2 \xrightarrow{\delta_2} P_1 \xrightarrow{\delta_1} P_o \xrightarrow{\delta_o} M \longrightarrow 0$$

is a projective resolution for $M \in \underline{\underline{M}}^f_R$, we apply $-\otimes_R N$ to it to obtain the complex

$$P \otimes_R N : \dots \longrightarrow P_2 \otimes_R N \xrightarrow{\delta_2 \otimes 1_N} P_1 \otimes_R N \xrightarrow{\delta_1 \otimes 1_N} P_o \otimes_R N \xrightarrow{\delta_o \otimes 1_N} M \otimes_R N \longrightarrow 0.$$

Since $-\otimes_R N$ is right exact, (cf. (1.10)), $\operatorname{Im}(\delta_1 \otimes 1_N) = \operatorname{Ker}(\delta_o \otimes 1_N)$, and one considers only the homology groups of the complex

$$(P \otimes_R N)' : \dots \longrightarrow P_3 \otimes_R N \xrightarrow{\delta_3 \otimes 1_N} P_2 \otimes_R N \xrightarrow{\delta_2 \otimes 1_N} P_1 \otimes_R N \xrightarrow{\delta_1 \otimes 1_N}$$
$$P_o \otimes_R N \longrightarrow 0 .$$

The homology groups of $(P \otimes_R N)'$ are denoted by

$$H_i((P \otimes_R N)', \delta \otimes 1_N) = \operatorname{Tor}^R_i(M,N)_P = \operatorname{Ker}(\delta_i \otimes 1_N)/\operatorname{Im}(\delta_{i+1} \otimes 1_N)$$

$$i = 0, 1, 2, \dots,\ \delta_o' \otimes 1 : P_o \otimes_R N \longrightarrow 0.$$

The map $\operatorname{Tor}^R_i(-,N)_P : \underline{\underline{M}}^f_R \longrightarrow \underline{\underline{A}},\ M \longmapsto \operatorname{Tor}^R_i(M,N)_P$, $i = 0,1,2,\dots$ is called the __i-th left derived functor of__ $-\otimes_R N$.

3.5 __Theorem__: The i-th right derived functor of $\hom(-,N)$ is an additive contravariant functor ${}_{\underline{\underline{R}}}\underline{\underline{M}}^f \longrightarrow \underline{\underline{A}}$, and the i-th left derived functor of $-\otimes_R N$ is a covariant functor $\underline{\underline{M}}^f_{\underline{\underline{R}}} \longrightarrow \underline{\underline{A}}$ (cf. (3.4)).

The proof is done in several steps:

First, we show how to define

$$\mathrm{ext}^i_R(\varphi,N)_P : \mathrm{Ext}^i_R(M,N)_P \longrightarrow \mathrm{Ext}^i_R(M',N)_P \quad \text{and}$$

$$\mathrm{tor}^R_i(\varphi,N)_P : \mathrm{Tor}^R_i(M',N)_P \longrightarrow \mathrm{Tor}^R_i(M,N)_P,$$

for $\varphi : M' \longrightarrow M$, M', M of finite type.

3.6 Lemma: Let $M', M \in {}_{\underline{\underline{R}}}\underline{\underline{M}}^f$ and $\varphi \in \mathrm{Hom}_R(M',M)$. If P' and P are projective resolutions for M' and M respectively, then there exists a chain map $\Psi : P' \longrightarrow P$ such that $\Psi_{-1} = \varphi$. Without too much abuse of the notation, we write $\varphi : P' \longrightarrow P$.

Proof: Since $P'_o \in {}_{\underline{\underline{R}}}\underline{\underline{P}}^f$, we can complete the following diagram commutatively:

$$\begin{array}{ccccc} & & P'_o & & \\ & \swarrow^{\Psi_o} & \downarrow{\scriptstyle \delta'_o} & & \\ & & M' & & \\ & & \downarrow{\scriptstyle \varphi} & & \\ P_o & \xrightarrow{\delta_o} & M & \longrightarrow & 0 \end{array} \quad ; \quad \text{i.e.,} \quad \begin{array}{ccccc} P'_o & \xrightarrow{\delta'_o} & M' & \longrightarrow & 0 \\ {\scriptstyle \Psi_o}\downarrow & & \downarrow{\scriptstyle \varphi} & & \\ P_o & \xrightarrow{\delta_o} & M & \longrightarrow & 0 \end{array}$$

is commutative. Now we define $\Psi_i : P'_i \longrightarrow P_i$ recursively: We have the situation

$$D: \quad \begin{array}{ccccc} P'_{i+1} & \xrightarrow{\delta'_{i+1}} & P'_i & \xrightarrow{\delta'_i} & P'_{i-1} \\ {\scriptstyle \Psi_{i+1}}\downarrow & & {\scriptstyle \Psi_i}\downarrow & & \downarrow{\scriptstyle \Psi_{i-1}} \\ P_{i+1} & \xrightarrow{\delta_{i+1}} & P_i & \xrightarrow{\delta_i} & P_{i-1}, \end{array}$$

where, a priori, the right square is commutative. Thus, $\delta'_{i+1}\Psi_i\delta_i = \delta'_{i+1}\delta'_i\Psi_{i-1} = 0$; i.e., $\operatorname{Im}\delta'_{i+1}\Psi_i \subset \operatorname{Ker}\delta_i = \operatorname{Im}\delta_{i+1}$, and consequently, the following diagram can be completed commutatively:

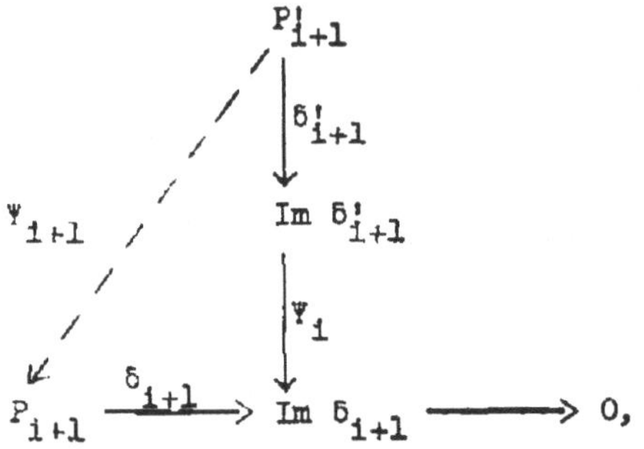

where we have identified the restriction of Ψ_i to $\operatorname{Im}\delta'_{i+1}$ with Ψ_i. Evidently, Ψ_{i+1} also completes the diagram D commutatively (or, more precisely, composed with the injection $\operatorname{Im}\delta_{i+1} \longrightarrow P_i$). #

<u>Definition of</u> $\operatorname{ext}^i_R(\varphi,N)$.

The chain map $\varphi : P' \longrightarrow P$ induces a chain map $\operatorname{hom}(\varphi,1_N) : \operatorname{hom}(P,N)' \longrightarrow \operatorname{hom}(P',N)'$, (cf. (3.4), because of the functoriality of $\operatorname{hom}_R(-,N)$. This in turn induces a family of maps: $\operatorname{ext}^i_R(\varphi,N) : \operatorname{Ext}^i_R(M,N)_P \longrightarrow \operatorname{Ext}^i_R(M',N)_{P'}$, (cf. (2.2)). Similarly $\operatorname{tor}^R_i(\varphi,N) : \operatorname{Tor}^R_i(M',N)_{P'} \longrightarrow \operatorname{Tor}^R_i(M,N)_P$ is defined.

<u>Remark</u>: For a particular projective resolution P of M, $\operatorname{Ext}^i_R(M,N)_P$ and $\operatorname{Tor}^R_i(M,N)_P$ are well defined. However, the chain map $\varphi : P' \longrightarrow P$ (cf. (3.6)) is not uniquely determined by this construction.

3.7 <u>Lemma</u>: If P' and P are projective resolutions of the R-modules of finite type M' and M resp., and if $\varphi_1,\varphi_2 : P' \longrightarrow P$ are two chain maps induced by $\varphi \in \operatorname{Hom}_R(M',M)$, (cf. (3.6)), then $\operatorname{ext}^i_R(\varphi_1,N) = \operatorname{ext}^i_R(\varphi_2,N)$, $\operatorname{tor}^R_i(\varphi_1,N) = \operatorname{tor}^R_i(\varphi_2,N)$.

<u>Proof</u>: We shall show that φ_1 and φ_2 are homotopic (cf. (2.4)).

For then $\hom(\varphi_1, 1_N)$ and $\hom(\varphi_2, 1_N)$ are homotopic (cf. (2.9)) and from (2.5) it will follow that $\mathrm{ext}^i_R(\varphi_1, N) = \mathrm{ext}^i_R(\varphi_2, N)$ and $\mathrm{tor}^R_i(\varphi_1, N) = \mathrm{tor}^R_i(\varphi_2, N)$. Thus we have to show that there exists a map $\rho : P' \longrightarrow P$, $\rho_i : P'_{i-1} \longrightarrow P_i$, such that the following diagram

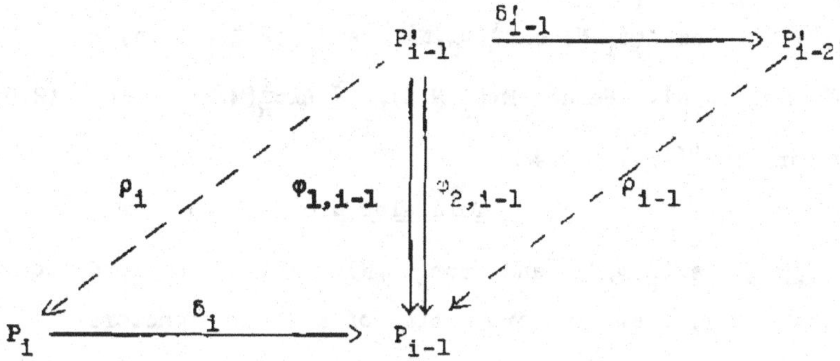

implies $\varphi_{1,i-1} - \varphi_{2,i-1} = \delta'_{i-1}\delta_{i-1} + \rho_i\delta_i$. We define ρ_i recursively: $\rho_o : M' \longrightarrow P_o$ as $\rho_o = 0$. Now, if $\rho_i : P'_{i-1} \longrightarrow P_i$ is constructed, we define ρ_{i+1} by:

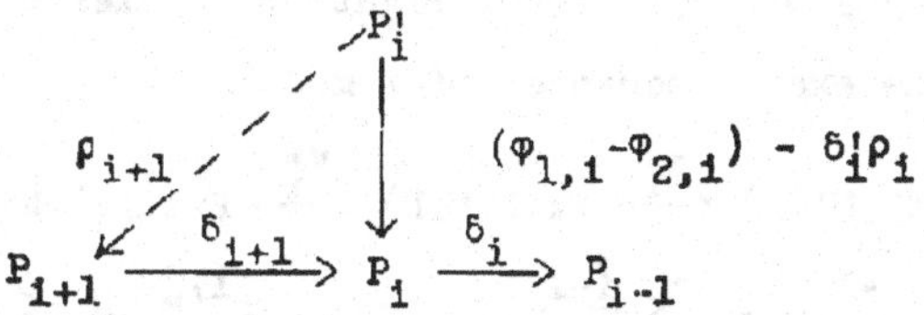

.

This is possible since

$$P'_i(\varphi_{1,i} - \varphi_{2,i} - \delta'_i\rho_i) \subset \mathrm{Im}\ \delta_{i+1} .$$

Now: $\varphi_{1i} - \varphi_{2i} - \delta'_i\rho_i = \rho_{i+1}\delta_{i+1}$. Thus $\varphi_1 \simeq \varphi_2$. #

3.8 Lemma: For $M \in {}_R\underline{\underline{M}}^f$, $M' \in \underline{\underline{M}}^f_R$ and $N \in {}_R\underline{\underline{M}}$, $\mathrm{Ext}^i_R(M,N)_P$ and $\mathrm{Tor}^R_i(M',N)_P$ are independent-up to isomorphism-of the chosen projective resolution, and thus, we shall omit the index P.

Proof: Given two projective resolutions P_1 and P_2 of M. Then the map $1_M : M \longrightarrow M$ induces two chain maps

$$\varphi_1 : P_1 \longrightarrow P_2 \quad \text{and}$$
$$\varphi_2 : P_2 \longrightarrow P_1 \quad (\text{cf. } (3.6))$$

such that $\varphi_1\varphi_2 \simeq 1_{P_2}$ and $\varphi_2\varphi_1 \simeq 1_{P_1}$. Thus, by (3.7), $\mathrm{ext}^i_R(\varphi_2\varphi_1,N) = \mathrm{ext}^i_R(1_{P_1},N)$ and $\mathrm{ext}^i_R(\varphi_1\varphi_2,N) = \mathrm{ext}^i_R(1_{P_2},N)$. But $\mathrm{ext}^i_R(\varphi_2\varphi_1,N) = \mathrm{ext}^i_R(\varphi_1,N)\ \mathrm{ext}^i_R(\varphi_2,N)$, and $\mathrm{ext}^i_R(\varphi_1\varphi_2,N) = \mathrm{ext}^i_R(\varphi_2,N)\ \mathrm{ext}^i_R(\varphi_1,N)$. Hence $\mathrm{Ext}^i_R(M,N)_{P_1} \cong \mathrm{Ext}^i_R(M,N)_{P_2}$ (cf. (2.5)). Similarly for $\mathrm{Tor}^R_i(-,N)$. #

<u>This also proves (3.5)</u>. #

3.9 <u>Lemma</u>: $\mathrm{ext}^i_R(\varphi,N)$ and $\mathrm{tor}^i_R(\varphi,N)$ satisfy the conditions (1.3, i, ii, iii,); i.e., the properties of additive functors.

We leave the <u>verification</u> as an exercise (cf. Ex. 3,1). #

3.10 <u>Theorem</u>: Let $N \in {}_R\underline{\underline{M}}$ and let

$$0 \longrightarrow M' \xrightarrow{\varphi} M \xrightarrow{\psi} M'' \longrightarrow 0$$

be an exact sequence of left (resp. right) R-modules of finite type. Then we obtain the exact sequences of Z-modules:

$$\begin{aligned} \cdots &\longleftarrow \mathrm{Ext}^i_R(M',N) \xleftarrow{\varphi^*_i} \mathrm{Ext}^i_R(M,N) \xleftarrow{\psi^*_i} \mathrm{Ext}^i_R(M'',N) \xleftarrow{\Delta_i} \\ &\longleftarrow \mathrm{Ext}^{i-1}_R(M',N) \xleftarrow{\varphi^*_{i-1}} \cdots \longleftarrow \mathrm{Ext}^1_R(M'',N) \xleftarrow{\Delta_1} \\ &\longleftarrow \mathrm{Hom}_R(M',N) \xleftarrow{\varphi^*} \mathrm{Hom}_R(M,N) \xleftarrow{\psi^*} \mathrm{Hom}_R(M'',N) \longleftarrow 0. \end{aligned}$$

In particular, for each n, $X \in {}_R\underline{\underline{M}}^f$, $\mathrm{Ext}^n_R(-,X)$ is a contravariant half exact functor and

$$\cdots \mathrm{Tor}^R_i(M',N) \xrightarrow{\varphi_i} \mathrm{Tor}^R_i(M,N) \xrightarrow{\psi_i} \mathrm{Tor}^R_i(M'',N) \xrightarrow{\Delta_i} \mathrm{Tor}^R_{i-1}(M',N) \longrightarrow \cdots$$
$$\cdots \mathrm{Tor}^R_1(M'',N) \xrightarrow{\Delta_1} M' \otimes_R N \xrightarrow{\varphi\otimes 1_N} M \otimes_R N \xrightarrow{\psi\otimes 1_N} M'' \otimes_R N \longrightarrow 0.$$

<u>Proof</u>: We first shall show: Given

$$0 \longrightarrow M' \xrightarrow{\varphi} M \xrightarrow{\psi} M'' \longrightarrow 0;$$

then we can find projective resolutions

$$\begin{array}{l} P' \longrightarrow M' \longrightarrow 0 \\ P \longrightarrow M \longrightarrow 0 \\ P'' \longrightarrow M'' \longrightarrow 0, \end{array}$$

such that

$$\begin{array}{ccccccccc} 0 & \longrightarrow & P' & \xrightarrow{\varphi} & P & \xrightarrow{\psi} & P'' & \longrightarrow & 0 \\ & & \downarrow & & \downarrow & & \downarrow & & \\ 0 & \longrightarrow & M' & \xrightarrow{\varphi} & M & \longrightarrow & M'' & \longrightarrow & 0 \\ & & \downarrow & & \downarrow & & \downarrow & & \\ & & 0 & & 0 & & 0 & & \end{array}$$

is an exact sequence of graded complexes (cf. (2.9)).

Let

$$\begin{array}{l} (P',\delta') \longrightarrow M' \longrightarrow 0 \quad \text{and} \\ (P'',\delta'') \longrightarrow M'' \longrightarrow 0 \end{array}$$

be projective resolutions of M' and M'' resp. We have to fill in the following diagram commutatively:

$$\begin{array}{cccccccccccc} & & 0 & & 0 & & 0 & & 0 & & 0 & \\ & & \downarrow & & \downarrow & & \downarrow & & \downarrow & & \downarrow & \\ & \longrightarrow & P'_i & \xrightarrow{\delta'_i} & P'_{i-1} & \longrightarrow \cdots \longrightarrow & P'_1 & \xrightarrow{\delta'_1} & P'_o & \xrightarrow{\delta'_o} & M' & \longrightarrow 0 \\ & & \downarrow \iota'_i & & \downarrow \iota'_{i-1} & & \downarrow \iota'_1 & & \downarrow \iota'_o & & \downarrow \varphi & \\ D: & \dashrightarrow & P_i & \xrightarrow{\delta_i} & P_{i-1} & \dashrightarrow \cdots \dashrightarrow & P_1 & \xrightarrow{\delta_1} & P_o & \xrightarrow{\delta_o} & M & \longrightarrow 0 \\ & & \downarrow \pi''_i & & \downarrow \pi''_{i-1} & & \downarrow \pi''_1 & & \downarrow \pi''_o & & \downarrow \psi & \\ & \longrightarrow & P''_i & \xrightarrow{\delta''_i} & P''_{i-1} & \longrightarrow \cdots \longrightarrow & P''_1 & \xrightarrow{\delta''_1} & P''_o & \xrightarrow{\delta''_o} & M'' & \longrightarrow 0, \\ & & \downarrow & & \downarrow & & \downarrow & & \downarrow & & \downarrow & \\ & & 0 & & 0 & & 0 & & 0 & & 0 & \end{array}$$

where $P_i = P'_i \oplus P''_i$ and $\iota_i : P'_i \longrightarrow P$ and $\pi''_i : P_i \longrightarrow P''_i$ are the corresponding injections and projections.

With the same method as in the proof of (3.6) we can fill in the following diagram commutatively:

$$\begin{array}{ccccccccccc}
\cdots & \longrightarrow & P''_2 & \xrightarrow{\delta''_2} & P''_1 & \xrightarrow{\delta''_1} & P''_o & \xrightarrow{\delta''_o} & M'' & \longrightarrow & 0 \\
 & & \Big\downarrow{\scriptstyle \sigma_2} & & \Big\downarrow{\scriptstyle \sigma_1} & & \Big\downarrow{\scriptstyle \sigma_o} & & \Big\downarrow{\scriptstyle 1_{M''}} & & \\
\cdots & \longrightarrow & P'_1 & \xrightarrow{\delta'_1} & P'_o & \xrightarrow{\delta'_o\varphi} & M & \xrightarrow{\Psi} & M'' & \longrightarrow & 0
\end{array}$$

(observe that in the proof of (3.6) we have only used that the toprow was a projective resolution and the bottom row was exact (cf. Ex. 3,2)). We have to show that the bottom row is exact: $\operatorname{Ker}\Psi = \operatorname{Im}\varphi = \operatorname{Im}\delta_o\varphi$ since δ_o is epic; and $\operatorname{Im}\delta'_1 = \operatorname{Ker}\delta'_o = \operatorname{Ker}\delta'_o\varphi$ since φ is monic. Now we can define $\delta_i : P_i \longrightarrow P_{i-1}$,

$$* \quad \begin{cases} i > 0, \ \delta_i = \pi'_i\delta'_i\iota'_{i-1} + \pi''_i\delta''_i\iota''_{i-1} + (-1)^i\pi''_i\sigma_i\iota'_{i-1} \\[2ex] \delta_o = \pi'_o\delta'_o\varphi + \pi''_o\delta''_o\tau_o . \end{cases}$$

This definition makes the diagram D commutative, with exact columns and rows (cf. Ex. 3,4). ⌜ ⌝ we have the exact sequence of graded complexes

$$\begin{array}{ccccccccc}
0 & \longrightarrow & P' & \longrightarrow & P & \longrightarrow & P'' & \longrightarrow & 0 \\
 & & \downarrow & & \downarrow & & \downarrow & & \\
0 & \longrightarrow & M' & \longrightarrow & M & \longrightarrow & M'' & \longrightarrow & 0,
\end{array}$$

where the upper row is a split exact sequence of chain maps, each P''_i being projective. Applying $\hom(-,N)$ to this exact complex gives the exact complex with split exact upper row (cf. (1.6))

$$\begin{array}{ccccccccc}
0 & \longrightarrow & \hom(P'',N) & \longrightarrow & \hom(P,N) & \longrightarrow & \hom(P',N) & \longrightarrow & 0 \\
 & & \uparrow & & \uparrow & & \uparrow & & \\
0 & \longrightarrow & \operatorname{Hom}_R(M'',N) & \longrightarrow & \operatorname{Hom}_R(M,N) & \longrightarrow & \operatorname{Hom}_R(M',N) & & \\
 & & \uparrow & & \uparrow & & \uparrow & & \\
 & & 0 & & 0 & & 0 & &
\end{array}$$

To compute $\mathrm{Ext}^n_R(-,N)$, we have to replace the middle row by zeros (cf. (3.4)).

If we apply the exact triangle theorem (cf. (2.10)) to this exact sequence of graded complexes, we obtain the desired result, if we can show that $\mathrm{Ext}^0_R(-,N) \sim \mathrm{Hom}_R(-,N)$ (cf. (1.11).

Similarly the theorem is proved for $\mathrm{Tor}^R_1(-,N)$, once it is shown that $\mathrm{Tor}^R_0(-,N) \sim - \otimes_R N$. This is done in the next lemma.

3.11 <u>Lemma</u>: We have a natural equivalence between the functors: $\mathrm{Ext}^0_R(-,N) \sim \mathrm{Hom}_R(-,N)$, $\mathrm{Tor}^R_0(-,N) \sim - \otimes_R N$.

<u>Proof</u>: Per definition (cf. (3.4)), we have, using the left exactness of $\mathrm{Hom}_R(-,N)$,
$\mathrm{Ext}^0_R(M,N) = \mathrm{Ker}(\hom(\delta_1,1_N)) = \mathrm{Im}(\hom(\delta_0,1_N)) \cong \mathrm{Hom}_R(M,N)$; and similarly $\mathrm{Tor}^R_0(M,N) \simeq M \otimes_R N$.

It remains to show that these are natural transformations; but this is an immediate consequence of (3.6) and (3.5): A map $\varphi : M \longrightarrow M'$ gives rise to the commutative diagram

$$\begin{array}{ccccccccc}
0 & \longrightarrow & \mathrm{Hom}_R(M,N) & \longrightarrow & \mathrm{Hom}_R(P_0,N) & \longrightarrow & \mathrm{Hom}_R(P_1,N) & \longrightarrow & \dots \\
 & & \uparrow{\scriptstyle \varphi_0^*} & & \uparrow{\scriptstyle \varphi_1^*} & & \uparrow{\scriptstyle \varphi_2^*} & & \\
0 & \longrightarrow & \mathrm{Hom}_R(M',N) & \longrightarrow & \mathrm{Hom}_R(P_0',N) & \longrightarrow & \mathrm{Hom}_R(P_1',N) & \longrightarrow & \dots ,
\end{array}$$

whence

$$\begin{array}{ccc}
\mathrm{Ext}^0_R(M,N) & \xrightarrow{\simeq} & \mathrm{Hom}_R(M,N) \\
\uparrow{\scriptstyle \hat{\varphi}} & & \uparrow{\scriptstyle \varphi_0^*} \\
\mathrm{Ext}^0_R(M',N) & \xrightarrow{\simeq} & \mathrm{Hom}_R(M',N)
\end{array}$$

is commutative. Similarly for $\mathrm{Tor}^R_0(-,N) \simeq - \otimes_R N$, and for maps. #

3.12 <u>Theorem</u>: Let

$$\begin{array}{ccccccccc}
0 & \longrightarrow & M' & \xrightarrow{\varphi} & M & \xrightarrow{\psi} & M'' & \longrightarrow & 0 \\
 & & \downarrow{\scriptstyle\alpha} & & \downarrow{\scriptstyle\beta} & & \downarrow{\scriptstyle\gamma} & & \\
0 & \longrightarrow & L' & \xrightarrow{\sigma} & L & \xrightarrow{\tau} & L'' & \longrightarrow & 0
\end{array}$$

be a commutative diagram with exact rows of left (resp. right) R-modules of finite type. Then for $N \in {}_R\underline{M}$ the following diagrams are commutative with exact rows.

$$\begin{array}{ccccccccccc}
\cdots \longrightarrow & \mathrm{Ext}^i_R(M'',N) & \xrightarrow{\psi^*_i} & \mathrm{Ext}^i_R(M,N) & \xrightarrow{\varphi^*_i} & \mathrm{Ext}^i_R(M',N) & \xrightarrow{\Delta_i} & \mathrm{Ext}^{i+1}_R(M'',N) & \longrightarrow \cdots \\
 & \uparrow{\scriptstyle\alpha^*_i} & & \uparrow{\scriptstyle\beta^*_i} & & \uparrow{\scriptstyle\gamma^*_i} & & \uparrow{\scriptstyle\alpha^*_{i+1}} & \\
\cdots \longrightarrow & \mathrm{Ext}^i_R(L'',N) & \xrightarrow{\tau^*_i} & \mathrm{Ext}^i_R(L,N) & \xrightarrow{\sigma^*_i} & \mathrm{Ext}^i_R(L',N) & \xrightarrow{\Delta_i} & \mathrm{Ext}^{i+1}_R(L'',N) & \longrightarrow \cdots
\end{array}$$

and

$$\begin{array}{ccccccccccc}
\cdots \longrightarrow & \mathrm{Tor}^R_i(M',N) & \xrightarrow{\tilde{\varphi}_i} & \mathrm{Tor}^R_i(M,N) & \xrightarrow{\tilde{\psi}_i} & \mathrm{Tor}^R_i(M'',N) & \xrightarrow{\Delta_i} & \mathrm{Tor}^R_{i-1}(M',N) & \longrightarrow \cdots \\
 & \downarrow{\scriptstyle\tilde{\alpha}_i} & & \downarrow{\scriptstyle\tilde{\beta}_i} & & \downarrow{\scriptstyle\tilde{\gamma}_i} & & \downarrow{\scriptstyle\tilde{\alpha}_{i-1}} & \\
\cdots \longrightarrow & \mathrm{Tor}^R_i(L',N) & \xrightarrow{\tilde{\sigma}_i} & \mathrm{Tor}^R_i(L,N) & \xrightarrow{\tilde{\tau}_i} & \mathrm{Tor}^R_i(L'',N) & \xrightarrow{\Delta_i} & \mathrm{Tor}^R_{i-1}(L',N) & \longrightarrow \cdots
\end{array}$$

<u>Proof</u>: We can find projective resolutions and chain maps such that the following diagram is commutative with split exact columns of graded complexes on the left (cf. (3.6) and the proof of (3.10)):

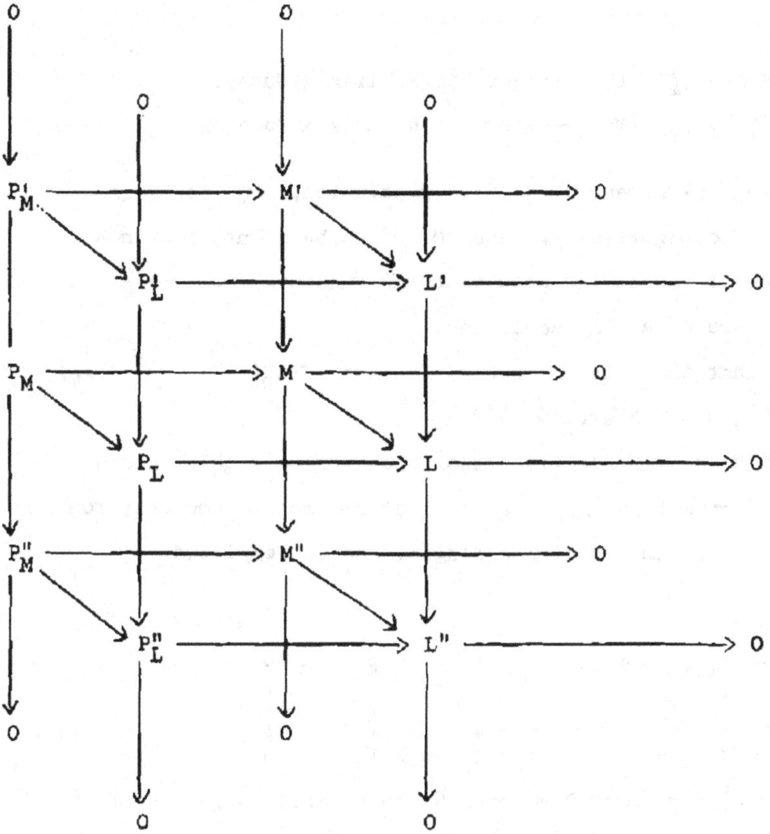

The desired result follows now from theorem (2.11). #

Exercises §3:

1.) Prove (3.9). (It should be observed that there are two things to be shown:

(i) For every $M \in {}_R\underline{\underline{M}}^f$, we select a particular projective resolution P_M, and form the category $\underline{\underline{M}}_P$, where $ob(\underline{\underline{M}}_P)$ are pairs $(M, \underline{\underline{P}}_M)$, $M \in {}_R\underline{\underline{M}}^f$ and the morphisms are homotopy classes $[\varphi]$ of chain maps φ, induced by $\varphi \in Hom_R(M, M')$. Then it is to be shown that

$\underline{\underline{E}}_P^N : \underline{\underline{M}}_P \longrightarrow \underline{\underline{A}}$ ($\underline{\underline{A}}$ = the category of abelian groups),
$(M, P_M) \longmapsto Ext_R^i(M,N)_P$, $[\varphi] \longmapsto ext_R^i(\varphi,N)_P$ is a functor.

(ii) If we choose a different projective resolution, P'_M for each $M \in {}_R\underline{\underline{M}}^f$, then the categories $\underline{\underline{M}}_P$ and $\underline{\underline{M}}_{P'}$ can be identified in a natural way. Now it remains to show that the functors $\underline{\underline{M}}_P \longrightarrow \underline{\underline{A}}$ and $\underline{\underline{M}}_{P'} \longrightarrow \underline{\underline{A}}$ are naturally equivalent.

It follows that there is an induced functor $Ext_R^i(-,N) : {}_R\underline{\underline{M}}^f \longrightarrow \underline{\underline{A}}$, $M \longmapsto Ext_R^i(M,N)$, $\varphi \longmapsto ext_R^i(\varphi,N)$.)

2.) Let $P \longrightarrow M$ be a projective resolution for $M \in {}_R\underline{\underline{M}}^f$, where R is a ring. If $X \longrightarrow M' \longrightarrow 0$ is an exact sequence, show that for every $\varphi \in Hom_R(M,M')$ the following diagram can be completed commutatively:

$$\begin{array}{ccccccccccccc} \cdots P_i & \longrightarrow & P_{i-1} & \longrightarrow & \cdots & \longrightarrow & P_1 & \longrightarrow & P_o & \longrightarrow & M & \longrightarrow & 0 \\ \downarrow & & \downarrow & & & & \downarrow & & \downarrow & & \downarrow \varphi & & \\ \cdots X_i & \longrightarrow & X_{i-1} & \longrightarrow & \cdots & \longrightarrow & X_1 & \longrightarrow & X_o & \longrightarrow & M' & \longrightarrow & 0 \end{array} .$$

3.) Let $0 \longrightarrow M' \longrightarrow M \longrightarrow M'' \longrightarrow 0$ be an exact sequence of left R-modules of finite type. Describe the connecting homomorphism $Ext_R^i(M',N) \xrightarrow{\Delta_i} Ext_R^{i+1}(M'',N)$ explicitly.

4.) Show that the diagram D in the proof of (3.10) becomes commutative if one defines the δ_i as in (*).

§4. Homological dimension.

The "change of rings theorem" for homological dimensions is proved, and the connections between the homological dimensions of the modules in a short exact sequence are derived.

In this section again let R be a noetherian ring.

4.1 Definition: Let $M \in {}_{R}\underline{M}^f$; M has homological dimension n (notation $hd_R(M) = n$) if

(i) there exists a projective resolution of M of length n; i.e., an exact sequence

$$0 \longrightarrow P_n \longrightarrow P_{n-1} \longrightarrow \cdots \longrightarrow P_1 \longrightarrow P_o \longrightarrow M \longrightarrow 0$$

with $P_i \in {}_{R}\underline{P}^f$, $0 \leq i \leq n$, and

(ii) there does not exist a projective resolution of M of length $\leq n - 1$.

4.2 Lemma: Let $M \in {}_{R}\underline{M}^f$. The following conditions are equivalent

(i) $M \in {}_{R}\underline{P}^f$,

(ii) $hd_R(M) = 0$

(iii) $Ext^1_R(M,X) = 0$, $\forall X \in {}_{R}\underline{M}$.

Proof: (i) $\Longrightarrow$ (ii) is trivial. (ii) $\Longrightarrow$ (iii): We have the projective resolution $P : \ldots \longrightarrow 0 \longrightarrow 0 \longrightarrow P_o \longrightarrow M \longrightarrow 0$; i.e., $P_o \simeq M$. This gives rise to the graded complex (cf. (3.4)):

$$hom(P,N)' : 0 \longrightarrow Hom_R(P_o,X) \xrightarrow{\delta_1^*} Hom_R(0,X) \xrightarrow{\delta_2^*} Hom_R(0,X) \longrightarrow \cdots$$

and $Ext^1_R(M,X) = Ker\ \delta_2^*/Im\ \delta_1^* = 0$.

(iii) $\Longrightarrow$ (i): From (3.10), we obtain for every exact sequence of left R-modules of finite type $E : 0 \longrightarrow M_1 \xrightarrow{\varphi} M_2 \xrightarrow{\psi} M \longrightarrow 0$, the exact sequence of $\underline{\underline{Z}}$-modules

$$0 \longrightarrow \mathrm{Hom}_R(M,M_1) \xrightarrow{\psi^*} \mathrm{Hom}_R(M_2,M_1) \xrightarrow{\varphi^*} \mathrm{Hom}_R(M_1,M_1) \longrightarrow 0;$$

i.e., $1_{M_1} = \varphi^*(\sigma)$ for some $\sigma \in \mathrm{Hom}_R(M_2,M_1)$. Hence $1_{M_1} = \varphi\sigma$, and the sequence E is split by (I,(2.2)). Thus $M \in {}_{\underline{\underline{R}}}\underline{P}^f$ by (I,(2.9)).#

4.3 <u>Theorem</u>: Let $M \in {}_{\underline{\underline{R}}}\underline{M}^f$. Then

$$\mathrm{hd}_R(M) < n \iff \mathrm{Ext}^n_R(M,X) = 0, \quad \forall\, X \in {}_{\underline{\underline{R}}}\underline{M}.$$

<u>Proof</u>: " $\Longrightarrow$ ": This direction is as obvious as (ii) $\Longrightarrow$ (iii) of (4.2) and is left as an exercise (Ex. 4,1).

Conversely, from the first part it follows in particular that $P \in {}_{\underline{\underline{R}}}\underline{P}^f$ implies $\mathrm{Ext}^n_R(P,X) = 0$, $\forall\, X \in {}_{\underline{\underline{R}}}\underline{M}$, $\forall\, n \geq 1$. Now, any projective resolution P of M gives rise to the following diagram

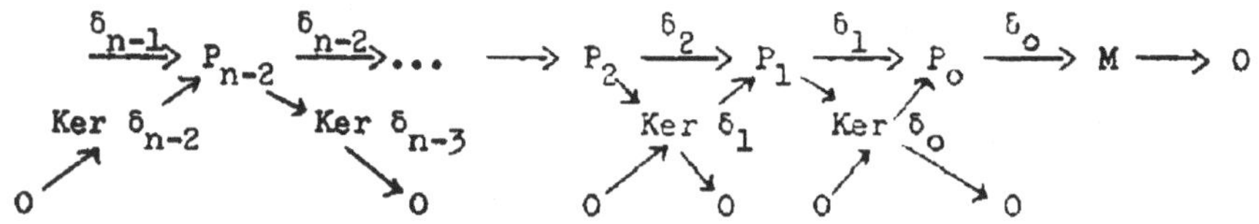

where the sequences $0 \longrightarrow \mathrm{Ker}\ \delta_i \longrightarrow P_i \longrightarrow \mathrm{Ker}\ \delta_{i-1} \longrightarrow 0$ are exact. By (3.10) these sequences induce, for each $X \in {}_{\underline{\underline{R}}}\underline{M}$, an exact sequence

$$0 = \mathrm{Ext}^k_R(P_i,X) \longrightarrow \mathrm{Ext}^k_R(\mathrm{Ker}\ \delta_i,X) \longrightarrow \mathrm{Ext}^{k+1}_R(\mathrm{Ker}\ \delta_{i-1},X) \longrightarrow$$
$$\longrightarrow \mathrm{Ext}^{k+1}_R(P_i,X) = 0$$

from which it follows that $\mathrm{Ext}^k_R(\mathrm{Ker}\ \delta_i,X) \cong \mathrm{Ext}^{k+1}_R(\mathrm{Ker}\ \delta_{i-1},X)$, $k,i > 0$. From this, in turn, we conclude by induction that

$$\mathrm{Ext}^k_R(\mathrm{Ker}\ \delta_i,X) \cong \mathrm{Ext}^{k+h}_R(\mathrm{Ker}\ \delta_{i-h},X), \qquad \forall\, h \leq i+1;$$

hence, $\mathrm{Ext}^1_R(\mathrm{Ker}\ \delta_{n-2}, X) \cong \mathrm{Ext}^n_R(M,X)$. Thus, if $\mathrm{Ext}^n_R(M,X) = 0$, then $\ker \delta_{n-2} \in {}_R\underline{\underline{P}}^f$, by (4.2), and we have obtained a projective resolution of length $(n-1)$ for M, i.e., $hd_R(M) < n$. #

4.4 <u>Lemma</u>: Let $0 \longrightarrow M' \longrightarrow M \longrightarrow M'' \longrightarrow 0$ be an exact sequence of left R-modules, of finite type. If $hd_R(M) < \infty$, then $hd_R(M) \leq \max(hd_R(M'), hd_R(M''))$.

<u>Proof</u>: If $hd_R(M') = \infty$ or $hd_R(M'') = \infty$, the above formula is obviously true. Thus we may assume $hd_R(M') = n' < \infty$, $hd_R(M'')=n''< \infty$, $hd_R(M) = n < \infty$ and set $n_o = \max(n',n'')$. But then $\mathrm{Ext}_R^{n_o}(M'',X) = \mathrm{Ext}_R^{n_o}(M',X) = 0$ by (4.3), which, according to (3.10), implies $\mathrm{Ext}_R^{n_o}(M,X) = 0$. This in turn, using (4.3) once more, shows that $hd_R(M) \leq \max(n',n'')$. #

4.5 <u>Lemma</u>: Let $M', M'' \in {}_R\underline{\underline{M}}^f$, $P \in {}_R\underline{\underline{P}}^f$ and assume that $0 \longrightarrow M' \longrightarrow P \longrightarrow M'' \longrightarrow 0$ is an exact sequence and $M'' \notin {}_R\underline{\underline{P}}^f$. Then $hd_R(M'') = 1 + hd_R(M')$.

<u>Proof</u>: From (3.10), we obtain for every $N \in {}_R\underline{\underline{M}}$: $\mathrm{Ext}^n_R(M',N) \simeq \mathrm{Ext}^{n+1}_R(M'',N)$, $n \geq 1$. If $M'' \notin {}_R\underline{\underline{P}}^f$, $hd_R(M'') \geq 1$, and thus the above formula follows. #

4.6 <u>Lemma</u>: Let R, S be noetherian rings and φ a homomorphism of rings $R \longrightarrow S$, with $\varphi(1) = 1$, and such that S is an R-module of finite type. If $M \in {}_S\underline{\underline{M}}^f$, then $hd_R(M) \leq hd_R({}_RS) + hd_S(M)$.

<u>Proof</u>: Since ${}_RS \in {}_R\underline{\underline{M}}^f$, it follows that $M \in {}_R\underline{\underline{M}}^f$. The theorem is obviously true if $hd_R({}_RS)$ or $hd_S(M)$ is infinite. Thus we may assume that $hd_R({}_RS) < \infty$ and $hd_S(M) < \infty$, and we shall prove the theorem by induction on $hd_S(M)$.

If $hd_S(M) = 0$, then by (4.2), $M \in {}_S\underline{\underline{P}}^f$; i.e., $M \oplus X \simeq ({}_SS)^{(n)}$ for some positive integer n (cf. I, (2.9)). Thus, if $hd_R({}_RS) = s$, then for every $N \in {}_R\underline{\underline{M}}$

$\text{Ext}_R^{s+1}(M,N) \oplus \text{Ext}_R^{s+1}(X,N) \simeq \text{Ext}_R^{s+1}(({}_RS)^{(n)},N) \simeq \oplus_1^n \text{Ext}_R^{s+1}({}_RS,N) = 0$ by (1.6) and (3.5); i.e., $\text{Ext}_R^{s+1}(M,N) = 0$. From (4.3) we get $\text{hd}_R(M) \leq s = \text{hd}_R({}_RS) + \text{hd}_S(M)$. Assume now that $\text{hd}_S(M) = m$, $m > 0$. Then there exists an exact sequence of left S-modules of finite type, $E : 0 \longrightarrow M' \longrightarrow P \longrightarrow M \longrightarrow 0$, where $P \in {}_S\underline{\underline{P}}^f$. Thus, by (4.5), $\text{hd}_S(M) = 1 + \text{hd}_S(M')$; i.e., $\text{hd}_S(M') = m - 1$. By the induction hypothesis, $\text{hd}_R(M') \leq \text{hd}_R({}_RS) + m - 1$. From the case $m = 0$, we obtain $\text{hd}_R(P) \leq s$. Thus $\text{hd}_R(M'), \text{hd}_R(P) < s + m$; and since E is also an exact sequence of R-modules, we get from (4.4):

$$\text{hd}_R(M) \leq s + m = \text{hd}_R({}_RS) + \text{hd}_S(M). \quad \#$$

Exercises §4:

1.) Let $M \in {}_R\underline{\underline{M}}^f$, where R is a ring. Show: $\text{hd}_R(M) < n$ implies $\text{Ext}_R^n(M,X) = 0, \quad \forall X \in {}_R\underline{\underline{M}}$.

2.) Show that $P \in {}_R\underline{\underline{P}}^f$ if and only if $\text{Ext}_R^1(P,X) = 0, \quad \forall X \in {}_R\underline{\underline{M}}^f$.

§5. Description of $\operatorname{Ext}^1_R(M,N)$ in terms of exact sequences.

We prove that $\operatorname{Ext}^1_R(M,N)$ is naturally equivalent to the group of equivalence classes of extensions of N by M under the Baer sum.

In this section R is a noetherian ring and all modules are left modules of finite type.

5.1 Definition: Let $M', M'' \in {}_{R}\underline{\underline{M}}^f$. An extension of M' by M'' is a short exact sequence

$$E : 0 \longrightarrow M' \xrightarrow{\varphi} X \xrightarrow{\psi} M'' \longrightarrow 0,$$

where $X \in {}_{R}\underline{\underline{M}}$ (then automatically $X \in {}_{R}\underline{\underline{M}}^f$ (cf. I, (2.3)), and φ, ψ are R-homomorphisms. On the set $\tilde{E}_R(M'',M')$ of all extensions of M' by M'' we introduce the relation ρ: If

$E : 0 \longrightarrow M' \xrightarrow{\varphi} X \xrightarrow{\psi} M'' \longrightarrow 0 \in \tilde{E}_R(M'',M')$ and

$E' : 0 \longrightarrow M' \xrightarrow{\varphi'} X' \xrightarrow{\psi'} M'' \longrightarrow 0 \in \tilde{E}_R(M'',M')$, then $E \rho E'$ if and only if $\exists\ \sigma \in \operatorname{Hom}_R(X,X')$, such that the following diagram is commutative:

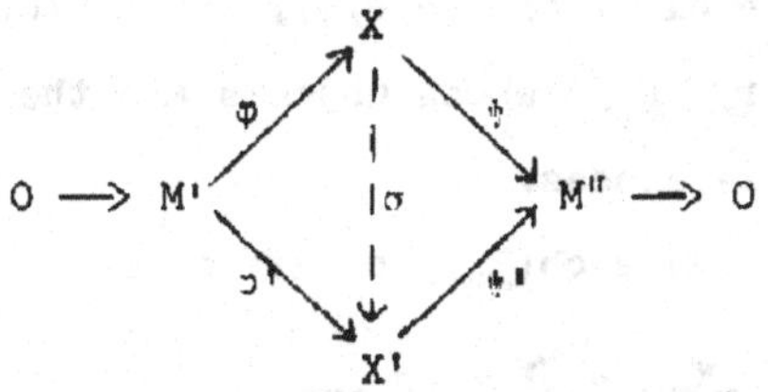

We leave it as an exercise to show that σ is necessarily an isomorphism and that ρ is an equivalence relation. It should be noted that $E : 0 \longrightarrow M' \longrightarrow X \longrightarrow M'' \longrightarrow 0$ and $E' : 0 \longrightarrow M' \longrightarrow X' \longrightarrow M'' \longrightarrow 0$ with $X \simeq X'$ does not necessarily imply $E \rho E'$ (cf. Ex. 5,1). By $E_R(M'',M')$ we denote the set theoretic quotient $\tilde{E}_R(M'',M')/\rho$. By $[E]$ we denote the image of $E \in \tilde{E}_R(M'',M')$ in $E_R(M'',M')$. If $E: 0 \to M' \xrightarrow{\varphi} M \xrightarrow{\psi} M'' \to 0$ and $E': 0 \to N' \xrightarrow{\varphi'} N \xrightarrow{\psi'} N'' \to 0$ are two exact sequences of left R-modules and R-homomorphisms, then a <u>morphism $E \to E'$</u> is a triple (α,β,γ) of R-homomorphisms, such that the following diagram is commutative:

$$\begin{array}{ccccccccccc} E: & 0 & \longrightarrow & M' & \xrightarrow{\varphi} & M & \xrightarrow{\psi} & M'' & \longrightarrow & 0 \\ {\scriptstyle(\alpha,\beta,\gamma)}\downarrow & & & \downarrow{\scriptstyle\alpha} & & \downarrow{\scriptstyle\beta} & & \downarrow{\scriptstyle\gamma} & & \\ E': & 0 & \longrightarrow & N' & \xrightarrow{\varphi'} & N & \xrightarrow{\psi'} & N'' & \longrightarrow & 0 \end{array} .$$

In this way we obtain a category $\tilde{\underline{\underline{E}}}$ where the objects are short exact sequences and the morphisms, written on the right, are triples (α,β,γ). For $E, E' \in \tilde{E}_R(M'',M')$, $E \rho E'$ if and only if $(1_{M'},\sigma,1_{M''})E = E'$ for some $\sigma \in \mathrm{Hom}_R(M,N)$.

5.2 <u>Definition</u>: If we now define an equivalence relation ρ' among the morphisms (α,β,γ) in $\tilde{\underline{\underline{E}}}_R$ by $(\alpha,\beta,\gamma)\,\rho'\,(\alpha',\beta',\gamma')$ if there exists $\sigma \in \mathrm{Hom}_R(M_1,M)$ and $\sigma' \in \mathrm{Hom}_R(N,N_1)$ such that $(1_{M'},\sigma,1_{M''})(\alpha,\beta,\gamma)(1_{N'},\sigma',1_{N''}) = (\alpha',\beta',\gamma')$ and if we denote by $[\alpha,\beta,\gamma]$ the equivalence class of (α,β,γ), then $[\alpha,\beta,\gamma]$ are the morphisms in the <u>category $\underline{\underline{E}}_R$</u>, whose objects are the equivalence classes of short exact sequences.

5.3 <u>Theorem</u>: Let $[E] \in \mathrm{ob}(\underline{\underline{E}}_R)$ be given, say

$$E: 0 \longrightarrow M' \xrightarrow{\varphi} X \xrightarrow{\psi} M'' \longrightarrow 0$$

for every $\alpha \in \mathrm{Hom}_R(M',N')$, there exists a unique $[E'] \in \mathrm{ob}(\underline{\underline{E}}_R)$, denoted by $[E]\alpha$, and a unique morphism $[\alpha,\beta,1_{M''}] \in \mathrm{morph}_{\underline{\underline{E}}_R}([E],[E'])$ such that $[E][\alpha,\beta,1_{M''}] = [E']$.

<u>Proof</u>: According to (1.14) and (1.15) we can complete the diagram

$$\begin{array}{lccccccccc}
E: & 0 & \longrightarrow & M' & \xrightarrow{\varphi} & X & \xrightarrow{\psi} & M'' & \longrightarrow & 0 \\
 & & & \downarrow{\scriptstyle\alpha} & & \downarrow{\scriptstyle\beta} & & \downarrow{\scriptstyle 1_M} & & \\
E': & 0 & \dashrightarrow & N' & \xrightarrow{\varphi'} & Q & \xrightarrow{\psi'} & M'' & \dashrightarrow & 0
\end{array}$$

where $(Q;\beta,\varphi')$ is the fiber coproduct of $(M';\varphi,\alpha)$. Since

$$\begin{array}{ccc}
M' & \xrightarrow{\varphi} & X \\
{\scriptstyle\alpha}\downarrow & & \downarrow{\scriptstyle\psi} \\
N' & \xrightarrow{0} & M''
\end{array}$$

is a commutative diagram, there exists a unique homomorphism $\psi' : Q \longrightarrow M''$, completing the second square commutatively and such that $\varphi'\psi' = 0$. From (1.16) we conclude that φ' is monic. It remains to show that $\psi' = \operatorname{coker} \varphi'$ (cf. Ex. §1).

Since $\varphi'\psi' = 0$, we can complete the following diagram uniquely

$$\begin{array}{ccccc}
N' & \xrightarrow{\varphi'} & Q & \xrightarrow{\operatorname{coker}\varphi'} & \operatorname{Coker}\varphi' \\
 & {\scriptstyle 0}\searrow & \downarrow{\scriptstyle\psi'} & \swarrow{\scriptstyle \exists!\tau} & \\
 & & M'' & &
\end{array}$$

and since $\varphi\beta \operatorname{coker} \varphi' = \alpha\varphi' \operatorname{coker} \varphi' = 0$, we can complete the diagram

$$\begin{array}{ccccc}
M' & \xrightarrow{\varphi} & X & \xrightarrow{\operatorname{coker}\varphi} & M'' = \operatorname{Coker}\varphi \\
 & & \downarrow{\scriptstyle\beta} & \nearrow{\scriptstyle\psi'} & \\
 & {\scriptstyle 0}\searrow & Q & \swarrow{\scriptstyle\exists!\sigma} & \\
 & & \downarrow{\scriptstyle\operatorname{coker}\varphi'} & & \\
 & & \operatorname{Coker}\varphi' & &
\end{array}$$

By the universality of $\operatorname{coker} \varphi'$, we get $\psi' = \operatorname{coker} \varphi'$, i.e.,

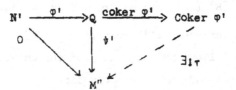

$M'' \cong Q/\mathrm{Im}\,\varphi'$. Now, if also $E(\alpha,\beta',1_{M''}) = E''$, then it follows from the universal property of the fiber coproduct (cf. (1.13)), that $E'' \rho E'$. Thus, we obtain a unique map

$$[E] \longrightarrow [E][\alpha,\beta,1_{M''}] = [E]\alpha.^{*)} \qquad \#$$

5.4 <u>Theorem</u>: Let $[E] \in \mathrm{ob}(\underline{\underline{E}}_R)$, say

$$E : 0 \longrightarrow M' \longrightarrow X \longrightarrow M'' \longrightarrow 0,$$

for every $\gamma \in \mathrm{Hom}_R(N'',M'')$ there exists a unique $[E'] \in \mathrm{ob}(\underline{\underline{E}}_R)$, denoted by $\gamma[E]$, and a unique $[1_{M'},\beta,\gamma] \in \mathrm{morph}_{\underline{\underline{E}}_R}([E'],[E])$ such that $[E'][1_{M'},\alpha,\gamma] = [E]$.

The <u>proof</u> is dual to that of (5.3), using the properties of the fiber product.$^{**)}$ #

5.5 <u>Theorem</u> (<u>Universal property of</u> $[E]\alpha$ <u>and</u> $\gamma[E']$): $[\alpha,\beta,\gamma] \in \mathrm{morph}_{\underline{\underline{E}}_R}([E],[E'])$ is determined by α and γ; namely: $[E][\alpha,\beta,\gamma]$ is the unique $[E']$ such that $\gamma[E'] = [E]\alpha$.

<u>Proof</u>: This follows immediately from the universal properties of the fiber product and the fiber coproduct.

5.6 <u>Corollary</u>: (i) $\gamma(\gamma'[E]) = (\gamma\gamma')[E]$,

(ii) $([E]\alpha)\alpha' = [E](\alpha\alpha')$,

(iii) $(\gamma[E])\alpha = \gamma([E]\alpha)$,

(iv) $(\gamma[E])[\alpha,\beta,\gamma] = \gamma([E][\alpha,\beta,\gamma]) = [E]\alpha$

<u>Proof</u>: These identities are an immediate consequence of (5.2)-(5.5); e.g., $[E'] = ([E])[\alpha,\beta,\gamma]$ is uniquely determined by the identity $\gamma[E'] = \gamma[E]\alpha$, but $[E]\alpha$ also satisfies this condition. #

Next we shall define an additive structure on $E_R(M'',M')$ - the so-called "Baer sum"-, which makes $E_R(M'',M')$ into an $(\mathrm{End}_R(M''),\mathrm{End}_R(M'))$-bimodule.

---

*) The sequence E' constructed in this proof is denoted by $E\alpha$.

**) By γE we denote the sequence $E' \in \gamma[E]$ constructed with the help of the fiber product.

5.7 **Theorem**: $E_R(M'',M)$ is an abelian group under the Baer sum, defined below. In addition to this, the following formulae are satisfied, for $\alpha,\alpha_1,\alpha_2 \in \mathrm{Hom}_R(M',N')$ and $\gamma,\gamma_1,\gamma_2 \in \mathrm{Hom}_R(M'',M'')$:

I. $\gamma([E] + [E']) = \gamma[E] + \gamma[E']$

II. $([E] + [E'])\alpha = [E]\alpha + [E']\alpha$

III. $(\gamma + \gamma')[E] = \gamma[E] + \gamma'[E]$

IV. $[E](\alpha + \alpha') = [E]\alpha + [E]\alpha'$.

Proof: To define the Baer sum, let $E_1, E_2 \in \tilde{E}_R(M'',M')$ be given:

$$E_1 : 0 \longrightarrow M' \xrightarrow{\varphi_1} X \xrightarrow{\psi_1} M'' \longrightarrow 0$$

$$E_2 : 0 \longrightarrow M' \xrightarrow{\varphi_2} X' \xrightarrow{\psi_2} M'' \longrightarrow 0.$$

By (I, Ex. 2,1a)

$$E_1 \oplus E_2 : 0 \longrightarrow M' \oplus M' \xrightarrow{\varphi_1 \oplus \varphi_2} X \oplus X' \xrightarrow{\psi_1 \oplus \psi_2} M'' \oplus M'' \longrightarrow 0$$

is an exact sequence and it is readily verified that $[E_1] = [E_1']$ and $[E_2] = [E_2']$ implies $[E_1 \oplus E_2] = [E_1' \oplus E_2']$. Therefore we may define $[E_1] \oplus [E_2] = [E_1 \oplus E_2]$. In (Ex. 1,7) we defined the diagonal and the codiagonal maps: $\Delta : M'' \longrightarrow M'' \oplus M''$; $m'' \longmapsto (m'',m'')$; $\nabla : M' \oplus M' \longrightarrow M'$; $(m_1',m_2') \longmapsto m_1' + m_2'$. Then $\Delta[E_1 \oplus E_2]\nabla \in E_R(M'',M')$.

Now we define the Baer sum

$$[E_1] + [E_2] = \Delta[E_1 \oplus E_2]\nabla \; .$$

(5.5) ensures the consistency of this definition.

We observe that for $\tau,\sigma \in \mathrm{Hom}_R(X,Y)$ we have

$$(\sigma + \tau) = \Delta_X(\sigma \oplus \tau)\nabla_Y \quad \text{(cf. Ex. 1,7)}.$$

Moreover it is easy to verify, using (5.5), that

$$[E_1 \oplus E_2](\alpha_1 \oplus \alpha_2) = [E_1]\alpha_1 \oplus [E_2]\alpha_2$$
$$(\gamma_1 \oplus \gamma_2)[E_1 \oplus E_2] = \gamma_1[E_1] \oplus \gamma_2[E_2],$$

and $\gamma\Delta = \Delta(\gamma \oplus \gamma)$, $\alpha = (\alpha \oplus \alpha)$, (cf. Ex. 1,7).

<u>To prove I</u>: $\gamma([E] + [E']) = \gamma\Delta[E \oplus E']\nabla = \Delta(\gamma \oplus \gamma)[E \oplus E']\nabla$
$= \Delta(\gamma[E] \oplus \gamma[E'])\nabla = \gamma[E] + \gamma[E']$.

II is proved similarly.

<u>To prove III</u>: We have up to some abuse of notation $[E \oplus E] = [E(\Delta, \Delta, \Delta)]$, more precisely, $[E \oplus E] = [E(\Delta_{M'}, \Delta_X, \Delta_{M''})]$ as is easily seen; i.e., with (5.5), $[\Delta(E \oplus E)] = [E(\Delta, \tau, 1_{M''})] = [E]\Delta$. Similarly, $\nabla[E] = [E \oplus E]\nabla$. Thus: $(\gamma + \gamma')[E] = \Delta(\gamma \oplus \gamma')\nabla[E]$
$= \Delta(\gamma \oplus \gamma')[E \oplus E]\nabla = \Delta(\gamma[E] \oplus \gamma'[E])\nabla$
$= \gamma[E] + \gamma'[E]$.

Similarly, for IV.

It remains to show that this makes $\underline{\underline{E}}_R(M'', M')$ into an abelian group.

(i) Associativity:

$$[E_1] + ([E_2] + [E_3]) = [E_1] + \Delta[E_2 \oplus E_3]\nabla = \Delta([E_1] \oplus \Delta[E_2 \oplus E_3]\nabla)\nabla$$
$$= \Delta((1_{M''} \oplus \Delta)[E_1 \oplus (E_2 \oplus E_3)](1_{M'} \oplus \nabla))\nabla$$
$$= \Delta((\Delta \oplus 1_{M''})[(E_1 \oplus E_2) \oplus E_3](\nabla \oplus 1_{M'}))\nabla$$
$$= ([E_1] + [E_2]) + [E_3] \quad \text{(cf. Ex. 1,7).}$$

(ii) The class $[E_o]$ of the split exact sequence is the zero element of $E_R(M'', M')$; in fact, for every $E \in \tilde{E}_R(M'', M')$, we have $[E_o] = [E]0_{M'}$ (cf. (1.15)) and hence $[E] = [E](1 + 0_M) = [E] + [E_o]$ by the distributive law.

(iii) Similarly one shows that for $-1_{M'} : M' \longrightarrow M'$, $-1_{M'} : m' \longmapsto -m'$, we have $[E] + [E](-1_{M'}) = [E_o]$.

(iv) For the proof of $[E_1] + [E_2] = [E_2] + [E_1]$, let $\tau : X \oplus Y \xrightarrow{\sim} Y \oplus X$ be the natural isomorphism (cf. Ex. 1,7). Then $(\tau, \tau, \tau) : E_1 \oplus E_2 \longrightarrow E_2 \oplus E_1$ - with some more abuse of notation - shows $[E_2 \oplus E_1] = \tau[E_1 \oplus E_2]\tau$, (cf.(5.6)) and since $\Delta\tau = \Delta$ and $\tau\nabla = \nabla$, we obtain $[E_1] + [E_2] = [E_1 \oplus E_2]$ $= \Delta\tau[E_1 \oplus E_2]\tau\nabla = \Delta[E_2 \oplus E_1]\nabla = [E_2] + [E_1]$. #

5.8 <u>Corollary</u>:

(i) $E_R(M'',-)$ is a covariant functor

$E_R(M'',-) : {}_R\underline{\underline{M}}^f \longrightarrow \underline{\underline{A}};\ M' \longmapsto E_R(M'',M')$

$\mathrm{Hom}_R(M',N') \longrightarrow \mathrm{morph}_{\underline{\underline{A}}}(E_R(M'',M'),E_R(M'',N'))$;

$\varphi \longmapsto \varphi_1^* : [E] \longmapsto [E]\varphi$.

(ii) $E_R(-,M')$ is a contravariant functor

$E_R(-,M') : {}_R\underline{\underline{M}}^f \longrightarrow \underline{\underline{A}};\ M'' \longmapsto E_R(M'',M')$,

$\mathrm{Hom}_R(N'',M'') \longrightarrow \mathrm{morph}_{\underline{\underline{A}}}(E_R(M'',M'),\ E_R(N'',M'))$;

$\varphi \longmapsto \varphi_1^* : [E] \longmapsto \varphi[E]$.

<u>Proof</u>: This is an immediate consequence of the previous theorems. #

5.9 <u>Theorem</u>: There is a natural equivalence $\mathrm{Ext}_R^1(-,M') \sim E_R(-,M')$.

Once this result is established, we have a one-to-one correspondence between the homomorphisms $\alpha : E_R(M'',M') \longrightarrow E_R(N'',M')$, $[E] \longrightarrow \alpha[E]$ and $\mathrm{ext}_R^1(\alpha,M') : \mathrm{Ext}_R^1(M'',M') \longrightarrow \mathrm{Ext}_R^1(N'',M')$, for $\alpha \in \mathrm{Hom}_R(N'',M')$. We shall use the abbreviations $\alpha\,\mathrm{Ext}_R^1(M'',M')$ for $\mathrm{Im}(\mathrm{ext}_R^1(\alpha,M'))$, and $\mathrm{Ext}_R^1(M'',M')\alpha$ for $\mathrm{Im}(\mathrm{ext}_R^1(M'',\alpha))$.

<u>Proof</u>: We construct a map $\Phi : \mathrm{Ext}_R^1(M'',M') \longrightarrow E_R(M'',M')$. Since $M'' \in {}_R\underline{\underline{M}}^f$, there exists an exact sequence

$$E_1 : 0 \longrightarrow Y \xrightarrow{\ x\ } P \xrightarrow{\ \lambda\ } M'' \longrightarrow 0$$

with $P \in {}_R\underline{\underline{P}}^f$. From (3.10) we obtain the exact sequence

$$0 \longrightarrow \mathrm{Hom}_R(M'',M') \xrightarrow{\lambda^*} \mathrm{Hom}_R(P,M') \xrightarrow{x^*} \mathrm{Hom}_R(Y,M') \xrightarrow{\Delta_1}$$
$$\longrightarrow \mathrm{Ext}_R^1(M'',M') \longrightarrow 0,$$

since $\operatorname{Ext}^1_R(P,M') = 0$ by (4.2).

Thus, we obtain an isomorphism $\chi : \operatorname{Ext}^1_R(M'',M') \xrightarrow{\sim} \operatorname{Hom}_R(Y,M')/\operatorname{Im}\kappa^*$. Now, to define Φ, let $\alpha + \operatorname{Im}\kappa^* \in \operatorname{Hom}_R(Y,M')/\operatorname{Im}\kappa^*$ be given. Then $[E_1]\alpha \in E_R(M'',M')$, and we define $\Phi : \operatorname{Ext}^1_R(M'',M') \longrightarrow E_R(M'',M')$, $\Phi : a \longmapsto \chi(a) = \alpha + \operatorname{Im}\kappa^* \longmapsto [E_1]\alpha$.

(i) <u>Φ is well defined</u>; i.e., we have to show: if $\alpha \in \operatorname{Im}\kappa^*$, then $[E_1]\alpha = 0$, i.e., $[E_1]\alpha$ contains a split exact sequence. But for $\alpha \in \operatorname{Im}\kappa^*$, $\alpha = \kappa\beta$, for some $\beta \in \operatorname{Hom}_R(P,M')$, and $[E_1]\alpha = [E_1](\kappa\beta) = ([E_1]\kappa)\beta$. But from the commutative diagram

$$\begin{array}{lccccccccc}
E_1 : & 0 & \longrightarrow & Y & \xrightarrow{\kappa} & P & \xrightarrow{\lambda} & M'' & \longrightarrow & 0 \\
 & & & \Big\downarrow{\scriptstyle \kappa} & & \Big\downarrow{\scriptstyle \iota_1 + \lambda\iota_2} & & \Big\downarrow{\scriptstyle 1_{M''}} & & \\
 & 0 & \longrightarrow & P & \xrightarrow{\iota_1} & P \oplus M'' & \xrightarrow{\pi_2} & M'' & \longrightarrow & 0
\end{array}$$

we conclude that $[E_1]\kappa = [E_0]$ and consequently $[E_1]\alpha = [E_0]$; i.e., Φ is well defined.

(ii) <u>Φ is additive</u>: Given $\alpha + \operatorname{Im}\kappa^*$ and $\alpha' + \operatorname{Im}\kappa^*$ in $\operatorname{Hom}_R(Y,M')/\operatorname{Im}\kappa^*$, then $[E_1](\alpha + \alpha') = [E_1]\alpha + [E_1]\alpha'$; i.e., Φ is additive.

(iii) <u>To show that Φ is an isomorphism</u>, we construct a map $\Psi : E_R(M'',M') \longrightarrow \operatorname{Ext}^1_R(M'',M')$. Given $E \in \tilde{E}_R(M'',M')$, $E : 0 \longrightarrow M' \xrightarrow{\varphi} X \xrightarrow{\psi} M'' \longrightarrow 0$, then we can complete the diagram

D:

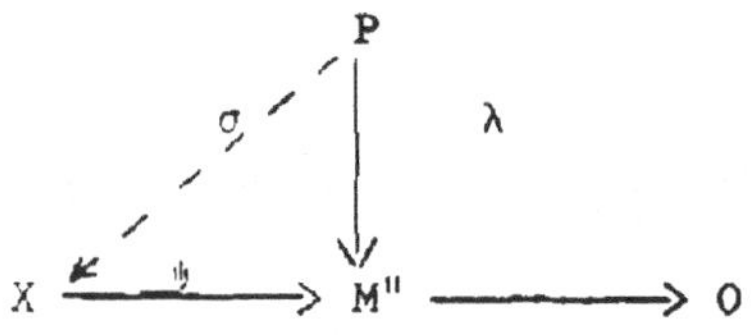

and it remains to fill in the following diagram

commutatively:

$$D' : \quad \begin{array}{lccccccccc} E_1 & : 0 & \longrightarrow & Y & \xrightarrow{\kappa} & P & \xrightarrow{\lambda} & M'' & \longrightarrow & 0 \\ & & & \downarrow \alpha & & \downarrow \sigma & & \downarrow 1_{M''} & & \\ E & : 0 & \longrightarrow & M' & \xrightarrow{\varphi} & X & \xrightarrow{\psi} & M'' & \longrightarrow & 0 \; . \end{array}$$

We put $\alpha : Y \longrightarrow M'$ $\alpha : y \longmapsto y\kappa\sigma\tilde{\varphi}$, $\tilde{\varphi} : m'\varphi \longmapsto m'$, where $\tilde{\varphi} : \mathrm{Im}\,\varphi \longrightarrow M'$ exists, since φ is monic. Then α is well-defined; indeed, $(y\kappa\sigma)\psi = y\kappa\lambda = 0$; i.e., $y\kappa\sigma \in \mathrm{Im}\,\varphi$.

Now we set

$$\Psi : E_R(M'',M') \longrightarrow \mathrm{Ext}^1_R(M'',M'); \; [E] \longmapsto \chi^{-1}(\alpha + \mathrm{Im}\,\kappa^*)$$

(i) Ψ is well-defined, for, if $[E]\alpha = [E_0]$, then $E_1\alpha$ splits, and so α factors through κ (cf. D') and $\alpha \in \mathrm{Im}\,\kappa^*$.

(ii) $\Psi\Phi = 1_{\mathrm{Ext}^1_R(M'',M')}$ and $\Phi\Psi = 1_{E_R(M'',M')}$. We have $\Phi\Psi([E]) = \Phi(\chi^{-1}(\alpha + \mathrm{Im}\,\kappa^*)) = [E_1]\alpha = [E]$, by the universal property of $E_1\alpha$ (cf. (5.5)). Conversely, $\Psi\Phi(a) = \Psi([E_1]\alpha) = \chi^{-1}(\alpha + \mathrm{Im}\,\kappa^*)$ where $\chi(a) = (\alpha + \mathrm{Im}\,\kappa^*)$. This shows that $E_R(M'',M') \simeq \mathrm{Ext}^1_R(M'',M')$.

Next we show that this is a natural equivalence; i.e., given $\sigma : N'' \longrightarrow M''$, we show that the following diagram is commutative:

$$\begin{array}{ccccc} \mathrm{Ext}^1_R(M'',M') & \xrightarrow{\Phi_{M''}} & E_R(M'',M') & & [E] \\ \downarrow \sigma_1^* & & \downarrow \sigma & & \downarrow \\ \mathrm{Ext}^1_R(N'',M') & \xrightarrow{\Phi_{N''}} & E_R(N'',M') & & \sigma[E] \; . \end{array}$$

Let P and P' be projective resolutions of M'' and N'' resp. From (3.6) we obtain the commutative diagram

$$\begin{array}{ccccc} P_o & \xrightarrow{\delta_o} & M'' & \longrightarrow & 0 \\ {\scriptstyle\sigma'}\uparrow & & \uparrow{\scriptstyle\sigma} & & \\ P'_o & \xrightarrow{\delta'_o} & N'' & \longrightarrow & 0 \end{array} .$$

And if we define $\rho : \operatorname{Ker} \delta'_o \longrightarrow \operatorname{Ker} \delta_o$, $\rho : x \longmapsto x\sigma_1$, then we obtain the commutative diagram

$$\begin{array}{ccccccccc} E_1 : 0 & \longrightarrow & \operatorname{Ker} \delta_o & \xrightarrow{\kappa} & P_o & \xrightarrow{\lambda} & M'' & \longrightarrow & 0 \\ & & {\scriptstyle\rho}\uparrow & & {\scriptstyle\sigma'}\uparrow & & {\scriptstyle\sigma}\uparrow & & \\ E'_1 : 0 & \longrightarrow & \operatorname{Ker} \delta'_o & \xrightarrow{\kappa'} & P'_o & \xrightarrow{\lambda'} & N'' & \longrightarrow & 0 . \end{array}$$

This in turn induces the commutative diagram with exact rows

$$\begin{array}{ccccccccc} 0 \longrightarrow & \operatorname{Hom}_R(M'',M') & \longrightarrow & \operatorname{Hom}_R(P_o,M') & \xrightarrow{\kappa^*} & \operatorname{Hom}_R(\operatorname{Ker} \delta_o,M') & \longrightarrow & \operatorname{Ext}^1_R(M'',M') & \longrightarrow 0 \\ & {\scriptstyle\sigma^*}\downarrow & & {\scriptstyle\sigma'^*}\downarrow & & {\scriptstyle\rho^*}\downarrow & & \downarrow{\scriptstyle\sigma^*_1} & \\ 0 \longrightarrow & \operatorname{Hom}_R(N'',M') & \longrightarrow & \operatorname{Hom}_R(P'_o,M') & \xrightarrow{\kappa'^*} & \operatorname{Hom}_R(\operatorname{Ker} \delta'_o,M') & \longrightarrow & \operatorname{Ext}^1_R(N'',M') & \longrightarrow 0. \end{array}$$

It is now obvious that the isomorphism χ is natural, and it remains to show that

$$\begin{array}{ccc} \operatorname{Hom}_R(\operatorname{Ker} \delta_o,M')/\operatorname{Im} \kappa^* & \xrightarrow{\mathfrak{F}_{M''}} & E_R(M'',M') \\ \tilde{\sigma} & & \sigma \\ \operatorname{Hom}_R(\operatorname{Ker} \delta'_o,M')/\operatorname{Im} \kappa'^* & \xrightarrow{\mathfrak{F}_{N''}} & E_R(N'',M') \end{array}$$

is a commutative diagram, where

$$\alpha + \operatorname{Im} \kappa^* \overset{\tilde{\sigma}}{\longmapsto} \rho\alpha + \operatorname{Im} \kappa'^* \overset{\mathfrak{F}_{N''}}{\longmapsto} [E'_1]\rho\alpha$$

$$\alpha + \operatorname{Im} \kappa^* \overset{\mathfrak{F}_{M''}}{\longmapsto} [E_1]\alpha \longmapsto \sigma([E^1_1]\alpha) \quad .$$

But from (5.5) it follows that $\sigma([E_1]\alpha) = (\sigma[E_1])\alpha = (E^1_1]\rho)\alpha = [E^1_1]\rho\alpha$;

hence the desired result. #

Exercises §5.

1.) (a) Show that ρ in (5.1) and (5.2) is an equivalence relation.

(b) Let $E, E' \in \underline{\underline{E}}_R(M'', M')$, where R is a ring and $M'', M' \in {}_R\underline{\underline{M}}^f$:

$$E : 0 \longrightarrow M' \longrightarrow X \longrightarrow M'' \longrightarrow 0$$

$$E : 0 \longrightarrow M' \longrightarrow X \longrightarrow M'' \longrightarrow 0 \quad .$$

Show: $E \rho E' \Longrightarrow X \simeq X'$.

(c) Construct two exact sequences E, E' such that $X \simeq X'$ but not $E \rho E'$.

2.) Show that the Baer sum is well-defined.

3.) Show that $[(E_1 \oplus E_2)](\alpha_1 \oplus \alpha_2) = [E_1]\alpha_1 \oplus [E_2]\alpha_2$.

Chapter III

MORITA THEOREMS AND SEPARABLE ALGEBRAS

In this chapter, all rings are assumed to be left and right noetherian.

§1. Projective modules and generators

If S is a commutative ring, B is a left noetherian S-algebra and C is an S-flat S-algebra, then

$$C \otimes_S \operatorname{Ext}_B^n(M,N) \overset{\text{nat}}{\cong} \operatorname{Ext}^n_{C\otimes_S B}(C \otimes_S M, C \otimes_S N)$$

for all $M,N \in {}_{\underline{\underline{B}}}\underline{\underline{M}}^f$. We derive the basic properties of the maps $\mu_{M,N} : M^* \otimes_S N \longrightarrow \operatorname{Hom}_S(M,N)$ and $\tau_M : M \otimes_{\operatorname{End}_S(M)} M^* \longrightarrow S$, and we prove five properties of modules equivalent to "being a generator." A faithful exact functor preserves projective modules and generators.

1.1 <u>Notation</u>: Let S and T be rings; then ${}_{\underline{\underline{S}}}\underline{\underline{M}}$ = category of left S-modules, $\underline{\underline{M}}_{\underline{\underline{S}}}$ = category of right S-modules, ${}_{\underline{\underline{S}}}\underline{\underline{M}}^f$ = category of finitely generated left S-modules, ${}_{\underline{\underline{S}}}\underline{\underline{P}}^f$ = category of finitely generated projective left S-modules, ${}_{\underline{\underline{S}}}\underline{\underline{M}}_{\underline{\underline{T}}}$ = category of (S,T)-bimodules.

1.2 <u>Theorem</u> (Auslander-Goldman [1]): Let S be a commutative ring, B and C S-algebras. Moreover, assume that B is left noetherian and that C is S-flat; i.e., that $C \otimes_S -$ is an exact functor on ${}_{\underline{\underline{S}}}\underline{\underline{M}}$. If $M \in {}_{\underline{\underline{B}}}\underline{\underline{M}}^f$, then

$$C \otimes_S \operatorname{Ext}_B^n(M,N) \overset{\text{nat}}{\cong} \operatorname{Ext}^n_{C \otimes_S B}(C \otimes_S M, C \otimes_S N), \ n = 0,1,2,\ldots$$

for every $N \in {}_{\underline{\underline{B}}}\underline{\underline{M}}^f$.

<u>Proof</u>: $B \otimes_S C$ is an S-algebra, and by (Ex. 1,3) $\operatorname{Ext}^n_B(M,N) \in {}_S\underline{\underline{M}}$. Define

$$\alpha : C \otimes_S \operatorname{Hom}_B(M,N) \longrightarrow \operatorname{Hom}_{C \otimes_S B}(C \otimes_S M,\ C \otimes_S N)$$

by $\alpha : c \otimes \varphi \longmapsto (c \otimes \varphi)^{\alpha}$, where $(c' \otimes m)(c \otimes \varphi)^{\alpha} = c'c \otimes m\varphi$; $c, c' \in C$, $m \in M$, $\varphi \in \operatorname{Hom}_B(M,N)$, and extend α $\underline{\underline{Z}}$-linearly. Then α is a natural homomorphism, as is easily seen.

<u>Claim</u>: α is an isomorphism.

If M is B-free; i.e., $M \cong {}_B B^{(t)}$, then we have the commutative diagram

$$\begin{array}{ccc} C \otimes_S \operatorname{Hom}_B({}_B B^{(t)}, N) & \xrightarrow{\ \alpha\ } & \operatorname{Hom}_{C \otimes_S B}(C \otimes_S {}_B B^{(t)}, C \otimes_S N) \\ \downarrow \wr & & \downarrow \wr \\ C \otimes_S N^{(t)} & \xrightarrow{\ 1_{C \otimes_S N^{(t)}}\ } & C \otimes_S N^{(t)} \end{array} ,$$

where the vertical maps are natural isomorphisms and the bottom map is the identity. Thus α is a natural isomorphism for finitely generated free left B-modules M. Now, if $M \in {}_B\underline{\underline{M}}^f$, choose $F = {}_B B^{(t)} \in {}_B\underline{\underline{M}}^f$ such that $F \xrightarrow{\sigma} M \longrightarrow 0$ is a B-exact sequence. Moreover, B is left noetherian and hence $\operatorname{Ker} \sigma \in {}_B\underline{\underline{M}}^f$, and we can find $F' = {}_B B^{(s)} \in {}_B\underline{\underline{M}}^f$ such that $F' \longrightarrow F \longrightarrow M \longrightarrow 0$ is an exact sequence of left B-modules.

This sequence gives rise to the commutative diagram with exact rows (C is S-flat),

$$\begin{array}{ccccccc} 0 \longrightarrow & C \otimes_S \operatorname{Hom}_B(M,N) & \longrightarrow & C \otimes_S \operatorname{Hom}_B(F,N) & \longrightarrow & C \otimes_S \operatorname{Hom}_B(F',N) \\ & \downarrow \alpha'' & & \downarrow \alpha & & \downarrow \alpha' \\ 0 \longrightarrow & \operatorname{Hom}_{C \otimes_S B}(C \otimes_S M,\ C \otimes_S N) & \longrightarrow & \operatorname{Hom}_{C \otimes_S B}(C \otimes_S F, C \otimes_S N) & \longrightarrow & \operatorname{Hom}_{C \otimes_S B}(C \otimes_S F',\ C \otimes_S N) . \end{array}$$

Since α and α' are isomorphisms, so is α''. <u>This proves the claim</u>. Now let

$$Y : \ldots \longrightarrow P_n \longrightarrow P_{n-1} \longrightarrow \ldots \longrightarrow P_1 \longrightarrow M \longrightarrow 0$$

be a projective resolution of M such that $P_i \in {}_{\underline{\underline{B}}}\underline{M}^f$ (B is left noetherian). Since $P_i \in {}_{\underline{\underline{B}}}\underline{P}^f$, it follows from Ex. 1,1 that $C \otimes_S P_i \in {}_{C \otimes_S \underline{\underline{B}}}\underline{P}^f$; and since C is S-flat, we obtain a projective resolution of $C \otimes_S M \in {}_{C\otimes_S \underline{\underline{B}}}\underline{M}^f$:

$$C \otimes_S Y : \ldots \longrightarrow C \otimes_S P_n \longrightarrow C \otimes_S P_{n-1} \longrightarrow \ldots \longrightarrow C \otimes_S P_1 \longrightarrow C \otimes_S M \longrightarrow 0.$$

Y, $C \otimes_S Y$ and α give rise to the commutative diagram

$$\begin{array}{lccccc}
X_1: & 0 \longrightarrow & C \otimes_S \mathrm{Hom}_B(M,N) & \longrightarrow & C \otimes_S \mathrm{Hom}_B(P_1,N) & \longrightarrow \ldots \\
\alpha^* \downarrow & & \alpha_0 \downarrow & & \alpha_1 \downarrow & \\
X_2: & 0 \longrightarrow & \mathrm{Hom}_{C \otimes_S B}(C \otimes_S M, C \otimes_S N) & \longrightarrow & \mathrm{Hom}_{C \otimes_S B}(C \otimes_S P_1, C \otimes_S N) & \longrightarrow \ldots
\end{array}$$

where α^* is an isomorphism of chaincomplexes (cf. II, (2.1)). Consequently X_1 and X_2 have isomorphic homology groups (cf. II, (2.10)); i.e.,

$$H_n(\alpha) : H_n(C \otimes_S \mathrm{Hom}_B(Y,N)) \xrightarrow{\text{nat}} H_n(\mathrm{Hom}_{C \otimes_S B}(C \otimes_S Y, C \otimes_S N)).$$

The latter homology group is $\mathrm{Ext}^n_{C \otimes_S B}(C \otimes_S M, C \otimes_S N)$ (cf. II, (3.4)). Since C is S-flat, $H_n(C \otimes_S \mathrm{Hom}_B(Y,N) \overset{\text{nat}}{=} C \otimes_S H_n(\mathrm{Hom}_B(Y,N))$ (cf. II, (2.12)). Hence

$$\mathrm{Ext}^n_{C \otimes_S B}(C \otimes_S M, C \otimes_S N) \overset{\text{nat}}{\cong} C \otimes_S \mathrm{Ext}^n_B(M,N), \quad n = 0,1,2,\ldots \quad \#$$

1.3 <u>Remarks</u>: Let S be a ring; set, for $M \in {}_S\underline{M}^f$, $\Omega(M) = \mathrm{End}_S(M)$. Then $M \in {}_{\underline{\underline{S}}}\underline{M}_{\Omega(M)}$; moreover, we put $\mathrm{Hom}_S(M,S) = {}_SM^*$, the dual of $M \in {}_{\underline{\underline{S}}}\underline{M}^f$ and $\mathrm{Hom}_{\Omega(M)}(M,\Omega(M)) = M^*_{\Omega(M)}$.

For $\varphi \in {}_{\underline{\underline{S}}}\underline{M}^*$, we define $m(\varphi s) = (m\varphi)s$, $m \in M$, $s \in S$, $m(\omega\varphi) = (m\omega)\varphi$, $m \in M$, $\omega \in \Omega(M)$, and for $\psi \in M^*_{\Omega(M)}$, $(\omega\psi)m = \omega(\psi m)$,

$m \in M$, $\omega \in \Omega(M)$, $(\psi s)m = \psi(sm)$, $m \in M$, $s \in S$; then ${}_S M^*$, $M^*_{\Omega(M)} \in {}_{\Omega(M)}\underline{\underline{M}}_S$. In (Ex. I, 3, 5) it has been shown that ${}_S M^* \otimes_S M$ is a ring. The above definitions show that ${}_S M^* \otimes_S M$ is also an $(\Omega(M), \Omega(M))$-bimodule. We shall generally write bimodule homomorphisms as exponents.

1.4 Definitions: For $M, N \in {}_{\underline{\underline{S}}}\underline{M}^f$, we define

(i) $\underline{\mu_{M,N}} : {}_S M^* \otimes_S N \longrightarrow \mathrm{Hom}_S(M,N)$, $m(\varphi \otimes n)^{\mu_{M,N}} = (m\varphi)n$, $\varphi \in {}_S M^*$, $n \in N$;

(ii) $\underline{\tau_M} : M \otimes_{\Omega(M)} {}_S M^* \longrightarrow S$, $(m \otimes \varphi)^{\tau_M} = m\varphi$, $m \in M$, $\varphi \in {}_S M^*$, or, more generally, $\tau_{M,N} : M \otimes_{\Omega(M)} \mathrm{Hom}_S(M,N) \longrightarrow N$, $m \otimes \varphi \longmapsto m\varphi$.

$\mu_{M,N}$ is a natural homomorphism of $(\Omega(M), \Omega(N))$-bimodules (cf. I, Ex. 3,5). Similarly one shows that τ_M is a natural homomorphism of (S,S)-bimodules. Thus $\mathrm{Im}\,\mu_{M,N} \in {}_{\Omega(M)}\underline{\underline{M}}_{\Omega(N)}$ and $\mathrm{Im}\,\tau_M \in {}_{\underline{\underline{S}}}\underline{M}_S$. Since τ_M is also a ring homomorphism, $\mathrm{Im}\,\tau_M$ is a two-sided ideal in S. In particular, if $M = N$, we write μ_M, and $\mathrm{Im}\,\mu_M$ is a two-sided $\Omega(M)$-ideal.

1.5 Lemma: Let $M \in {}_{\underline{\underline{S}}}\underline{M}^f$. Then M is projective if and only if μ_M is epic. Moreover, $\mu_{M,N}$ is an isomorphism for every $N \in {}_{\underline{\underline{S}}}\underline{M}^f$, if μ_M is epic.

Proof: (i) If $M \in {}_{\underline{\underline{S}}}\underline{P}^f$, then M is the epimorphic image of a free left S-module $F \cong {}_S S^{(n)}$, $F \xrightarrow{\sigma} M \longrightarrow 0$. Let $\{e_i\}_{1 \leq i \leq n}$ be a basis of F; then the set $\{m_i : m_i = e_i\sigma\}_{1 \leq i \leq n}$ is a system of generators for M. Since $M \in {}_{\underline{\underline{S}}}\underline{P}^f$, there exists $\rho \in \mathrm{Hom}_S(M,F)$, such that $\rho\sigma = 1_M$. If we write $m\rho = \Sigma_{i=1}^n s_i(m)e_i$, then it is easily checked that the $\varphi_i : M \longrightarrow S$, $m \longmapsto s_i(m)$, $1 \leq i \leq n$, belong to ${}_S M^*$. Moreover, for every $m \in M$, we have $m = \Sigma_{i=1}^n (m\varphi_i)m_i$; i.e., $1_M = (\Sigma_{i=1}^n \varphi_i \otimes m_i)^{\mu_M}$, and μ_M is epic, since it is an $\Omega(M)$-homomorphism.

(ii) Conversely, if μ_M is epic, then there exists $\Sigma_{i=1}^{n} \varphi_i \otimes m_i \in M^* \otimes_S M$ such that $1_M = (\Sigma_{i=1}^{n} \varphi_i \otimes m_i)^{\mu_M}$. Obviously, the elements $\{m_i\}_{1\leq i\leq n}$ form a system of generators for M. Let F be a free left S-module with a basis $\{e_i\}_{1\leq i\leq n}$ and define the epimorphism $\sigma : F \longrightarrow M$ by $e_i \longmapsto m_i$, $1 \leq i \leq n$. Now, $\rho : M \longrightarrow F$, $m \longmapsto \Sigma_{i=1}^{n} (m)\varphi_i e_i$ is an S-homomorphism such that $\rho\sigma = 1_M$; i.e., $M \in {}_{\underline{S}}\underline{P}^f$.

To prove the second part of (1.5), let $\Sigma_{i=1}^{n} \varphi_i \otimes m_i \in {}_S M^* \otimes_S M$ be such that $(\Sigma_{i=1}^{n} \varphi_i \otimes m_i)^{\mu_M} = 1_M$. That $\mu_{M,N}$ is an isomorphism for every $N \in {}_{\underline{S}}\underline{M}$, is established by the map

$$\nu_{M,N} : \mathrm{Hom}_S(M,N) \longrightarrow {}_S M^* \otimes_S N;\ \psi \longmapsto \Sigma_{i=1}^{n} \varphi_i \otimes m_i\psi .$$

For, $\nu_{M,N}\mu_{M,N} = 1_{\mathrm{Hom}_S(M,N)}$ and $\mu_{M,N}\nu_{M,N} = 1_{{}_S M^* \otimes_S N}$, as is easily seen. In addition, $\nu_{M,N}$ is an $(\Omega(M),\Omega(N))$-homomorphism. #

1.6 <u>Lemma</u>: If τ_M is epic, then τ_M is an isomorphism; in fact, $\tau_{M,N}$ is an isomorphism for all $N \in {}_{\underline{S}}\underline{M}^f$.

<u>Proof</u>: If τ_M is epic, then there are elements $m_i \in M$, $\varphi_i \in {}_S M^*$, $1 \leq i \leq n$, such that $1 = \Sigma_{i=1}^{n} m_i\varphi_i$. We now put $\sigma_M : S \longrightarrow M \otimes_{\Omega(M)} {}_S M^*$; $s \longmapsto \Sigma_{i=1}^{n} sm_i \otimes \varphi_i$. It follows easily that $\sigma_M \tau_M = 1_S$ and $\tau_M \sigma_M = 1_{M \otimes_{\Omega(M)} {}_S M^*}$. (Observe that

$$n\psi \cdot m \otimes \varphi = n\cdot(\psi \otimes m)^{\mu_M} \otimes \varphi = n \otimes (\psi \otimes m)^{\mu_M} . \varphi = n \otimes \psi \cdot m\varphi \ .)$$

To prove the last statement show that $\rho_{M,N} : N \longrightarrow M \otimes_{\Omega(M)} \mathrm{Hom}_S(M,N)$, $n \longmapsto \Sigma_{i=1}^{n} m_i \otimes \varphi_i n^{\sigma}$, - where σ is the canonical isomorphism $N \longrightarrow \mathrm{Hom}_S(S,N)$, - is an inverse for $\tau_{M,N}$. #

1.7 <u>Lemma</u>: If $M \in {}_{\underline{S}}\underline{P}^f$, then $(\mathrm{Im}\ \tau_M)M = M$.

<u>Proof</u>: By (1.5), if M is projective, then $\exists\ m_i \in M$, $\varphi_i \in {}_S M^*$, $i = 1,\ldots,n$, such that $m = \Sigma_{i=1}^{n}(m\varphi_i)m_i$, $\forall\, m \in M$. But $m\varphi_i \in \mathrm{Im}\ \tau_M$, $i = 1,\ldots,n$. Hence $M = (\mathrm{Im}\ \tau_M)M$. #

1.8 Remark: It is in general not true that $\mathrm{Im}\,\tau_M = S$ if $M \in {}_{\underline{S}}\underline{P}^f$. For example, let $S = Ke \oplus Ke'$, $e^2 = e$, $(e')^2 = e'$, $ee' = e'e = 0$ where K is a field. Then S is a ring, and $Ke \in {}_{\underline{S}}\underline{P}^f$. But $(Ke \otimes_{\Omega(Ke)} (Ke)^*)^{\tau_{Ke}} = Ke$.

1.9 Definition: $M \in {}_{\underline{S}}\underline{M}^f$ is called a generator, if $\mathrm{Im}\,\tau_M = S$, $M \in {}_{\underline{S}}\underline{P}^f$ is called a progenerator, if $\mathrm{Im}\,\tau_M = S$. A progenerator is sometimes also called faithfully projective (Strooker [1]) or a projective completely faithful module (Endo [1]).

1.10 Theorem (Strooker [1], Cohn [1]): For $M \in {}_{\underline{S}}\underline{M}^f$ the following statements are equivalent:

(i) $\mathrm{Im}\,\tau_M = S$.

(ii) $\exists\, X \in {}_{\underline{S}}\underline{M}^f$, such that $X \oplus {}_S S \cong M^{(n)}$ for some positive integer n.

(iii) $\mathrm{Hom}_S(M,-)$ is a faithful functor on ${}_{\underline{S}}\underline{M}^f$.

(iv) Every $X \in {}_{\underline{S}}\underline{M}^f$ is the homomorphic image of $M^{(n)}$ for some positive integer n.

(v) $(\mathrm{Im}\,\tau_M)M = M$, and for every maximal right ideal I in S, $IM \neq M$.

Proof: (i) $\Longrightarrow$ (ii): Since $\mathrm{Im}\,\tau_M = S$, $\exists\, m_i \in M$, $\varphi_i \in {}_S M^*$, $i = 1,\dots,n$, such that $\Sigma_{i=1}^n m_i\varphi_i = 1$. Then we have an epimorphism $\rho : M^{(n)} \longrightarrow {}_S S$, $(x_i)_{1\leq n} \longrightarrow \Sigma_{i=1}^{i=n} x_i \varphi_i$, and we obtain (ii).

(ii) $\Longrightarrow$ (iii): It suffices to show that $\hom(1_M, \psi) \neq 0$, whenever $\psi \in \mathrm{Hom}_S(X,X')$ is not zero. According to (ii) we have the exact sequence

$$M^{(n)} \longrightarrow {}_S S \longrightarrow 0,$$

which induces the exact commutative diagram

$$\begin{array}{ccccc} 0 \longrightarrow & \mathrm{Hom}_S({}_S S, X) & \longrightarrow & \mathrm{Hom}_S(M^{(n)}, X) \\ & \big\downarrow \mathrm{hom}\,(1_S, \psi) & & \big\downarrow \mathrm{hom}(1_{M^{(n)}}, \psi) \\ 0 \longrightarrow & \mathrm{Hom}_S({}_S S, X') & \longrightarrow & \mathrm{Hom}_S(M^{(n)}, X') \end{array} .$$

Since $\mathrm{Hom}_S({}_S S, X') \cong X$, $\mathrm{hom}(1_S, \psi) \neq 0$, but then $\mathrm{hom}(1_{M^{(n)}}, \psi) \neq 0$, and hence $\mathrm{hom}(1_M, \psi) \neq 0$.

(iii) $\Longrightarrow$ (i). Assume that $\mathrm{Im}\,\tau_M \neq S$. Then the canonical map $\varphi : S \longrightarrow S/\mathrm{Im}\,\tau_M$ is non-zero. However, since $\mathrm{Im}\,\psi \subset \mathrm{Im}\,\tau_M$ for all $\psi \in \mathrm{Hom}_S(M, S)$, $\psi\varphi = 0$, and $\mathrm{hom}(1, \varphi) = 0$; i.e., (iii) also fails.

(ii) $\Longleftrightarrow$ (iv). If $X \in {}_{S}\underline{\underline{M}}^f$ then there exists an exact sequence of the form $S^{(m)} \longrightarrow X \longrightarrow 0$. But if $X \oplus S \simeq M^{(n)}$, then this gives rise to the epimorphism $M^{(nm)} \longrightarrow X \longrightarrow 0$. Conversely if (iv) holds then $M^{(n)} \longrightarrow S \longrightarrow 0$ is exact for some n, but since this sequence splits, (ii) holds.

(i) $\Longleftrightarrow$ (v): Trivially, (v) $\Longrightarrow$ (i), since $\mathrm{Im}\,\tau_M$ is a right ideal in S. Conversely, let I be a right ideal in S such that $IM = M$. Then $0 = M/IM \cong S/I \otimes_S M$ (cf. I, (3.18)). Thus $0 = (S/I \otimes_S M)^{(n)} \cong S/I \otimes_S M^{(n)} \cong S/I \otimes_S S \oplus S/I \otimes_S X$, for some $X \in {}_{S}\underline{\underline{M}}^f$, by (ii); but (i) $\Longrightarrow$ (ii). Thus $S/I \otimes_S S \cong S/I = 0$; i.e., $S = I$. #

1.11 <u>Lemma</u>: Let S be a commutative ring, B an S-algebra which is faithfully flat as an S-module; i.e., $B \otimes_S -$ is a faithful exact functor, and C an S-algebra. Then $M \in {}_{C}\underline{\underline{M}}^f$ is projective (a generator) if and only if $B \otimes_S M \in {}_{C}\underline{\underline{M}}^f$ is projective (a generator).

<u>Proof</u>: "$\Longrightarrow$". This direction is obvious, since $B \otimes_S -$ is an additive functor carrying free modules into free modules. Conversely, if $B \otimes_S M \in {}_{B \otimes_S C}\underline{\underline{P}}^f$, then

$\mathrm{Ext}^1_{B \otimes_S C}(B \otimes_S M, B \otimes_S X) = 0, \quad \forall X \in {}_{\underline{\underline{C}}}\underline{\underline{M}}$. By (1.2), $0 = B \otimes_S \mathrm{Ext}^1_C(M,X)$. Since $B \otimes_S -$ is a faithful functor, $\mathrm{Ext}^1_C(M,X) = 0, \quad \forall X \in {}_{\underline{\underline{C}}}\underline{\underline{M}}$; i.e., $M \in {}_{\underline{\underline{C}}}\underline{\underline{P}}^f$.

Let now $B \otimes_S M$ be a generator. If $\mathrm{hom}(1_M, \psi) = 0$ for some $\psi \in \mathrm{Hom}_C(X,X')$ then $0 = 1_B \otimes \mathrm{hom}(1_M,\psi) \cong \mathrm{hom}(1_B \otimes 1_M, 1_B \otimes \psi)$, and $1_B \otimes \psi = 0$, since $B \otimes_S M$ is a generator. But $B \otimes_C -$ is a faithful functor; hence $\psi = 0$. #

__Exercises §1__:

1.) Let S be a commutative ring, A and B S-algebras. If $P \in {}_{\underline{\underline{A}}}\underline{\underline{P}}^f$, show that $B \otimes_S P \in {}_{B \otimes_S A}\underline{\underline{P}}^f$.

2.) Show that the map σ_M defined in the proof of (1.6) is a ring - and an (S,S)-homomorphism such that $\sigma_M \tau_M = 1_S$ and $\tau_M \sigma_M = 1_M \otimes_{\Omega(M)} {}_S M^*$.

3.) Finish the proof of (1.7).

4.) Let S be a commutative ring and B an S-algebra. If $M,N \in {}_{\underline{\underline{B}}}\underline{\underline{M}}$, show that $\mathrm{Ext}^n_B(M,N) \in {}_{\underline{\underline{S}}}\underline{\underline{M}}$. If, in addition, B is a finite S-algebra and if $M,N \in {}_{\underline{\underline{B}}}\underline{\underline{M}}^f$, show that $\mathrm{Ext}^n_B(M,N) \in {}_{\underline{\underline{S}}}\underline{\underline{M}}^f$.

§2 Morita equivalence:

The Morita theorems are proved: If $E \in {}_S\underline{\underline{M}}^f$ is a progenerator, there exists a categorical equivalence between ${}_S\underline{\underline{M}}^f$ and ${}_{\mathrm{End}_S(M)}\underline{\underline{M}}^f$. Various natural isomorphisms are derived. As general references we list: Auslander-Goldman [1], Bass [2], Cohn [1], Morita [1].

2.1 Theorem (Morita [1]): Let S be a ring and $E \in {}_S\underline{\underline{P}}^f$ a progenerator, and write $\Omega = \Omega(E) = \mathrm{End}_S(E)$. Then there exists a categorical equivalence between ${}_S\underline{\underline{M}}^f$ and ${}_\Omega\underline{\underline{M}}^f$:

$$h^E : {}_S\underline{\underline{M}}^f \longrightarrow {}_\Omega\underline{\underline{M}}^f,\ X \longmapsto \mathrm{Hom}_S(E,X),$$
$$h^E : \mathrm{Hom}_S(X,X') \longrightarrow \mathrm{Hom}_\Omega(\mathrm{Hom}_S(E,X),\mathrm{Hom}_S(E,X')),$$
$$\varphi \longmapsto \mathrm{hom}(1_E,\varphi).$$

This categorical equivalence is called a Morita equivalence between ${}_S\underline{\underline{M}}^f$ and ${}_\Omega\underline{\underline{M}}^f$. Moreover, it is an order isomorphism; in particular, the S-submodules of E correspond to the left ideals in Ω, and the (S,Ω)-submodules of E correspond to the two-sided ideals of Ω.

For greater lucidity, we shall postpone the proof for a moment.

2.2 Lemma: (i) Let $E \in {}_S\underline{\underline{P}}^f$. Then

α) ${}_SE^* = \mathrm{Hom}_S(E,S)\ {}_\Omega \overset{\mathrm{nat}}{\cong}{}_S\ \mathrm{Hom}_\Omega(E,\Omega) = E^*_\Omega$ where $\Omega = \mathrm{End}_S(E)$.

β) $E^*_\Omega,\ {}_SE^* \in \underline{\underline{P}}^f{}_S$.

γ) E and ${}_SE^*$ are generators in $\underline{\underline{M}}^f_\Omega$ and ${}_\Omega\underline{\underline{M}}^f$ respectively (cf. (1.9)).

(ii) Let $E \in {}_S\underline{\underline{M}}^f$ be a generator. Then

α) ${}_SE^*$ is a generator in $\underline{\underline{M}}^f_S$.

β) $E \in \underline{\underline{P}}^f_\Omega$, ${}_SE^* \in {}_\Omega\underline{\underline{P}}^f$.

Note that we indicate by attaching subscripts to which category an isomorphism belongs, e.g., $\cong_S$ denotes an isomorphism of right

S-modules.

<u>Proof</u>: (i) α) We have the following chain of natural isomorphisms of bimodules:

$${}_S E^* \ {}_\Omega{\cong}_S \ \mathrm{Hom}_S(E \otimes_\Omega ({}_S E^* \otimes_S E), S) \ {}_\Omega{\cong}_S \ \mathrm{Hom}_S(E \otimes_\Omega {}_S E^*, \mathrm{Hom}_S(E,S))$$
$${}_\Omega{\cong}_S \ \mathrm{Hom}_\Omega(E, \mathrm{Hom}_S({}_S E^*, {}_S E^*)) \ {}_\Omega{\cong}_S \ \mathrm{Hom}_\Omega(E, {}_S E^* \otimes_S ({}_S E^*)_S{}^*)$$
$${}_\Omega{\cong}_S \ \mathrm{Hom}_\Omega(E, {}_S E^* \otimes_S E) \ {}_\Omega{\cong}_S \ \mathrm{Hom}_\Omega(E,\Omega) = E^*_\Omega \quad (1.5).$$

β) By assumption, $E \oplus X \cong_S S^{(n)}$ for some X and n. But then ${}_S E^* \oplus {}_S X^* \cong_S \mathrm{Hom}_S(S^{(n)}, S) \cong_S (\mathrm{Hom}_S(S,S))^{(n)} \cong_S (S)^{(n)}$ and ${}_S E^*$ is right S-projective. The same holds for E^*_Ω by α).

γ) If $E \oplus X \cong {}_S S^{(n)}$ for some $X \in {}_S\underline{\underline{P}}^f$ and some natural number n, then

$$E^{(n)} \cong_\Omega \mathrm{Hom}_S({}_S S^{(n)}, E) \cong_\Omega \mathrm{Hom}_S(E,E) \oplus \mathrm{Hom}_S(X,E);$$

i.e., E is a generator in $\underline{\underline{M}}^f_\Omega$ (cf. (1.10)). Similarly it follows from β) that ${}_S E^*$ and E^*_Ω are generators in ${}_\Omega\underline{\underline{M}}^f$.

(ii) α) Since E is a left S-generator, there are $X \in {}_S\underline{\underline{M}}^f$ and $n \in N$ such that $E^{(n)} {}_S{\cong} X \oplus_S S$. Therefore $({}_S E^*)^{(n)} \cong_S {}_S X^* \oplus {}_S S^* = {}_S X^* \oplus S$ and ${}_S E^*$ is indeed a generator.

β) By assumption $E^{(n)} {}_S{\cong} X \oplus {}_S S$. Thus

$$\Omega^{(n)} \cong_\Omega \mathrm{Hom}_S(E^{(n)}, E) \cong_\Omega \mathrm{Hom}_S({}_S S, E) \oplus \mathrm{Hom}_S(X,E) \cong_\Omega E \oplus \mathrm{Hom}_S(X,E);$$

i.e., $E \in \underline{\underline{P}}^f_\Omega$. Similarly, $E^* \in {}_\Omega\underline{\underline{P}}^f$ is established. #

2.3 <u>Remark</u>: I. If $E \in {}_S\underline{\underline{P}}^f$, we have the following natural isomorphisms:

(i) ${}_S E^* \ {}_\Omega{\cong}_S \ E^*_\Omega, \ \Omega = \mathrm{End}_S(E)$,

(ii) $\Omega = \mathrm{End}_S(E) \overset{\text{ring}}{\cong} {}_S E^* \otimes_S E$ (cf. (1.4)),

(iii) $E \otimes_\Omega E^*_\Omega \overset{\text{ring}}{\cong} \mathrm{End}_\Omega (E)$ (cf. (2.2), (1.10)),

(iv) $E \ {}_S{\cong}_\Omega \ {}_\Omega({}_S E^*)^*$.

II. If $E \in {}_S\underline{\underline{M}}^f$ is a generator, we have the following natural isomorphisms:

(i) $S \overset{\text{ring}}{\cong} E \otimes_{\Omega} {}_S E^*$,

(ii) $E \otimes_{\Omega} E^*_{\Omega} \overset{\text{ring}}{\cong} \mathrm{End}_{\Omega}(E)$.

III. Combining these isomorphisms, we obtain for a progenerator $E \in {}_S\underline{\underline{M}}^f$ the following natural isomorphisms:

(i) ${}_S E^* \; {}_{\Omega}\!\cong_S \; E^*_{\Omega}$,

(ii) $\Omega = \mathrm{End}_S(E) \overset{\text{ring}}{\cong} {}_S E^* \otimes_S E$,

(iii) $S \overset{\text{ring}}{\cong} E \otimes_{\Omega} {}_S E^* \overset{\text{ring}}{\cong} \mathrm{End}_{\Omega}(E)$,

(iv) $E \; {}_S\!\cong_{\Omega} \; {}_{\Omega}({}_S E^*)^* \; {}_S\!\cong_{\Omega} \; (E^*_{\Omega})^*_S$, and

(v) $\forall\, N \in {}_S\underline{\underline{M}}^f$: ${}_S E^* \otimes_S N \; {}_{\Omega(E)}\!\cong_{\Omega(N)} \; \mathrm{Hom}_S(E,N)$, where $\Omega(E) = \mathrm{Hom}_S(M,M)$ and $\Omega(N) = \mathrm{Hom}_S(N,N)$ (cf. (1.5)).

Now we turn to the <u>proof of (2.1)</u>: Let $E \in {}_S P^f$ be a generator. From (1.10) and (2.2) it follows that the following functors are faithful:

$$h^E:\ {}_S\underline{\underline{M}}^f \longrightarrow {}_{\Omega}M^f;\ N \longmapsto \mathrm{Hom}_S(E,N),$$

$$\mathrm{Hom}_S(N,N') \longrightarrow \mathrm{Hom}_{\Omega}(\mathrm{Hom}_S(E,N),\mathrm{Hom}_S(E,N')),\ \varphi \longmapsto \mathrm{hom}_S(1_E,\varphi)$$

and

$$t^E:\ {}_{\Omega}\underline{\underline{M}}^f \longrightarrow {}_S\underline{\underline{M}}^f;\ M \longmapsto \mathrm{Hom}_{\Omega}({}_S E^*,Y),$$

$$\mathrm{Hom}_{\Omega}(M,M') \longrightarrow \mathrm{Hom}_S(\mathrm{Hom}_{\Omega}({}_S E^*,M),\mathrm{Hom}_{\Omega}({}_S E^*,M')),\ \psi \longmapsto \mathrm{hom}_{\Omega}(1_{{}_S E^*},\psi).$$

We shall show that $h^E t^E \sim 1_{{}_S\underline{\underline{M}}^f}$ and $t^E h^E \sim 1_{{}_{\Omega}\underline{\underline{M}}^f}$ (cf. II, (1.10)):

$$h^E t^E:\ N \overset{h^E}{\longmapsto} \mathrm{Hom}_S(E,N) \overset{t^E}{\longmapsto} \mathrm{Hom}_{\Omega}({}_S E^*,\mathrm{Hom}_S(E,N))$$
$$\overset{\text{nat}}{{}_S\cong} \mathrm{Hom}_S(E \otimes_{\Omega} {}_S E^*,N) \overset{\text{nat}}{{}_S\cong} \mathrm{Hom}_S(S,N) \overset{\text{nat}}{{}_S\cong} N,$$

$$t^E h^E:\ M \overset{t^E}{\longmapsto} \mathrm{Hom}_{\Omega}({}_S E^*,M) \overset{h^E}{\longmapsto} \mathrm{Hom}_S(E,\mathrm{Hom}_{\Omega}({}_S E^*,M))$$
$$\overset{\text{nat}}{{}_{\Omega}\cong} \mathrm{Hom}_{\Omega}({}_S E^* \otimes_S E,M) \overset{\text{nat}}{{}_{\Omega}\cong} \mathrm{Hom}_{\Omega}(\Omega,M) \overset{\text{nat}}{{}_{\Omega}\cong} M.$$

It should be observed that, in order to show $h^E t^E(N) \overset{\text{nat}}{\cong} N$, we have only used the fact that E is a generator in ${}_S\underline{\underline{M}}^f$, whereas for $t^E h^E(M) \overset{\text{nat}}{\cong} M$ we have used that $E \in {}_S\underline{\underline{P}}^f$. For the homomorphisms we have

$$\begin{aligned} h^E(\mathrm{Hom}_S(N,N')) &= \mathrm{Hom}_\Omega(\mathrm{Hom}_S(E,N),\mathrm{Hom}_S(E,N')) \\ &\overset{\text{nat}}{\cong} \mathrm{Hom}_\Omega({}_S E^* \otimes_S N,\ \mathrm{Hom}_S(E,N') \\ &\overset{\text{nat}}{\cong} \mathrm{Hom}_S(E \otimes_S N),N') \overset{\text{nat}}{\cong} \mathrm{Hom}_S(N,N'). \end{aligned}$$

This shows that the functor $h^E = \mathrm{Hom}_S(E,-)$ is a categorical isomorphism. Now, let $M' \subset_S M$; i.e., let M' be an S-submodule of M. Then the exact sequence $0 \longrightarrow M' \longrightarrow M \longrightarrow M/M' \longrightarrow 0$ induces the exactness of the sequence of left Ω-modules $0 \longrightarrow \mathrm{Hom}_S(E,M') \longrightarrow \mathrm{Hom}_S(E,M)$; and $h^E(M')$ is a left Ω-submodule of $h^E(M)$. In particular, if $M = E$, then $h^E(M')$ is a left Ω-ideal and if, in addition, $M \in {}_{\underline{\underline{S}}}\underline{\underline{M}}^f_\Omega$, it is clear that $h^E(M)$ is a two-sided ideal in Ω. #

2.4 __Remark__: It should be observed that t^E from the proof of (2.1) is also a Morita equivalence (cf. (2.2)).

2.5 __Remark__: (2.1) also holds for ${}_{\underline{\underline{S}}}\underline{\underline{M}}$ and ${}_{\underline{\underline{\Omega}}}\underline{\underline{M}}$ since t^E and h^E preserve injective limits (cf. Cohn [1]).

2.6 __Remark__: The Morita equivalence $h^E : {}_\rho\underline{\underline{M}}^f \longrightarrow {}_\Omega\underline{\underline{M}}^f$, $M \longrightarrow \mathrm{Hom}_S(E,M)$, - with $E \in {}_{\underline{\underline{S}}}\underline{\underline{M}}^f$ a progenerator and $\Omega = \mathrm{Hom}_\rho(E,E)$, - preserves projectives, generators and faithful modules.

__Exercises §2__:

1.) Let S be a ring and let $E \in {}_{\underline{\underline{S}}}\underline{\underline{M}}^f$ be a progenerator. Show that the following two pairs of functors are naturally equivalent

(i) $\mathrm{Hom}_S(E,-)$ and $E^* \otimes_S -$,

(ii) $\mathrm{Hom}\ (E^*,-)$ and $E \otimes_\Omega -$, where $\Omega = \mathrm{End}_S(E)$.

2.) Let S be a ring and $E \in {}_{\underline{\underline{S}}}\underline{\underline{P}}^f$. Then E is a progenerator if and only if $- \otimes_S E$ is a faithful functor on $\underline{\underline{M}}^f_{\underline{\underline{S}}}$ (cf. (1.9)), if and only if $X \otimes_S E = 0$ implies $X = 0$, $\forall X \in {}_{\underline{\underline{S}}}\underline{\underline{M}}^f$.

3.) Let S be a ring and $E \in {}_S\underline{\underline{M}}^f$ a progenerator. Show that for every $M \in {}_S\underline{\underline{M}}^f$, $\mathrm{Hom}_S(E,M) \in {}_\Omega\underline{\underline{M}}^f$, where $\Omega = \mathrm{End}_S(E)$.

4.) Let $M \in {}_S\underline{\underline{M}}^f$, and show that

(i) if M is projective, then $\mathrm{Hom}_S(M,-): {}_S\underline{\underline{M}}^f \longrightarrow {}_{\mathrm{End}_S(M)}\underline{\underline{M}}^f$, preserves generators and faithfulness,

(ii) if M is a generator, then $\mathrm{Hom}_S(M,-)$ preserves projectives.

§3 Norm and trace

This section is a survey of trace, norm, discriminant and dual bases of finite dimensional algebras over a field. K denotes a field and A a finite dimensional K-algebra; i.e., A is a ring, which is at the same time a finite dimensional K-vectorspace.

3.1 Definitions: Let $a \in A$. Then $\varphi_a : A \longrightarrow A$, $x \longmapsto ax$, is a linear transformation of A as a K-module. Let $(a_{ij})_{1\leq i,j\leq n}$ be the matrix of φ_a relative to some fixed K-basis $\{w_i\}_{1\leq i\leq n}$ of A and let X be an indeterminate over K. Then we define $Pc_{A/K}(a,X) = \det(X\underline{E}_n - (a_{ij}))$, where $\underline{E}_n$ is the $(n \times n)$ identity matrix, to be the characteristic polynomial of $a \in A$ with respect to K. $Tr_{A/K}(a) = tr(a_{ij}) = \Sigma_{i=1}^n a_{ii}$ is called the trace of $a \in A$ over K, and $N_{A/K}(a) = \det(a_{ij})$ is the norm of $a \in A$ over K.

The matrix $(Tr_{A/K}(w_i w_j))_{1\leq i,j\leq n} = \underline{D}_{A/K}(w_1,\ldots,w_n)$ is called the discriminant matrix of the basis $\{w_i\}_{1\leq i\leq n}$ of A over K, and $D_{A/K}(w_1,\ldots,w_n) = \det(\underline{D}_{A/K}(w_1,\ldots,w_n))$ is called the discriminant of the basis $\{w_i\}_{1\leq i\leq n}$ of A over K.

Since A is finite over K, A is integral over K (cf. I, (5.4)) and thus $a \in A$ satisfies a monic polynomial with coefficients in K; because $K[X]$ is a Euclidean domain there is a unique monic polynomial $\min_{A/K}(a,X)$ of minimal degree in $K[X]$, which has $a \in A$ as a root, the minimum polynomial of $a \in A$ over K.

3.2 Lemma: For $a \in A$, $Pc_{A/K}(a,X)$, $Tr_{A/K}(a)$ and $N_{A/K}(a)$ are independent of the chosen basis. If $\{w_i\}_{1\leq i\leq n}$ and $\{w_i'\}_{1\leq i\leq n}$ are two bases of A over K, and if $\underline{B}$ is the matrix of the linear transformation $w_i \longmapsto w_i'$, $1 \leq i \leq n$, relative to the basis

$\{w_i\}_{1\le i\le n}$. Then

$$\underline{\underline{B}}\underline{\underline{D}}_{A/K}(w_1,\ldots,w_n)\underline{\underline{B}}^t = \underline{\underline{D}}_{A/K}(w_1',\ldots,w_n'),$$

where $\underline{\underline{B}}^t$ is the transposed matrix of $\underline{\underline{B}}$. Thus the vanishing of the discriminant of any basis is a property of the algebra A over K alone.

Proof: Let $w_i' = \Sigma_{j=1}^{n} \beta_{ij} w_j$. If $\underline{\underline{A}}$ and $\underline{\underline{A}}'$ are the matrices of φ_a for $a \in A$, relative to the bases $\{w_i\}_{1\le i\le n}$ and $\{w_i'\}_{1\le i\le n}$ resp., then $\underline{\underline{A}}' = \underline{\underline{B}}\,\underline{\underline{A}}\,\underline{\underline{B}}^{-1}$. Thus $Pc_{A/K}(a,X)_{\{w_i\}} = \det(X\underline{\underline{E}}_n - \underline{\underline{A}})$ $= \det(\underline{\underline{B}}(X\underline{\underline{E}}_n - \underline{\underline{A}})\underline{\underline{B}}^{-1}) = \det(X\underline{\underline{E}}_n - \underline{\underline{A}}') = Pc_{A/K}(a,X)_{\{w_i'\}}$.

If $Pc_{A/K}(a,X) = X^n + k_{n-1} X^{n-1} + \ldots + k_o$, then it is easily seen that

3.2' $$Tr_{A/K}(a) = -k_{n-1} \quad\text{and}\quad N_{A/K}(a) = (-1)^n k_o.$$

Thus the norm and trace are independent of the particular underlying basis. The formula for the discriminant is verified by a straight-forward computation. #

3.3 Lemma (Cayley-Hamilton): Let $a \in A$; then $Pc_{A/K}(a,a) = 0$.

Proof: Since we have an isomorphism $A \xrightarrow{\sim} A_L = Hom_A(A_A,A_A)$, $a \longrightarrow \varphi_a$, it suffices to show that $Pc_{A/K}(a,(a_{ij})_{1\le i,j\le n}) = 0$ where $(a_{ij})_{1\le i,j\le n}$ is the matrix of the linear transformation φ_a relative to some fixed basis. Let $\underline{\underline{A}} = (a_{ij})_{1\le i,j\le n}$. The following formula is easily checked by "multiplying out":

$$X^i\underline{\underline{E}}_n - \underline{\underline{A}}^i = (X^{i-1}\underline{\underline{E}}_n + X^{i-2}\underline{\underline{A}} + \ldots + \underline{\underline{A}}^{i-1})(X\underline{\underline{E}}_n - \underline{\underline{A}}),$$

for every natural number i. We write this as $X^i\underline{\underline{E}}_n - \underline{\underline{A}}^i = \underline{\underline{B}}_i(X)(X\underline{\underline{E}}_n - \underline{\underline{A}})$. Now $Pc(\underline{\underline{A}},X) = \det(X\underline{\underline{E}}_n - \underline{\underline{A}}) = \Sigma_{i=1}^{n} k_i X^i$, $k_n = \pm 1$. Let $\underline{\underline{A}}(X)$ be the matrix of the cofactors of $X\underline{\underline{E}}_n - \underline{\underline{A}}$ (cf. proof of I,(5.2)). Then $\underline{\underline{A}}(X)(X\underline{\underline{E}}_n - \underline{\underline{A}}) = Pc(\underline{\underline{A}},X)\cdot\underline{\underline{E}}_n$. Thus

$$Pc(\underline{\underline{A}},\underline{\underline{A}}) = \Sigma_{i=1}^{n} k_i \underline{\underline{A}}^i = \Sigma_{i=1}^{n} k_i X^i \underline{\underline{E}}_n - \Sigma_{i=1}^{n} k_i \underline{\underline{B}}_i(X)(X^i \underline{\underline{E}}_n - \underline{\underline{A}}).$$

But $(\Sigma_{i=1}^{n} k_i X^i)\underline{\underline{E}}_n = Pc(\underline{\underline{A}},X) \cdot \underline{\underline{E}}_n = \underline{\underline{A}}(X)(X\underline{\underline{E}}_n - \underline{\underline{A}})$. Hence

$$Pc(\underline{\underline{A}},\underline{\underline{A}}) = \underline{\underline{A}}(X)(X\underline{\underline{E}}_n - \underline{\underline{A}}) - \Sigma_{i=1}^{n} k_i \underline{\underline{B}}_i(X)(X\underline{\underline{E}}_n - \underline{\underline{A}}),$$

$$Pc(\underline{\underline{A}},\underline{\underline{A}}) = (\underline{\underline{A}}(X) - \Sigma_{i=1}^{n} k_i \underline{\underline{B}}_i(X))(X\underline{\underline{E}}_n - \underline{\underline{A}}).$$

Since the left hand side is independent of X, the right hand side has to be zero. #

3.4 Corollary: For $a \in A$, $\min_{A/K}(a,X)$ divides $Pc_{A/K}(a,X)$.

Proof: This is an immediate consequence of (3.3). #

3.5 Lemma: If $(A : K) = n$, then, for $a \in A$, $Pc_{A/K}(a,X)$ divides $(\min_{A/K}(a,X))^n$.

Proof: Let $\underline{\underline{A}}$ be the matrix of φ_a relative to some fixed basis; then it suffices to show that $Pc(\underline{\underline{A}},X)$ divides $\min(\underline{\underline{A}},X)^n$. Let $\min(\underline{\underline{A}},X) = f(X) = \Sigma_{i=1}^{t} k_i X^i$. We define recursively the matrices

$$\begin{aligned} \underline{\underline{B}}_{t-1} &= k_t \cdot \underline{\underline{E}}_n \\ \underline{\underline{B}}_{t-2} &= k_{t-1} \cdot \underline{\underline{E}}_n + A\,\underline{\underline{B}}_{t-1} \\ &\vdots \\ \underline{\underline{B}}_o &= k_1 \cdot \underline{\underline{E}}_n + \underline{\underline{A}}\underline{\underline{B}}_1. \end{aligned}$$

Then $0 = f(\underline{\underline{A}}) = \Sigma_{i=1}^{t-1}(\underline{\underline{A}}^i \underline{\underline{B}}_{i-1} - \underline{\underline{A}}^{i+1}\underline{\underline{B}}_i) + \underline{\underline{A}}^t \underline{\underline{B}}_{t-1} + k_o \underline{\underline{E}}_n$. Thus $k_o \underline{\underline{E}}_n = -\underline{\underline{A}}\underline{\underline{B}}_o$. If we now put $\underline{\underline{B}}(X) = \Sigma_{i=o}^{t-1} \underline{\underline{B}}_i X^i$, then $(X\underline{\underline{E}}_n - \underline{\underline{A}})\underline{\underline{B}}(X) = f(X) \cdot \underline{\underline{E}}_n$, as one checks easily. Taking determinants of this matrix equation, we obtain $Pc(\underline{\underline{A}},X) \cdot \det(\underline{\underline{B}}(X)) = (f(X))^n$. #

3.6 Definition: Let A be a K-algebra. An (associative) bilinear form f is a map $f : A \times A \longrightarrow K$ such that $f(a+b,c) = f(a,c) + f(b,c)$, $f(a,b+c) = f(a,b) + f(a,c)$, $f(ka,b) = f(a,kb) = kf(a,b)$, $f(ab,c) = f(a,bc)$. f is said to be non-degenerate if $f(x,A) = 0$ implies $x = 0$. An example of such a bilinear form is $Tr_{A/K} : A \times A \longrightarrow K$, $Tr_{A/K} : (a,b) \longrightarrow Tr_{A/K}(ab)$. Tr is non-degenerate if and only if the discriminant of some K-basis of A is different from zero.

3.7 Lemma: Let A be a K-algebra, and $f : A \times A \longrightarrow K$ a non-degenerate bilinear form. If $\{w_i\}_{1\leq i\leq n}$ is a K-basis of A, then there exists a so-called dual basis with respect to f, $\{w_i^*\}_{1\leq i\leq n}$, satisfying

(i) $f(w_i, w_j^*) = \delta_{ij}$ and

(ii) if $w_i a = \Sigma_{i=1}^{n} k_{ij} w_j$, then $a w_i^* = \Sigma_{i=1}^{n} k_{ji} w_j^*$.

Moreover, the discriminant of $\{w_i\}$ relative to f $\det(f(w_i,w_j)) \neq 0$. Conversely, if f is a bilinear form such that, for some K-basis $\{w_i\}_{1\leq i\leq n}$ of A, the discriminant relative to f is different from zero, then the same is true for any K-basis of A.

Proof: Since f is non-degenerate,

$$\det(f(w_i,w_j)) \neq 0 \qquad \text{(cf. Ex. 3,6)},$$

and hence $(f(w_i,w_j))_{1\leq i,j\leq n}$ is invertible. Let $(k_{ij})_{1\leq i,j\leq n} = [(f(w_i,w_j))_{1\leq i,j\leq n}]^{-1}$ and set $w_i^* = \Sigma_{j=1}^{n} k_{ji}^{o} w_j$. Then $\{w_i^*\}_{1\leq i\leq n}$ is a K-basis for A and $f(w_i,w_j^*) = \delta_{ij}$. (ii) follows from an easy computation. The remainder is proved as in (3.2). #

3.8 Remark: The dual bases of a K-basis $\{w_i\}_{1\leq i\leq n}$ with respect to different non-degenerate bilinear forms may very well be different.

Example: Let G be a finite group of order n, and K a field such that $\operatorname{char}(K) \nmid n$. Then $\{g\}_{g\in G}$ and $\{1/n\cdot g^{-1}\}_{g\in G}$ are a pair of dual bases of KG with respect to the trace function. But also $f : KG \times KG \longrightarrow K$, $f : (g,g') \longmapsto \begin{cases} 0 & \text{if } g' \neq g^{-1} \\ 1 & \text{if } g' = g^{-1} \end{cases}$ is a non-degenerate bilinear form, and $\{g\}_{g\in G}$ and $\{g^{-1}\}_{g\in G}$ are a pair of dual bases with respect to f.

<u>Exercises §3</u>:

1.) Let A be an n-dimensional K-algebra. Show:

(i) $N_{A/K}(\alpha a\cdot b) = \alpha^n N_{A/K}(a) N_{A/K}(b)$, $\alpha \in K$, $a,b \in A$,

(ii) $Tr_{A/K}(\alpha a+b) = \alpha Tr(a) + Tr_{A/K}(b)$, $\alpha \in K$, $a,b \in A$.

2.) Let L be an extension field of K, and define $A^L = L \otimes_K A$. If A is an n-dimensional K-algebra, then A^L is an n-dimensional L-algebra. Show $Pc_{A/K}(a,X) = Pc_{A^L/L}(1 \otimes a, X)$, and thus, (cf. (3.2')), $Tr_{A/K}(a) = Tr_{A^L/L}(1 \otimes a)$ and $N_{A/K}(a) = N_{A^L/L}(1 \otimes a)$.

3.) Let A be a finite dimensional K-algebra, such that $A = \oplus_{i=1}^n A_i$ as K-algebra. Show: $Pc_{A/K}(a,X) = \Pi_{i=1}^n Pc_{A_i/K}(a_i,X)$, where $a = \Sigma_{i=1}^n a_i$, $a_i \in A_i$, $1 \le i \le n$. $Tr_{A/K}(a) = \Sigma_{i=1}^n Tr_{A_i/K}(a_i)$ $N_{A/K}(a) = \Pi_{i=1}^n N_{A_i/K}(a_i)$.

If $\{w_{ij}\}_{1\le j\le n_i}$ is a K-basis for A_i, $1 \le i \le n$,

$$D((w_{ij})_{\substack{1\le i\le n\\ 1\le j\le n_i}}) = \Pi_{i=1}^n D((w_{ij})_{1\le j\le n_i}).$$

4.) If A is a K-algebra and L a subfield of K such that $K \in {}_L\underline{M}^f$ show: $Tr_{A/L}(a) = Tr_{K/L}(Tr_{A/K}(a))$, $N_{A/L}(a) = N_{K/L}(N_{A/K}(a))$.

5.) Show that $Tr_{A/K} : A \times A \longrightarrow K$ is non-degenerate if and only if the discriminant of a K-basis of A is different from zero.

6.) Let $f : A \times A \longrightarrow K$ be a non-degenerate bilinear form. Show that $\det(f(w_i,w_j)) \neq 0$, where $(w_i)_{1\le i\le n}$ is a K-basis for A. Use this to prove (3.7).

7. A finite dimensional K-algebra A is called a <u>Frobenius algebra</u> if there exists a non-degenerate bilinear form $f : A \times A \longrightarrow K$. Let G be a finite group and K a field such that $\mathrm{char}(K) \mid |G|$.

Show that $A = KG$ is a Frobenius algebra. Find an example of a Frobenius algebra A which is not semi-simple. (Hint: Let $G = (g)$ be the cyclic group of order p and K a Galois-field with p elements. Show that $1 + g + \ldots + g^{p-1} \in \mathrm{rad}(KG)$. Generalize this idea and show that $\mathrm{rad}(KG) \neq 0$ if $\mathrm{char}(K) \mid |G|$.)

§4. The enveloping algebra

If B is a finite S-algebra, where S is a commutative ring, then

$$\mathrm{Ext}^n_{B^e}(B, \mathrm{Hom}_S(N,M)) \overset{\text{nat.}}{\cong} \mathrm{Ext}^n_B(N,M),$$

if $B^e = B \otimes_S B^{op}$ is the enveloping algebra and $M, N \in {}_{B}\underline{\underline{M}}^f$ are such that $\mathrm{Ext}^n_S(N,M) = 0$.

4.1 Definition: Let S be a commutative noetherian ring and B a finite S-algebra. Then B^{op} is also an S-algebra, and $B^e = B \otimes_S B^{op}$ is called the enveloping algebra of B. The correspondence $(x \otimes y^{op})m \longleftrightarrow xmy$ induces a bijection between ${}_{B^e}\underline{\underline{M}}$ and the subcategory ${}_{BSB}\underline{\underline{M}}$ of (B,B)-bimodules M such that $sm = ms$ for every $s \in S$, $m \in M$. In general, this is a proper subcategory of ${}_B\underline{\underline{M}}_B$.

4.2 Lemma: The B^e-map $\varepsilon : B^e \longrightarrow B$; $x \otimes y^{op} \longmapsto xy$ is called an augmentation of B^e and the augmentation ideal, Ker ε, is generated by $\{x \otimes 1^{op} - 1 \otimes x^{op} : x \in B\}$ as a left B^e-ideal.

Proof: Let $\Sigma_i\, x_i \otimes y_i^{op} \in \mathrm{Ker}\,\varepsilon$; i.e., $\Sigma_i\, x_i y_i = 0$; then $\Sigma_i\, x_i \otimes y_i^{op} = \Sigma_i[(x_i \otimes 1^{op})(1 \otimes y_i^{op} - y_i \otimes 1^{op})]$. ♣

4.3 Definition: Let $M \in {}_{B^e}\underline{\underline{M}}$; a derivation of M (crossed homomorphism) is an S-homomorphism $\varphi : B \longrightarrow M$, with $(xy)^\varphi = (x \otimes 1^{op})y^\varphi + (1 \otimes y^{op})x^\varphi$ or $(xy)^\varphi = x(y^\varphi) + (x^\varphi)y$. (Derivations are written as exponents.)

A derivation φ of M is called an inner derivation, if, for some $m \in M$, $x^\varphi = (x \otimes 1^{op})m - (1 \otimes x^{op})m$ (or $x^\varphi = xm - mx$), $x \in B$. The S-module of all derivations of M is denoted by Der(B,M), the S-module of all inner derivations by InDer(B,M).

4.4 <u>Lemma</u>: $\varphi_o : B \longrightarrow \mathrm{Ker}\,\varepsilon,\ x \longmapsto x \otimes 1^{op} - 1 \otimes x^{op}$ is a derivation.

<u>Proof</u>: Obviously, φ_o is an S-homomorphism. Moreover, $(xy)^{\varphi_o} = xy \otimes 1^{op} - 1 \otimes (xy)^{op} = xy \otimes 1^{op} - 1 \otimes y^{op}x^{op} = (x \otimes 1^{op})y^{\varphi_o} + (1 \otimes y^{op})x^{\varphi_o}$. #

4.5 <u>Theorem</u>: Let $M \in {}_{B^e}\underline{\underline{M}}$. Then we have a natural isomorphism $\Phi_M : \mathrm{Hom}_{B^e}(\mathrm{Ker}\,\varepsilon, M) \xrightarrow{\ \sim\ } \mathrm{Der}(B,M);\ \alpha \longmapsto \varphi_o\alpha$.

<u>Proof</u>: $\varphi_o\alpha : B \longrightarrow M$ is a derivation, since φ_o is one, and if $\varphi_o\alpha = \varphi_o\beta$; $\alpha,\beta \in \mathrm{Hom}_{B^e}(\mathrm{Ker}\,\varepsilon, M)$ then $(x \otimes 1^{op} - 1 \otimes x^{op})\alpha = (x \otimes 1^{op} - 1 \otimes x^{op})\beta$, $\forall x \in B$; hence $\alpha = \beta$ by (4.2). It remains to show that any given $\varphi \in \mathrm{Der}(B,M)$ can be factored through φ_o. Given φ, we define $\alpha : \mathrm{Ker}\,\varepsilon \longrightarrow M$, $\Sigma_i\, x_i \otimes y_i^{op} \longmapsto -\Sigma_i (x_i \otimes 1^{op})(y_i^{\varphi})$, for $\Sigma_i\, x_i y_i = 0$. This is obviously an S-homomorphism. But it is also B^e-linear, as is easily verified. Moreover, $x^{\varphi} - x^{\varphi_o\alpha} = x^{\varphi} - (x \otimes 1^{op} - 1 \otimes x^{op})\alpha = x^{\varphi} - x^{\varphi} = 0$; hence $\varphi = \varphi_o\alpha$. (Observe that a derivation takes value 0 on 1.)

That Φ is natural follows simply from the fact that Φ_M coincides with $\mathrm{hom}_S(\varphi_o, 1_M)$, with codomain restricted to the image. For $\sigma \in \mathrm{Hom}_{B^e}(M, M')$ we have $\mathrm{hom}_S(\varphi_o, 1_M)\mathrm{hom}_{B^e}(1_B, \sigma) = \mathrm{hom}_S(\varphi_o, \sigma) = \mathrm{hom}_{B^e}(1_{\mathrm{Ker}\,\varepsilon}, \sigma)\mathrm{hom}_S(\varphi_o, 1_{M'})$; hence the following diagram, where $\tilde{\sigma}$ denotes the appropriate restriction of $\mathrm{hom}_{B^e}(1_B, \sigma)$, commutes:

$$\begin{array}{ccc}
\mathrm{Hom}_{B^e}(\mathrm{Ker}\,\varepsilon, M) & \xrightarrow{\ \mathrm{hom}(1_{\mathrm{Ker}\,\varepsilon},\sigma)\ } & \mathrm{Hom}_{B^e}(\mathrm{Ker}\,\varepsilon, M') \\
\Big\downarrow \Phi_M & & \Big\downarrow \Phi_{M'} \\
\mathrm{Der}(B,M) & \xrightarrow{\qquad \tilde{\sigma} \qquad} & \mathrm{Der}(B,M')
\end{array}$$

#

4.6 <u>Lemma</u>: $\alpha \in \mathrm{Hom}_{B^e}(\mathrm{Ker}\ \varepsilon, M)$ can be extended to $\alpha' \in \mathrm{Hom}_{B^e}(B^e, M)$ if and only if $\varphi_o\alpha$ is an inner derivation; i.e., In Der(B,M) is isomorphic to the image of the natural restriction map $M \cong \mathrm{Hom}_{B^e}(B^e, M) \longrightarrow \mathrm{Hom}_{B^e}(\mathrm{Ker}\ \varepsilon, M)$.

<u>Proof</u>: Clearly $\alpha : \mathrm{Ker}\ \varepsilon \longrightarrow M$ can be extended to a B^e-map $\alpha' : B^e \longrightarrow M$ if and only if for some fixed $m \in M$, $(\Sigma_i\ x_i \otimes y_i^{op})\alpha = \Sigma_i (x_i \otimes y_i^{op})m$, $\forall\ \Sigma_i\ x_i \otimes y_i^{op} \in \mathrm{Ker}\ \varepsilon$. But this is equivalent to the condition that, for some $m \in M$, $\varphi_o\alpha : B \longrightarrow M$; $x \longmapsto (x \otimes 1^{op} - 1 \otimes x^{op})\alpha = (x \otimes 1^{op})m - (1 \otimes x^{op})m$; i.e., that $\varphi_o\alpha$ be an inner derivation. #

4.7 <u>Definition</u>: (i) B is called <u>separable</u>, if $B \in {}_{B^e}\underline{\underline{P}}^f$; i.e., if the exact B^e-sequence $E : 0 \longrightarrow \mathrm{Ker}\ \varepsilon \xrightarrow{\iota} B^e \xrightarrow{\varepsilon} B \longrightarrow 0$ splits. (Observe that $B \in {}_{B^e}\underline{\underline{M}}^f$, via $(x \otimes y^{op})b = xby$.)

(ii) For $M \in {}_{B^e}\underline{\underline{M}}^f$, we define the <u>n-th cohomology group of M</u> as $H^n(B,M) = \mathrm{Ext}^n_{B^e}(B,M)$, $n = 1,2,\ldots$ (cf. II,(3.4,i)), (cf. Hochschild [1]).

<u>Example</u>: Let S be a commutative ring and G a finite group. Let $SG = \oplus_{g\in G}\ Sg$. We make SG into an S-algebra by defining $g'\ \Sigma_{g\in G}\ s_g g = \Sigma_{g\in G} s_g(g'g)$, $s_g \in S$, $g' \in G$; and extending S-linearly. SG is called the <u>group algebra of G over S</u>.

4.8 <u>Theorem</u> (Higman [1]): If S is a commutative noetherian ring and G a finite group such that $|G|\cdot 1$ is a unit in S, then SG is a separable S-algebra.

<u>Proof</u>: To show that the sequence E in (4.7) splits, we define $\rho : SG \longrightarrow SG^e$ by $1 \longrightarrow (|G|\cdot 1)^{-1}(\Sigma_{g\in G}\ g^{-1} \otimes g^{op})$, and extend SG-linearly. It is now easily verified that ρ is an SG^e-map and that $\rho\varepsilon = 1_{SG}$. #

4.9 <u>Remark</u>: The exact sequence E of (4.7,i) yields, for each $M \in {}_{B^e}\underline{M}$, an exact sequence

$$\cdots \longrightarrow \mathrm{Hom}_{B^e}(B^e,M) \xrightarrow{\iota^*} \mathrm{Hom}_{B^e}(\mathrm{Ker}\ \varepsilon,M) \longrightarrow \mathrm{Ext}^1_{B^e}(B,M) \longrightarrow 0$$

(cf. II, (3.10) and (4.2)). Thus

$$\mathrm{Ext}^1_{B^e}(B,M) \overset{\text{nat.}}{\cong} \mathrm{Hom}_{B^e}(\mathrm{Ker}\ \varepsilon,M)/\mathrm{Im}\ \iota^* \overset{\text{nat}}{\cong} \mathrm{Der}(B,M)/\mathrm{InDer}(B,M)$$

(cf. II,(3.12)) and B is separable if and only if every derivation is inner (cf. (4.6)), if and only if $H^n(B,M) = 0$, $n = 1,2,\ldots$, $\forall\ M \in {}_{B^e}\underline{M}^f$ (cf. II,(4.2) and (4.3)).

4.10 Lemma: B is separable if and only if φ_o of (4.4) is an inner derivation.

Proof: Because of (4.9) it suffices to show that every derivation is inner if φ_o is inner. But if φ_o is inner then so is $\varphi_o\alpha$ for every $\alpha \in \mathrm{Hom}_{B^e}(\mathrm{Ker}\ \varepsilon,M)$, and hence, by (4.6) ι^* is epic and $\mathrm{InDer}(B,M) = \mathrm{Der}(B,M)$. #

Next we shall show that the cohomology groups $H^n(B,-)$ are closely related to $\mathrm{Ext}^n_B(-,-)$.

4.11 Theorem: Let $M,N \in {}_{B}\underline{M}^f$, such that $\mathrm{Ext}^n_S(N,M) = 0$. Make $\mathrm{Hom}_S(N,M)$ - with the morphisms written on the left - into a B^e-module by defining $(x \otimes y^{op})\rho(n) = x\rho(yn)$, for $x,y \in B$, $\rho \in \mathrm{Hom}_S(N,M)$. Then $H^n(B,\mathrm{Hom}_S(N,M)) \overset{\text{nat}}{\cong} \mathrm{Ext}^n_B(N,M)$, $n = 1,2,\ldots$.

Proof: The proof for arbitrary n may be found in Cartan-Eilenberg [1], Ch. IX, (4.4). We shall give a proof for $n = 1$, which is the most interesting case for our purpose. $\mathrm{Ext}^1_B(N,M)$ consists of congruence classes of short exact B-sequences
$E : 0 \longrightarrow M \longrightarrow M \longrightarrow X \longrightarrow N \longrightarrow 0$, which split over S, since $\mathrm{Ext}^1_S(N,M) = 0$ (cf. II(5.9)). We define a map
$\gamma : \mathrm{Der}(B,\mathrm{Hom}_S(N,M)) \longrightarrow \tilde{E}_B(N,M)$, $\varphi \longmapsto E_\varphi$, (cf. II,(5.1),

III, (4.3)), where $E_\varphi \in {}_{\underline{B}}\underline{M}$ is defined as follows: we put $X_\varphi = M \oplus N$ as S-module, and define the action of $b \in B$ by $b(m,n) = (bm + (b^\varphi)(n), bn)$. Since $\varphi \in Der(B, Hom_S(N,M))$, $X_\varphi \in {}_{\underline{B}}\underline{M}$, (cf. Ex. 4,1), and the canonical S-maps $M \xrightarrow{\iota} M \oplus N$ and $M \oplus N \xrightarrow{\pi} N$ are in fact B-homomorphisms. Thus $E_\varphi : 0 \longrightarrow M \longrightarrow X_\varphi \longrightarrow N \longrightarrow 0 \in \tilde{E}_B(N,M)$ (cf. Ex. 4,2). If $\varphi \in InDer(B, Hom_S(N,M))$, then $b^\varphi = -b\rho + \rho b$, for some $\rho \in Hom_S(N,M)$ (cf. (4.3)), and we define $\psi : N \longrightarrow X_\varphi$, $n \longmapsto (\rho(n), n)$. Then $b(n\psi) = b(\rho(n),n) = (b^\varphi(n) + b\rho(n), bn) = (\rho(bn), bn)$ and $(bn)\psi = (\rho(bn), bn)$; i.e., $\psi \in Hom_B(N, X_\varphi)$ such that $\psi\pi = 1_N$. Thus E_φ is a split exact sequence.

If, conversely, E_φ is a split exact sequence, then there exists $\chi \in Hom_B(N, X_\varphi)$ such that $\chi\pi = 1_N$. Hence, for every $n \in N$, $\chi : n \longmapsto (\rho(n), n)$, where $\rho(n)$ is easily seen to be an S-homomorphism from N to M. But χ is B-linear; i.e., $(\rho(bn), bn) = (b\rho(n) + b^\varphi(n), bn)$. Hence $b^\varphi(n) = \rho(bn) - b\rho(n)$, $n \in N$; i.e., φ is an inner derivation.

We leave it as an exercise to show that $\tilde{\Phi}$ is a $\underline{Z}$-homomorphism. From the above proved properties it follows that $\tilde{\Phi}$ induces a $\underline{Z}$-monomorphism

$$\Phi : Der(B, Hom_S(N,M))/InDer(B, Hom_S(N,M)) \longrightarrow E_B(N,M)$$

(cf. II, (5.1) and proof of II, (5.7)). To show that Φ is an epimorphism, let $E : 0 \longrightarrow M \xrightarrow{\sigma} X \xrightarrow{\tau} N \longrightarrow 0 \in \tilde{E}_B(N,M)$. There is an S-module map $\psi : X \xrightarrow{\sim} M \oplus N$. We use this to make $M \oplus N$ into a left B-module by defining $b(m,n) = \psi(b(\psi^{-1}(m,n)))$. Then the map $\psi : X \longrightarrow M \oplus N$ becomes a B-isomorphism. Since E is congruent to the split sequence in $E_S(N,M)$ (cf. II, (5.1)), the following diagram is commutative:

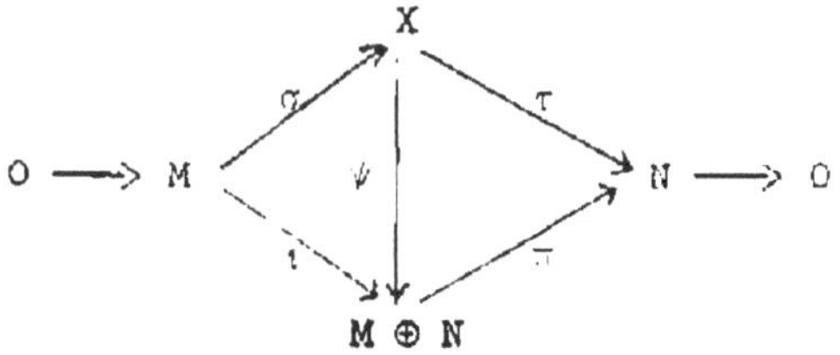

,

where ι and π are the S-injection and S-projection respectively. Under our definition of $M \oplus N \in {}_{B}\underline{\underline{M}}$ it is easily checked that ι and π are also B-homomorphisms. Hence the sequence E is equivalent to the exact sequence

$$E' : 0 \longrightarrow M \xrightarrow{\iota} M \oplus_S N \xrightarrow{\pi} N \longrightarrow 0 \in \mathscr{E}_B(N,M).$$

And, to finish the proof, it suffices to show that $E' = \Phi(\varphi)$ for some $\varphi \in \mathrm{Der}(B,\mathrm{Hom}_S(N,M))$. Since $M\iota$ is a B-submodule of $M \oplus_S N$, we must have $b(m,0) = (bm,0)$ in $M \oplus_S N$, hence $b(m,n) = (bm + \varphi(b,n), bn)$. From the B-module properties of $M \oplus_S N$, it follows now easily that for every $b \in B$, $\varphi(b,-) : N \longrightarrow M$ is an S-homomorphism, and that $b \longmapsto \varphi(b,-)$ is a derivation. This proves that Φ is a $\underline{Z}$-isomorphism. We leave the verification of the naturality of Φ as an exercise.

Thus by (II,(5.9)),

$$\mathrm{Ext}^1_B(N,M) \cong \mathrm{Der}(B,\mathrm{Hom}_S(N,M))/\mathrm{InDer}(B,\mathrm{Hom}_S(N,M)),$$

and by (4.9) $\mathrm{Ext}^1_B(N,M) \cong H^1(B,\mathrm{Hom}_S(N,M))$. #

4.12 Corollary: If B is separable, then $\mathrm{Ext}^1_B(N,M) = 0$ $M,N \in {}_{B}\underline{\underline{M}}^f$, when N is S-projective.

Exercises §4:

In the proof of (4.9) show:

1.) $X_\varphi \in {}_{B}\underline{\underline{M}}$.

2.) Show that the sequence $0 \longrightarrow M \xrightarrow{\iota} X \xrightarrow{\pi} N \longrightarrow 0 \in \tilde{E}_B(N,M)$.

3.) Show that Φ is a $\underline{\underline{Z}}$-homomorphism.

4.) Show that Φ is natural.

5.) Let G be a finite group and S a ring such that $|G| \cdot 1$ is invertible in SG. Show that $\{g\}_{g \in G}$ and $\{\frac{1}{|G|} g^{-1}\}_{g \in G}$ are dual bases with respect to the trace function, and that $\rho : SG \longrightarrow SG^e$; $x \longmapsto \frac{1}{|G|} \Sigma_{g \in G}\, xg^{-1} \otimes g^{op}$ is an SG^e-homomorphism.

§5. Separable algebras

Wedderburn's theorem is stated, and it is shown, that separable algebras are semi-simple and remain separable under extensions of the ground field.

In this section, K is a field and A a finite dimensional K-algebra.

5.1 Definitions: (i) A is said to be a semi-simple K-algebra, if $\text{rad}\, A = 0$.

(ii) A is said to be a separable K-algebra, if the sequence $E : 0 \longrightarrow \text{Ker}\,\varepsilon \longrightarrow A^e \xrightarrow{\varepsilon} A \longrightarrow 0$ is a split exact A^e-sequence.

5.2 Theorem: If A is semi-simple, then every $M \in {}_A\underline{\underline{M}}^f$ can be expressed uniquely up to isomorphism as a direct sum of simple left A-modules (cf. I, (4.5)). We shall show even more than that:

5.3 Theorem: Let S be a ring which is left artinian and left noetherian and such that $\text{rad}\, S = 0$. Then there exists only a finite number of non-isomorphic simple left S-modules; a complete set of them is given by the non-isomorphic minimal left ideals of S. Moreover, every $M \in {}_S\underline{\underline{M}}^f$ is projective and can be expressed uniquely up to isomorphism as a direct sum of simple left S-modules.

For the proof we shall show first

5.4 Lemma: If $N \in {}_S\underline{\underline{M}}^f$ is a direct sum of simple left S-modules, and if $X \subset N$, then X is a direct summand of N.

Proof: Let $X \subset N$, where $N = \oplus_{i=1}^n M_i$ is a direct sum of the simple left S-modules $\{M_i\}_{1 \leq i \leq n}$, and let $i_1, \ldots, i_\tau$ be a maximal subset of $1, \ldots, n$ such that the sum $X + \oplus_{j=1}^{\tau} M_{i_j}$ is direct. If for some $k \neq i_1, \ldots, i_\tau$, $(X \oplus (\oplus_{j=1}^{\tau} M_{i_j})) \cap M_k = 0$, then the set $i_1, \ldots, i_\tau$, would not be maximal (cf. I, (1.9)). Thus, for every

$k \neq i_1,\dots,i_\tau$, $(X \oplus (\oplus_{j=1}^{\tau} M_{i_j})) \cap M_k = M_k$. M_k being simple, hence $X \oplus \oplus_{j=1}^{\tau} M_{i_j} = N$. #

We now turn to the proof of (5.3): We shall show at first, that ${}_SS$ is a direct sum of a finite number of simple left S-modules; i.e., minimal left ideals. Since ${}_SS$ is artinian there exists a finite family $\{I_i\}_{1 \leq i \leq n}$ of maximal left ideals of S such that

$$0 = \text{rad } S = \cap \{I \subset {}_SS : I \text{ maximal}\} = \cap_{i=1}^{n} I_i \quad (\text{cf. Ex.5,12}).$$

Let $\varphi_i : {}_SS \longrightarrow {}_SS/I_i$, $1 \leq i \leq n$, be the canonical homomorphism. Then the map

$$\varphi : {}_SS \longrightarrow \oplus_{i=1}^{n} S/I_i, \quad s \longmapsto (s\varphi_i)_{1 \leq i \leq n}$$

is a monomorphism (cf. I, (2.4)). From (5.4) and the validity of the Krull-Schmidt-theorem (cf. I, (4.10)), we conclude, that ${}_SS$ is isomorphic to a direct sum of a finite number of simple left S-modules of the form S/I, where I is a maximal left ideal. Let ${}_SS = \oplus_{i=1}^{r} J_i$ be the decomposition of ${}_SS$ into minimal left ideals $\{J_i\}_{1 \leq i \leq r}$, and let $M \in {}_S\underline{\underline{M}}^f$ be simple. Then, for $0 \neq m \in M$, we have $M = Sm = \oplus_{i=1}^{r} J_i m$, and hence there exists exactly one i, $1 \leq i \leq r$, such that $M = J_i m$. Since M and J_i are simple left S-modules, $M \cong J_i$; i.e., every simple left S-module is isomorphic to a minimal left ideal of S, which is a direct summand of ${}_SS$, whence every simple left S-module is projective. Moreover, since the Jordan-Hölder theorem (I, (4.6), I, (4.7)) is valid for ${}_S\underline{\underline{M}}^f$, every $M \in {}_S\underline{\underline{M}}^f$ is projective, and it has a unique expression as a direct sum of simple left S-modules, as follows from the Krull-Schmidt theorem (I, (4.10)).

5.5 Corollary (Wedderburn's structure theorem): Let A be a semi-simple K-algebra. Then

(i) There is only a finite number of non-isomorphic simple left A-modules, $M_1,\dots,M_n$.

(ii) $A \underset{\text{ring}}{\cong} \oplus_{i=1}^n (D_i)_{n_i}$, where $D_i = \text{End}_A(M_i)$ is a skewfield, $1 \le i \le n$, and the $\{(D_i)_{n_i}\}_{1\le i\le n}$ are the only minimal two-sided ideals in A. Moreover, the D_i are uniquely determined by A, up to ring isomorphism, and ${}_A A \cong \oplus_{i=1}^n M_i^{(n_i)}$.

(iii) $(D_i)_{n_i}$ is a <u>simple K-algebra</u>, $1 \le i \le n$; i.e., a semi-simple K-algebra with only one simple left module. Conversely, every simple K-algebra is isomorphic to a full matrix ring over a finite dimensional skewfield over K.

<u>Proof:</u> (i) follows readily from (5.3).

(ii) We write
$$ {}_A A = \oplus_{i=1}^n M_i^{(n_i)} $$
where $\{M_i\}_{1\le i\le n}$ are the non-isomorphic minimal left ideals in A (cf. (5.3)). Then $M = \oplus_{i=1}^n M_i$ is a pro generator for ${}_{A}\underline{\underline{M}}^f$ (cf. (1.10)). Moreover, since $\text{Hom}_A(M_i,M_j) = 0$, if $i \ne j$, M_i and M_j being simple, $\text{End}_A(M) \overset{\text{nat}}{\underset{\text{ring}}{\cong}} \oplus_{i=1}^n \text{End}_A(M_i) = \oplus_{i=1}^n D_i$, where $D_i = \text{End}_A(M_i)$ is a skewfield of finite dimension over K (cf. I, Ex. 4,3), since A is a finite dimensional K-algebra. By (2.1), we have a Morita equivalence between ${}_{A}\underline{\underline{M}}^f$ and ${}_{D}\underline{\underline{M}}^f$, where $D = \oplus_{i=1}^n D_i$ as ring (cf. I, Ex. 4,7); in particular,
$$ A \underset{\text{ring}}{\cong} \text{End}_D(M) \qquad \text{(cf. (2.3))}. $$
But each M_i, $1 \le i \le n$, is in a natural way a right D-module: $md_i = m(d_1,\dots,d_n)$ $m \in M_i$. Hence $\text{End}_D(M) \overset{\text{nat}}{\cong} \oplus_{i,j=1}^n \text{Hom}_D(M_i,M_j) = \oplus_{i=1}^n \text{End}_{D_i}(M_i)$ (cf. Ex. 5,1). Since D_i is a skewfield,

$M_i \underset{D_i}{\cong} {}_{D_i}D_i^{(m_i)}$ (cf. Ex. 5,2) and $\text{End}_{D_i}(M_i) \underset{\text{ring}}{\cong} (D_i)_{m_i}$ (cf. Ex. 5,3).

Thus $A \underset{\text{ring}}{\overset{\text{nat.}}{\cong}} \oplus_{i=1}^{n} (D_i)_{m_i}$. Counting the dimensions over K, we find, if $s_i = (D_i : K)$, then $(M_i : K) = s_i \cdot m_i$. On the other hand, if $A = \oplus_{i=1}^{n} A_i$, where $A_i \cong (D_i)_{m_i}$, $1 \le i \le n$, then

$$A_i M_j = \begin{cases} 0 & \text{if } i \ne j \\ M_i & \text{if } i = j \end{cases}. \quad \text{Hence } {}_A A = \oplus_{i=1}^{n} M_i^{(n_i)}, \text{ implies}$$

${}_{A_i} A_i = M_i^{(n_i)}$ as left modules; i.e., $s_i m_i^2 = s_i m_i n_i$; hence $m_i = n_i$. Obviously, the $\{A_i\}_{1 \le i \le n}$ are two-sided ideals in A; and if I is another two-sided ideal in A, then I is isomorphic to a direct sum of some of the $\{A_i\}_{1 \le i \le n}$, since the only two-sided ideals in $\oplus_{i=1}^{n} (D_i)_{n_i}$ are direct sums of some of the $\{(D_i)_{n_i}\}_{1 \le i \le n}$ (cf. Ex. 5.4). If $A \underset{\text{ring}}{\cong} \oplus_{i=1}^{t} (D_i')_{k_i}$ where $\{D_i'\}_{1 \le i \le t}$ are finite dimensional skewfields, then ${}_A A = \oplus_{i=1}^{t} N_i^{(k_i)}$ (cf. Ex. 5.4), where the N_i are simple left A-modules. Thus, $t = n$, $k_i = n_i$ (if necessary after renumbering the $\{N_i\}_{1 \le i \le n}$) and $N_i \cong M_i$. Since $\mathrm{End}_A(N_i) \cong D_i'$ (cf. Ex. 5.4), the uniqueness of the decomposition is established. (cf. I,(4.10), the Krull-Schmidt theorem).

(iii) is now an immediate consequence of the proof of (ii). #

5.6 <u>Definitions:</u> Let S be a ring. An element $0 \ne e \in S$ is called an <u>idempotent</u>, if $e^2 = e$; e is called a <u>central idempotent</u>, if e lies in the center of S, e is said to be <u>primitive</u>, if $e = e_1 + e_2$, $e_1 e_2 = e_2 e_1 = 0$, where $e_i^2 = e_i$, $i = 1,2$, implies $e_1 = 0$ or $e_2 = 0$. Two idempotents e_1 and e_2 are called <u>equivalent</u> if $Se_1 \cong Se_2$ as left S-modules. A set $\{e_i\}_{1 \le i \le n}$ of idempotents is called a <u>complete set of primitive orthogonal idempotents</u>, if each e_i is primitive, $1 \le i \le n$, $1 = \sum_{i=1}^{n} e_i$ and $e_i e_j = 0$ if $i \ne j$. Similar definitions hold in a selfexplanatory way for central idempotents.

5.7 <u>Lemma:</u> Let A be a semi-simple K-algebra. An idempotent e in A is primitive if and only if Ae is a simple left A-module; conversely, every simple left A-module is isomorphic to Ae for some primitive idempotent e. There exists a unique decomposition $1 = \sum_{i=1}^{n} e_i$ into orthogonal primitive central idempotents in A. Moreover, $Ae_i \cong (D_i)_{n_i}$ is a simple K-algebra, $1 \leq i \leq n$, and if $e_i = \sum_{j=1}^{n_i} e_{ij}$, $1 \leq i \leq n$, is a decomposition of e_i into orthogonal primitive idempotents, then e_{ij} is equivalent to e_{kh} if and only if $i = k$.

<u>Proof:</u> From (5.5, ii)it follows, that $A = \oplus_{i=1}^{n} A_i$, where $A_i \cong (D_i)_{n_i}$ is a decomposition of A into minimal two-sided ideals in A. Let $\pi_i : A \longrightarrow A_i$, $1 \leq i \leq n$, be the corresponding ring projections (cf. I, Ex. 4,7) and put $e_i = \pi_i(1)$, $1 \leq i \leq n$. Then the $\{e_i\}_{1\leq i\leq n}$ form a complete set of orthogonal primitive central idempotents (cf. I, (1.10)). If also $\{e_i'\}_{1\leq i\leq m}$ were a complete set of orthogonal primitive central idempotents, then

$$1 = \sum_{i=1}^{n} e_i = \sum_{j=1}^{m} e_j', \quad \text{and} \quad e_1 = \sum_{i=1}^{n} e_1 e_i = \sum_{j=1}^{m} e_1 e_j'.$$

Since $e_1 e_j'$ lies in the center of A it is a central idempotent; and since e_1 is a primitive central idempotent, there exists exactly one j, say $j = 1$, such that $e_1 e_1' = e_1$ and $e_1 e_j' = 0$, $2 \leq j \leq m$. A similar argument shows that $e_1' = e_1 e_1'$; i.e., $e_1 = e_1'$. Continuing in this way, we conclude $n = m$ and $\{e_i\}_{1\leq i\leq n} = \{e_j'\}_{1\leq j\leq m}$.

It follows from (5.3), that every simple left A-module M is isomorphic to a direct summand M' of ${}_A A$. If $\pi : {}_A A \longrightarrow M'$ is the projection, then $(1)\pi = e$ is an idempotent, since $((1)\pi)\pi = e$ and $((1)\pi)\pi = (e)\pi = e(1)\pi = e^2$. Moreover, $M' \cong {}_A Ae$ and e is primitive since M' is simple. On the other hand, if e is a primitive idempotent and ${}_A Ae$ were not simple, then ${}_A Ae = M_1 \oplus M_2$, $M_i \neq 0$, $i = 1,2$, and thus, e could not be primitive. The rest of the statements follow now easily from the proof of (5.5), and the details are left as an exercise (cf. Ex. 5,4). #

5.8 Remark: Whereas the decomposition into orthogonal primitive central idempotents is unique, this is not necessarily true for the decomposition into orthogonal primitive idempotents.

5.9 Theorem: Let A be a separable K-algebra. Then A is semi-simple, and for every extension field L of K, $A^L = L \otimes_K A$ is a separable L-algebra.

Proof: From the definition (4.7) we conclude that $\mathrm{Ext}^1_{A^e}(A,X) = 0$, $\forall X \in {}_{A^e}\underline{\underline{M}}$ (cf. II, (4.2)). Hence, by (4.9)

$\mathrm{Ext}^1_A(M,N) = 0$, $\forall M,N \in {}_A\underline{\underline{M}}^f$, and thus $M \in {}_A\underline{\underline{P}}^f$ (cf. II, Ex. 4,2). This holds in particular for $A/\mathrm{rad}\, A$, $0 \to \mathrm{rad}\, A \to A \to A/\mathrm{rad}\, A \to 0$ then implies that ${}_A A = \mathrm{rad}\, A \oplus X$. From Nakayama's lemma (I,(4.18)) it follows that $X = A$; i.e., $\mathrm{rad}\, A = 0$ and A is semi-simple. If now L is an extension field of K, we obtain the split exact sequence (cf. II, (1.7)) $E^L : 0 \to L \otimes_K \mathrm{Ker}\, \epsilon \to L \otimes_K A^e \to A^L \to 0$ induced from the sequence E in (5.1). But $L \otimes_K A^e \cong A^L \otimes_L (A^L)^{op} = (A^L)^e$; i.e., $A^L \in {}_{(A^L)^e}\underline{\underline{P}}^f$, and thus A^L is separable. #

Exercises §5:

1.) Let A be semi-simple, $\{M_i\}_{1 \le i \le n}$ a complete set of non-isomorphic simple left A-modules. Let $D_i = \mathrm{End}_A(M_i)$ and $D = \bigoplus_{i=1}^{n} D_i$, as rings. Show, $\mathrm{Hom}_D(M_i,M_j) = 0$, if $i \neq j$.

2.) Let D be a skewfield, $M \in {}_D\underline{\underline{M}}^f$. Show, $M \cong {}_D D^{(n)}$ for some

$n \in \underline{\underline{N}}$.

3.) Let S be a ring, $M \in {}_S\underline{\underline{M}}$. If $\Omega = \mathrm{End}_S(M)$, show, $\mathrm{End}_S(M^{(n)}) \cong (\Omega)_n$ as a ring.

4.) Let D be a skewfield. Show,

(i) $(D)_n$ has no two-sided ideals

(ii) $(D)_n$ is simple. Describe the simple $(D)_n$-modules and find a complete set of orthogonal primitive idempotents in $(D)_n$.

5.) Show, $(K)_n \otimes_K (K)_m \overset{\text{ring}}{\cong} (K)_{nm}$.

6.) Let S be a noetherian and artinian ring. Show, $(\mathrm{rad}\ S)_n = \mathrm{rad}(S)_n$. (Hint: Show, that $(\mathrm{rad}\ S)_n \subset \mathrm{rad}(S)_n$. Then $(S)_n/(\mathrm{rad}\ S)_n \cong (S/\mathrm{rad}\ S)_n$. Hence it suffices to show: if S is semi-simple, so is $(S)_n$. But we have a Morita equivalence between S and $(S)_n$.)

7.) Let K_1 be a finite extension of K. Assume, that for every extension field of L of K, $L \otimes_K K_1$ is semi-simple. Show, that $\mathrm{min}_{K_1/K}(\alpha, X)$, where $\alpha \in K_1$, has no repeated linear factor over an algebraically closed field containing K_1. (Hint: $K_1 \otimes_K L$ is commutative and semi-simple. Show, that $K_1 \otimes_K L$ does not contain nilpotent elements. Now, for $\alpha \in K_1$, $K(\alpha)\otimes_K L$ is semi-simple implies that $K(\alpha)\otimes_K L \cong L(\alpha) = L[X]/(\mathrm{min}_{K_1/K}(\alpha, X))$: now, prove the statement by choosing L large enough.) This shows, that K_1 has to be a separable field extension of K.

8.) Let K be a field of characteristic zero. Show, that any extension field L of K is a separable extension field of K.

9.) Let $\underline{\underline{H}}$ be the quaternion algebra over the field of rational numbers: $\underline{\underline{H}} = \underline{\underline{Q}}(i, j, k)$, where $\underline{\underline{Q}}$ is the field of rational numbers and i,j,k satisfy the relations $i^2 = k^2 = j^2 = -1$, $ij = -ji = k$, $jk = -kj = i$, $ki = -ik = j$. Show that $\underline{\underline{H}}$ is a central division algebra.

10.) Let $A = (\underline{\underline{H}})_2$, where $\underline{\underline{H}}$ is the rational quaternion algebra (cf. Ex. 5,9). Show that there are different embeddings of $\underline{\underline{H}}$ into A.

11.) Give an example of a central simple K-algebra A, for which $1 \in A$ has two different decompositions into primitive orthogonal idempotens.

12.) Let S be an artinian ring and put $X = \bigcap_{i \in I} M_i$, where I is an index set and $M_i \subset M \in {}_S\underline{\underline{M}}^f$, $i \in I$. Show that X is the intersection of a finite subset of $\{M_i\}_{i \in I}$.

§6. Splitting Fields

Separable algebras over a field have finite dimensional splitting fields. Central simple algebras are separable. The center of a separable algebra is the image of the Gaschütz-Casimir operator.

In this section, K is a field and A is a finite dimensional K-algebra.

6.1 Definition: If A is semi-simple, an extension field L of K is called a splitting field for A, if $A^L \cong \oplus_{i=1}^k (L)_{n_i}$ (cf. (5.5)).

6.2 Lemma: There are no finite dimensional skewfields over an algebraically closed field.

Proof: We recall, that an algebraically closed field is one, over which every polynomial decomposes completely into linear factors (cf. Ex. 6,1). Now, let L be an algebraically closed field and D a finite dimensional skewfield over L. Then $d \in D$ is the root of a monic polynomial $f(X) \in L[X]$ (cf. I, (5.4)). Since L is algebraically closed, $f(X) = \prod_{i=1}^n (X - \alpha_i)$, $\alpha_i \in L$. But $0 = f(d) = \prod_{i=1}^n (d - \alpha_i)$ implies $d = \alpha_i$ for some $1 \leq i \leq n$, since D is skewfield; i.e., $D \subset L$. #

6.3 Theorem: Let A be a separable K-algebra. Then there exist splitting fields for A.

The proof follows from (5.5) and (6.2). In fact, every algebraically closed extension of K is a splitting field for A. #

6.4 Lemma: Let D be a finite dimensional skewfield over K. Then there is a natural isomorphism of rings $(D)_n \cong (K)_n \otimes_K D$.

<u>Proof:</u> Since D is K flat (cf. I, (3.16)), we obtain from (1.2), (Ex. 5,3) $(D)_n \underset{\text{ring}}{\overset{\text{nat}}{\cong}} \mathrm{End}_D(D^{(n)}) \underset{\text{ring}}{\overset{\text{nat}}{\cong}} \mathrm{End}_{K \otimes_K D}(K \otimes_K D^{(n)})$
$\underset{\text{ring}}{\overset{\text{nat}}{\cong}} \mathrm{End}_{K \otimes_K D}(D^{(n)} \otimes_K D) \underset{\text{ring}}{\overset{\text{nat}}{\cong}} \mathrm{End}_K(K^{(n)}) \otimes_K D \underset{\text{ring}}{\overset{\text{nat}}{\cong}} (K)_n \otimes_K D.$ ∎

6.5 <u>Theorem:</u> Let D be a finite dimensional central skewfield over K. Let L be a (commutative) subfield of D, with $(L:K) = s_1$. Then $L \otimes_K D \underset{\text{rings}}{\cong} (D_1)_{s_1}$, where D_1 is a central skewfield over L. Moreover, if L is a maximal subfield of D, then L is a splitting field for D and $(D : K) = (L : K)^2$.

<u>Proof:</u> We view $D \in {}_L\underline{\underline{M}}^f$, where the action of L on D is left multiplication; we write ${}_LD$ for this L-module, and define a map $\varphi : L \otimes_K D \longrightarrow \mathrm{End}_L(D)$, $\ell \otimes d \longmapsto (\ell \otimes d)^\varphi$, where $d'(\ell \otimes d)^\varphi = \ell d' d$ for $d' \in {}_LD$. It is easily checked that φ is a ring homomorphism, and $1 \in \mathrm{Im}\varphi$. Clearly, $\mathrm{Ker}\,\varphi = 0$; i.e., $\varphi : L \otimes_K D \xrightarrow{\sim} \mathrm{Im}\varphi$. We denote $\mathrm{Im}\varphi$ by S. Then $D \in \underline{\underline{M}}^f_S$, and we write D_S for this module. Since ${}_LD \in {}_L\underline{\underline{M}}^f$ is a progenerator for ${}_L\underline{\underline{M}}^f$ (cf. (1.10)), and since $S \subset \mathrm{End}_L({}_LD)$, $\mathrm{End}_S(D_S) \supset \mathrm{End}_{\mathrm{End}_L({}_LD)}({}_LD)$

$= L$ (cf. (2.4)). Moreover, $1 \in L$ implies $S \supset \mathrm{End}_D({}_DD) = D_r$ where D_r denotes right multiplication by the elements in D. Now, ${}_DD$ is a progenerator in ${}_D\underline{\underline{M}}$ and thus, $\mathrm{End}_S(D_S) \subset \mathrm{End}_D({}_DD) = D$. Altogether, we have the following chain of inclusions: $L \subset \mathrm{End}_S(D_S) \subset D$. Moreover, since $L \subset S$, we have $\alpha\ell = \ell\alpha$ for $\alpha \in \mathrm{End}_S(D_S)$, $\ell \in L$. Hence, for every $\alpha \in \mathrm{End}_S(D_S)$, $L(\alpha) \subset D$ is a commutative subfield of D; and since $\mathrm{End}_S(D_S) \in {}_L\underline{\underline{M}}^f$, $\mathrm{End}_S(D_S) = \sum_{\text{finite}} L(\alpha)$. Since $\mathrm{End}_S(D_S)$ is a finite dimensional L-algebra, it is a skewfield D_1 of finite dimension over L, for

D_S is a simple S-module, $S \supset D$.

If L is a maximal subfield of D, then $L(\alpha) \subset L$ for every $\alpha \in End_S(D_S)$. Thus $End_S(D_S) = L$ in this case.

Now, back to the general case: The map $\mu_D : D_S \otimes_S Hom_S(D_S,S) \longrightarrow End_S(D_S) = D_1$ (cf. (1.4)) is not zero, since $D \subset S$ and, since D_S is a simple S-module, it is an isomorphism and thus $D_S \in \underline{\underline{P}}_S^f$ (cf. (1.5)). Moreover, $S \subset End_L(D)$, a simple L-algebra and hence D_S is a faithful S-module. Since the Krull-Schmidt theorem is valid in $\underline{\underline{M}}_S^f$, D_S is a progenerator in $\underline{\underline{M}}_S^f$, and thus we have (cf. Ex. 6,8) $End_{D_1}(D) = S$, $End_S(D) = D_1$. Since D is a free left D_1-module, $S = End_{D_1}(D) = (D_1)_{n_1}$; i.e., $L \otimes_K D \cong (D_1)_{n_1}$. In case L is a maximal subfield of D, we obtain $L \otimes_K D = (L)_{n_1}$; i.e., L is a splitting field for D.

Now we count the dimensions: $(D:K) = (D:D_1)(D_1:L)(L:K) = (D^L:L) = n_1^2 \cdot (D_1:L)$. Since $n_1 = (D:D_1)$ we have $(L:K) = n_1$. In case L is a maximal subfield, this shows that $(D:K) = m^2 = (L:K)^2$. In this proof we have frequently identified structures which are naturally isomorphic, so as to avoid unnecessary complication. #

6.6 <u>Lemma:</u> Let D be a central skewfield over K and S_1, S_2 two maximal subfields of D, which are isomorphic as K-algebras, $\varphi : S_1 \xrightarrow{\sim} S_2$. Then there exists $d \in D$ such that $\varphi(s_1) = ds_1d^{-1}$, $s_1 \in S_1$.

<u>Proof:</u> Let S be a subfield of D which is isomorphic to S_1 and S_2 as K-algebras. Since S is also a maximal subfield of D $S \otimes_K D \underset{ring}{\cong} (S)_n$, where $n = (S:K)$ (cf. (6.7)). Now,

$S_1, S_2 \subset S \otimes_K D$ and hence we obtain two monomorphisms as K-algebras $\varphi_1 : S_1 \longrightarrow (S)_n$, $\varphi_2 : S_2 \longrightarrow (S)_n$. Since S is isomorphic to S_1 and S_2 as K-algebras, we obtain two monomorphisms, of K-algebras

$\psi_1 : S \longrightarrow S_1 \xrightarrow{\varphi_1} (S)_n$, $\psi_2 : S \longrightarrow S_2 \xrightarrow{\varphi_2} (S)_n$. Let L be the n-dimensional S-vectorspace, with a fixed basis $\{w_i\}_{1\leq i\leq n}$. We consider the K-submodules of L, M_1 and M_2, generated over K by the action of $\psi_1(s)$, and $\psi_2(s)$ resp., $s \in S$, on $\{w_i\}_{1\leq i\leq n}$. Then M_1 and M_2 are left S-modules of dimension n over K. Thus, they are isomorphic as S-modules (cf. (5.5)). Hence, this isomorphism can be extended to an automorphism of L, and thus, there exists a matrix $\underline{B} \in (S)_n$ such that $\psi_1(s)\underline{B} = \underline{B}\psi_2(s)$, $\forall s \in S$. Since M_1 and M_2 are isomorphic, $\underline{B}$ is invertible; i.e., $\underline{B}^{-1}\psi_1(s)\underline{B} = \psi_2(s)$. Consequently,

6.6' $$\underline{B}^{-1}\varphi_1(t\ )\underline{B} = \varphi_2(\varphi(t\ ))\ , \quad t \in S_1.$$

Under the isomorphism $S \otimes_K D \xrightarrow{\sim} (S)_n$, $\underline{B}$ corresponds to an element $\sum_{i=1}^{n'} s_i \otimes d_i \in S \otimes_K D$.

Here we can assume, that $\{s_i\}_{1\leq i\leq n'}$ are linearly independent over K since S is a free K-module. Now, the equation (6.6') reads $(\sum_{i=1}^{n'} s_i \otimes d_i)(1 \otimes \varphi(t)) = (1 \otimes t)(\sum_{i=1}^{n'} s_i \otimes d_i)$,

thus, $\sum_{i=1}^{n'} s_i \otimes d_i\ \varphi(t) = \sum_{i=1}^{n'} s_i \otimes s_1\ d_i\ ,\ \forall t \in S_1.$

Hence $d_i\ \varphi(t) = s_1\ d_i\ ,\ \forall t \in S_1\ ,\ 1 \leq i \leq n'$. If $d_1 \neq 0$ - this is the case for at least one d_i - then $\varphi(t) = d_1^{-1}\ t\ d_1, \forall t \in S_1$. #

6.7 <u>Theorem (Wedderburn):</u> Every finite skewfield is a field.

<u>Proof:</u> Let D be a skewfield with a finite number of elements. Then D is a central simple algebra over a Galois-field K. And all maximal subfields of D are isomorphic, since over a finite field there exists only one extension of a fixed degree. Since any element of D is contained in some maximal subfield of D, D is a

union of maximal subfields. On the other hand, if E_i and E_j are maximal subfields then, for some $d_i \in D$, $E_i = d_i^{-1} E_j d_i$ (cf.(6.9)). Thus the multiplicative group D^* of $D \setminus \{0\}$ is the union of conjugates of a proper subgroup E_o^*. However, this is impossible for a finite group.

6.8 Definition: A finite dimensional K-algebra is called a central simple K-algebra, if A is simple and center (A) = K.

6.9 Lemma: Let A be a central simple K-algebra. Then there exists a splitting field L for A with $(L : K) < \infty$.

Proof: Let $A = (D)_n$ (cf. (5.5)), where D is a central skewfield over K, and let L be a maximal subfield of D. Then, by (6.4) and (6.5), $L \otimes_K (D)_n \simeq L \otimes_K (K)_n \otimes_K D \simeq (K)_n \otimes_K L \otimes_K D$ $\simeq (K)_n \otimes_K (L)_m \simeq (K)_{nm} \otimes_K L \simeq (L)_{nm}$. #

6.10 Theorem: Let A be a separable K-algebra. Then there exists a splitting field L for A such that $(L : K) < \infty$.

Proof: By (5.5), $A = \oplus_{i=1}^n (D_i)_{n_i}$, and since, by (6.4) every finite extension field of a splitting field for an algebra is also a splitting field for this algebra, it obviously suffices to assume A to be simple. Thus, let $A = (D)_n$, K_1 = center (D) and let L be a maximal subfield of D. Then $L \otimes_K (D)_n \cong L \otimes_K D \otimes_K (K)_n$ $\cong L \otimes_{K_1} K_1 \otimes_K D \otimes_K (K)_n \cong L \otimes_{K_1} D \otimes_K K_1 \otimes_K (K)_n \cong (L)_m \otimes_K (K_1)_n$ $\cong (L)_{nm} \otimes_K K_1 \cong (L \otimes_K K_1)_{nm}$ (cf. proof of (6.4)). Since A is separable, $(D^L)_n$ is semi-simple (cf. (5.9)). In particular, $L \otimes_K K_1$ is semi-simple (cf. Ex. 5,6). The same argument shows, that, for every extension field L' of L, $L' \otimes_K K_1$ is semi-simple and now, the same technique as in (Ex. 5,7) shows, that K_1 is a separable extension of K. Since a finite separable extension of K is a simple extension, $K_1 = K(\alpha)$, for some $\alpha \in K_1$. Then $L \otimes_K K_1 \cong L(\alpha) \cong L[X]/(f(X))$, where $f(X) = \min_{K_1/K}(\alpha, X)$. Let L' be a finite extension field of L such that $f(X)$ decomposes completely into linear factors in L', say $f(X) = \Pi_{i=1}^{t}(\ell_i - X)$. Since α is separable over K, $\ell_i \neq \ell_j$ for $i \neq j$. Then $L' \otimes_L L \otimes_K K_1 \cong L' \otimes_K K_1 \cong L'[X]/f(X)) \cong L'[X]/\Pi_{i=1}^{t} (\ell_i - X)$.

Since the ideals $\{(\ell_i - X)\}_{1 \leq i \leq t}$ are maximal and different, a simple application of the Chinese remainder theorem (I, (7.7)), shows that the sequence

$$0 \longrightarrow \cap_{i=1}^{t} (\ell_i - X) \longrightarrow L'[X] \longrightarrow \oplus_{i=1}^{t} L'[X]/(\ell_i - X) \longrightarrow 0$$

(cf. I, (2.4)) is exact. Moreover, $\cap_{i=1}^{t} (\ell_i - X) = \Pi_{i=1}^{t} (\ell_i - X)$. Thus, $L' \otimes_L L \otimes_K K_1 \cong \oplus_{i=1}^{t} L'[X]/(\ell_i - X)$. But $L'[X]/(\ell_i - X) \cong L'$. Thus, $L' \otimes_K K_1 \cong L' \oplus \ldots \oplus L'$, t copies; i.e., $L' \otimes_K (D)_n = \oplus_1^t (L')_{n \cdot m}$.

We are now going to introduce the reduced characteristic polynomial, the reduced norm and the reduced trace of a separable K-algebra A. These are less complicated and more important then the characteristic polynomial, the norm and the trace; especially if char $K \neq 0$.

6.11 <u>Definitions:</u> Let A be a finite dimensional central simple K-algebra and L a finite dimensional splitting field for A (cf. (6.9)). If $A = (D)_n$ for some central skewfield D over K, we have $A^L = L \otimes_K A \xrightarrow[\sim]{\sigma} (L)_r$, where $r = n \cdot s$ with $[D : K] = s^2$. For $a \in A$, $(1 \otimes a)\sigma \in (L)_r$ is represented by an $(r \times r)$ - matrix with entries in L. We define the <u>reduced characteristic polynomial</u> of $a \in A$ relative to K as

$\mathrm{Pcrd}_{A/K}(a) = \det(X \cdot \underline{E}_r - (1 \otimes a)\sigma) \in L[X]$, the <u>reduced norm</u> of a as $\mathrm{Nrd}_{A/K}(a) = \det((1 \otimes a)\sigma) \in L$

and the reduced trace of a as

$$\mathrm{Trd}_{A/K}(a) = \mathrm{tr}((1 \otimes a)\sigma) \in L.$$

We then have

$$\mathrm{Pcrd}_{A/K}(a) = X^n - \mathrm{Trd}_{A/K}(a) + \ldots + (-1)^n \mathrm{Nrd}_{A/K}(a).$$

6.12 Lemma: For $a \in A$, $\mathrm{Pcrd}_{A/K}(a)$ is independent of the isomorphism $\sigma : A^L \longrightarrow (L)_r$ as well as of the choice of the splitting field L.

Proof: Let
$$\sigma_1 : L \otimes_K A \xrightarrow{\sim} (L)_r \quad \text{and}$$
$$\sigma_2 : L \otimes_K A \xrightarrow{\sim} (L)_r$$
be two algebra isomorphisms. Then one shows, as in the proof of (6.6), that the images of σ_1 and σ_2 are conjugate by a matrix in $(L)_r$; i.e., there exists a matrix $\underline{B} \in (L_r)$ such that

$$(1 \otimes a)\sigma_1 = \underline{B}^{-1}[(1 \otimes a)\sigma_2]\underline{B}.$$

But this shows that $\mathrm{Pcrd}_{A/K}(a)$ is independent of the chosen isomorphism. If now L_1 and L_2 are two finite dimensional splitting fields for A, then we choose a common extension field L of L_1 and L_2, and, using the previous result, we conclude that $\mathrm{Pcrd}_{A/K}(a)$ is independent of the chosen splitting field. #

Before showing that $\mathrm{Pcrd}_{A/K}(a) \in K[X]$ we have to derive some facts on central simple algebras which are of interest in themselves.

6.13 Lemma: Let A be a central simple K-algebra. Then there exists a separable extension field of K which splits A.

Proof: Because of (6.4) it suffices to assume that $A = D$ is a central skewfield over K. We shall show that D contains a maximal separable subfield. This will prove the assertion (cf. (6.5)). We claim that in $D \setminus K$ there exist separable elements. Assume to the contrary that every $d \in D$ satisfies an equation of minimal degree of the form $d^{p^s} = k \in K$, where $p > 0$ is the characteristic of K. This implies in particular, that the degree of $K(d)$ is a multiple of p, and that consequently, p^2 divides the degree of D over K. Let now L be a finite dimensional splitting field of D and consider an algebra homomorphism $\sigma : D \longrightarrow L \otimes_K D \xrightarrow{\sim} (L)_r$. Then, for $d \in D \setminus K$, $d\sigma$ also satisfies the equation $(d\sigma)^{p^s} = k\underline{E} \in K$, of minimal degree. But, since the minimum polynomial of $d\sigma$ divides the characteristic polynomial of $d\sigma$ (cf. (3.4)) we have $k(d\sigma) = 0$, for every $d \in D \setminus K$. And, since every element in $(L)_r$ has the form $\Sigma_i \ell_i (d_i \sigma)$ this implies $\operatorname{tr}(\underline{B}) = 0$ for every $B \in (\underline{L})_r \setminus L$, a contradiction.

Now we turn to the proof of (6.13). Let $d_1 \in D$ be separable of degree $m_1 > 1$ and $(D : K) = r^2$. Then $K(d_1) \otimes_K D \simeq (D_1)_{r/m_1}$, where D_1 is a central skewfield over $K(d_1)$. Now we continue the same construction with D_1 and $K_1 = K(d_1)$. After finitely many steps we get a separable extension $K(d_1, \ldots, d_t)$ of degree r over K which splits D. #

6.14 <u>Theorem</u>: Let A be a central simple K-algebra and $a \in A$. Then $\operatorname{Pcrd}_{A/K}(a) \in K[X]$; in particular, $\operatorname{Nrd}_{A/K}(a) \in K$ and $\operatorname{Trd}_{A/K}(a) \in K$.

<u>Proof</u>: According to (6.13) we can find a separable splitting field L of A. Extending it, if necessary, we may assume that L is a normal separable (i.e., Galois) extension of K. By G we denote the Galois group of L over K. To prove the theorem it suffices to show that $\operatorname{Pcrd}_{A/K}(a)$ is invariant under all $\rho \in G$. We set $(\rho)_r : (L)_r \longrightarrow (L)_r$, $(\ell_{ij}) \longmapsto (\ell_{ij}^{\rho})$ and fix an isomorphism $\sigma : L \otimes_K A \xrightarrow{\sim} (L)_r$. According to (6.12), $\operatorname{Pc}((1 \otimes a)\sigma) =$

$Pc((1\otimes a)\sigma(\rho)_r) = Pc((1\otimes a)\sigma)^\rho$, for all $\rho \in G$ and thus $Pcrd_{A/K}(a)$ is indeed invariant under the Galois group. #

6.15 Lemma: Let A be a central simple K-algebra. If $[A:K] = n^2$, then

(i) $[Pcrd_{A/K}(a)]^n = Pc_{A/K}(a)$

(ii) $[Nrd_{A/K}(a)]^n = N_{A/K}(a)$

(iii) $n \cdot Trd_{A/K}(a) = Tr_{A/K}(a)$, $\forall a \in A$.

Proof: It suffices to prove (i). Let L be a splitting field for A, then $M = \oplus_{i=1}^n L\underline{\underline{E}}_{i1}$ is a simple left A^L-module, if $\underline{\underline{E}}_{i1}$ is the matrix with 1 at the (i,1)-position and zeros elsewhere. A matrix $\underline{\underline{B}} = (b_{\rho\sigma}) \in A^L$ representing $a \in A$ acts on M by

$$\underline{\underline{B}}\,\underline{\underline{E}}_{i1} = \Sigma_{k=1}^n b_{ki}\,\underline{\underline{E}}_{k1};$$

and it follows that $Pc(\varphi_{1\otimes a}) = Pcrd_{A/K}(a)$, where $\varphi_{1\otimes a}$ denotes the matrix of the linear transformation of M induced by left multiplication with $(1\otimes a)$. Now, as left A^L-module, $A^L \simeq M^{(n)}$ and the result follows. #

6.16 Remark: (i) If A is a simple separable K-algebra with center L we define the reduced trace of A with respect to K by

$$Trd_{A/K}(a) = Tr_{L/K}(Trd_{A/L}(a)), \quad a \; A,$$

and it follows immediately that here too $r\cdot Trd_{A/K} = Tr_{A/K}(a)$, where $[A:L] = r^2$. For an arbitrary separable K-algebra the reduced trace is defined as the sum of the reduced traces of the simple components.
(ii) The trace function and the reduced trace function are symmetric, i.e., $Tr_{A/K}(ab) = Tr_{A/K}(ba)$, (cf. Ex. 6,10).

6.17 Theorem: A finite dimensional semi-simple K-algebra is separable if and only if there exists a finite dimensional splitting field for A.

Proof: Because of (6.10) we let L be a splitting field for A, say $A^L \simeq \oplus_{i=1}^n (L)_{r_i}$. We first show that A^L is separable. For this it suffices to show that $A' = (L)_r$ is separable; i.e., we have to show that the sequence

$$0 \longrightarrow \operatorname{Ker}\varepsilon \longrightarrow A'^e \xrightarrow{\varepsilon} A' \longrightarrow 0$$

splits over A'^e. For this it suffices to show that

$$\operatorname{hom}_{A'^e}(1_{A'},\varepsilon) = \varepsilon_* : \operatorname{Hom}_{A'^e}(A',A'^e) \longrightarrow \operatorname{End}_{A'^e}(A')$$

is an epimorphism; but ε_* is L-linear and $\operatorname{End}_{A'^e}(A') = \operatorname{center}(A') = L$, since $\varphi \in \operatorname{End}_{A'^e}(A')$ is uniquely determined by $(1)\varphi$. Thus it suffices to show that $\varepsilon_* \neq 0$. We choose a special basis $\{\underline{\underline{E}}_{ij}\}_{1\le i,j\le r}$ of A', where $\underline{\underline{E}}_{ij}$ is the matrix with 1 at the (i,j)-position and zeros elsewhere. It is easily checked that $\{\underline{\underline{E}}_{ij}^*\}$ with $\underline{\underline{E}}_{ij}^* = \underline{\underline{E}}_{ji}$ is a dual basis of $\{\underline{\underline{E}}_{ij}\}$ with respect to the reduced trace (cf. (6.11) and (3.7)). Now, we define, for every $b \in A'$, the map

$$\varphi_b : A' \longrightarrow A'^e,\ a \longmapsto \Sigma_{i,j=1}^r\, a\, E_{ij}^* b \otimes E_{ij}.$$

Then $\varphi_b \in \operatorname{Hom}_{A^e}(A,A^e)$, (cf. Ex. 6,2) and $\varepsilon_*(\varphi_b) = \varphi_b\varepsilon : A' \longrightarrow A'$, $a \longmapsto \Sigma_{i,j=1}^r\, a\, \underline{\underline{E}}_{ij}^*\, b\, \underline{\underline{E}}_{ji}$. If we choose $b = \underline{\underline{E}}_{11}$, then

$$(1)\varphi_{\underline{\underline{E}}_{11}} = \Sigma_{i,j=1}^n\, \underline{\underline{E}}_{ij}^*\, \underline{\underline{E}}_{11}\, \underline{\underline{E}}_{ij}$$

$$= \Sigma_{j=1}^n \underline{\underline{E}}_{1j}^* \underline{\underline{E}}_{1j} = \Sigma_{j=1}^n \underline{\underline{E}}_{j1}\underline{\underline{E}}_{1j} = \Sigma_{j=1}^n \underline{\underline{E}}_{jj} = 1.$$

Thus A' is separable. Hence we know that A^L is separable. But $L\otimes_K -$ is a faithful functor on ${}_K\underline{\underline{M}}^f$, and thus, $0 = \operatorname{Ext}^1_{A^{eL}}(A^L,X^L) = L \otimes_K \operatorname{Ext}^1_{A^e}(A,X)$ implies $\operatorname{Ext}^1_{A^e}(A,X) = 0$ for every $X \in {}_{A^e}\underline{\underline{M}}^f$, i.e., A is separable. #

6.18 <u>Corollary</u>: Let A be a separable finite dimensional K-algebra. Then the discriminant of every K-basis of A relative to the reduced trace function does not vanish and thus there exist dual bases relative to the reduced trace.

Proof: This is an immediate consequence of (3.7) and the proof of the previous theorem. #

6.19 Theorem: Let A be a central simple K-algebra. Then A is separable and it stays central simple under any extension of the ground field.

Proof: With (6.9) and (6.17) we conclude that A is separable, and it remains to show that for any extension field L of K, center $(A^L) = L$. But this follows from (1.2) since center $(A^L) = \mathrm{End}_{(A^L)^e}(A^L) \simeq L \otimes_K \mathrm{End}_{A^e}(A) \simeq L \otimes_K K \simeq L$. #

6.20 Theorem: Let A be a separable K-algebra and $f : A \times A \longrightarrow K$ a non-degenerate bilinear form. Let $\{w_i\}_{1\leq i\leq n}$ and $\{w_i^*\}_{1\leq i\leq n}$ be a pair of dual bases with respect to f. If $\varepsilon : A^e \longrightarrow A$ is the augmentation map, then $\mathrm{Im}\ \varepsilon_* = \hom_{A^e}(1_A, \varepsilon) = \{\Sigma_{i=1}^n w_i^* a w_i : a \in A\}$ $=$ center(A).

We observe that, since A is separable, there exists a non-degenerate bilinear form and a pair of dual bases relative to it, (cf. (3.7), (6.18) and Ex. 3,5), and thus the statement of the theorem is meaningful.

Proof: The map $\gamma : A \longrightarrow$ center(A), $a \longmapsto \Sigma_{i=1}^n w_i^* a w_i$ is called the Gaschütz-Casimir operator.

(i) We shall first show that $\mathrm{Im}\ \gamma$ is independent of the chosen basis. Let $w_j' = \Sigma_{i=1}^n \alpha_{ji} w_i$, $\alpha_{ji} \in K$ be another basis and put $\alpha_{ij}^* = (\alpha_{k\ell})_{ij}^{-1}$. Then the dual basis to $\{w_i'\}_{1\leq i\leq n}$ with respect to f is given by $w_j'^* = \Sigma_{i=1}^n \alpha_{ij}^* w_i^*$. If γ' is the Gaschutz-Casimir operator relative to the basis $\{w_i'\}_{1\leq i\leq n}$, then

$$\Sigma_{i=1}^n w_i'^* a w_i' = \Sigma_{i=1}^n \Sigma_{k=1}^n \alpha_{ki}^* w_k^* a \Sigma_{\ell=1}^n \alpha_{i\ell} w_\ell = \Sigma_{\ell=1}^n w_\ell^* a w_\ell ;$$

i.e., $\operatorname{Im} \gamma = \operatorname{Im} \gamma'$ as follows from symmetry, and, in fact, $\gamma = \gamma'$.
(ii) Next, we shall show that $\operatorname{Im} \gamma$ is independent of the chosen non-degenerate bilinear form. From (II, (1,12)) it follows that $\operatorname{Hom}_K({}_A A, K) \in \underline{\underline{M}}^f_A$. For the non-degenerate bilinear form $f : A \times A \longrightarrow K$, we define $\vartheta_f : {}_A A \longrightarrow \operatorname{Hom}_K(A_A, K)$, $a \longmapsto \varphi_a$, where $x^{\varphi_a} = f(x,a)$. Since f is non-degenerate, $\operatorname{Ker} \vartheta_f = 0$; moreover, if $\psi \in \operatorname{Hom}_K(A_A, K)$, then $\psi = \varphi_a$ with $a = \Sigma_{i=1}^n \psi(w_i) w_i^*$. Thus, ϑ_f is a K-isomorphism. Moreover, the relation $f(xy,z) = f(x,yz)$ implies that ϑ_f is an A-isomorphism. If now f_1 and f_2 are non-degenerate bilinear forms, then $\vartheta_{f_1} \vartheta_{f_2}^{-1} \in \operatorname{Hom}_A({}_A A, {}_A A)$ is an automorphism of ${}_A A$; i.e., right multiplication by some unit $a_o \in A$; and so $\vartheta_{f_1} = a_o \vartheta_{f_2}$. Thus, for every $x,y \in A$, $f_1(x,y) = x^{y \vartheta_{f_1}} = x^{y a_o \vartheta_{f_2}} = f_2(x, y a_o)$. If $\{w_i\}_{1 \le i \le n}$ and $\{w_i^*\}_{1 \le i \le n}$ are dual bases with respect to f_1, then $\{w_i\}_{1 \le i \le n}$ and $\{w_i^* a_o\}_{1 \le i \le n}$ are dual bases with respect to f_2. Since $a_o^{-1} A = A$, $\Sigma_{i=1}^n w_i^* a w_i = \Sigma_{i=1}^n w_i^* a_o (a_o^{-1} a) w_i$ shows that $\operatorname{Im} \gamma$ is independent of the chosen bilinear form.
(iii) Because of (i) we may choose the basis according to the simple components of A, and consequently, we may assume that A is simple. As a special non-degenerate bilinear form, we choose the reduced trace function (cf. (6,13)). Let $\{w_i\}_{1 \le i \le n}$ and $\{w_i^*\}_{1 \le i \le n}$ be a pair of dual bases relative to the reduced trace. Then, for every $b \in A$, the map $\psi_b : A \longrightarrow A^e$, $a \longmapsto \Sigma_{i=1}^n a w_i^* b \otimes w_i$ is an A^e-homomorphism (cf. Ex. 6,2) and $\gamma(b) = (1)\psi_b \epsilon$, where $\epsilon : A^e \longrightarrow A$ is the augmentation map. Hence $\gamma(b) = \epsilon_*(\psi_b) \in \operatorname{End}_{A^e}(A) = C = \operatorname{center}(A)$. Since C is a field (A was assumed to be simple) and γ is a C-homomorphism, it suffices to show that $\gamma(b) \neq 0$ for some $b \in A$. Let L be a finite dimensional splitting field for A. Now

one shows, as in the proof of (6.17), that there exists $\tilde{b} \in A^L$ such that $\gamma(\tilde{b}) \neq 0$; observe that $\{1 \otimes w_i\}_{1\leq i\leq n}$ and $\{1 \otimes w_i^*\}_{1\leq i\leq n}$ are dual bases of A^L relative to the trace function. If $\tilde{b} = \Sigma_k \ell_k \otimes a_k$, $\ell_k \in L$, $a_k \in A$, then $\gamma(\tilde{b}) = \Sigma_k \ell_k \otimes \Sigma_{i=1}^n w_i^* a_k w_i$; and hence at least one of the $\gamma(a_k) \neq 0$. #

Exercises §6:

1.) Show that every field K can be considered as a subfield of an algebraically closed field L; i.e., of a field L such that every irreducible polynomial $f(X) \in L[X]$ is of the form $x - \alpha$, $\alpha \in L$. (Hint: With each $f(X) \in K[X]$, degree $(f(X)) \geq 1$, associate a symbol X_f, and put $S = \{X_f : f(X) \in K[X], \text{degree } (f(X)) \geq 1\}$. Now, we form the polynomial ring $K[S]$; i.e., $K[S] = \{P(X_{f_1},\ldots,X_{f_n}) : P = \text{polynomial in } X_{f_i} \in S, 1 \leq i \leq n\}$, with the obvious addition and multiplication. Show that the ideal I, of $K[S]$ generated by $\{f(X_f) : X_f \in S\}$ is different from $K[S]$. Let M be a maximal ideal, $K[S] \supsetneqq M \supset I$, and let $\sigma : K[S] \longrightarrow K[S]/M$ be the canonical homomorphism. Show that for every $f(X) \in K[X]$, degree $(f(X)) \geq 1$, $\sigma(f)$ has a root in the field $L_1 = K[S]/M$. Now apply this same construction to L_1, etc. Thus, obtain a chain of fields $K \subset L_1 \subset L_2 \subset \ldots$, set $L = \bigcup_i L_i$, and show that L is an algebraically closed field containing K.)

2.) Let A be a separable K-algebra, and $\{w_i\}_{1\leq i\leq n}$ and $\{w_i^*\}_{1\leq i\leq n}$ a pair of dual bases with respect to some non-degenerate bilinear form. Show that for every $b \in A$, $\varphi_b : A \longmapsto A^e$, $\varphi_b : a \longmapsto \sum_{i=1}^n a\, w_i^*\, b \otimes w_i$ is an A^e-homomorphism.

3.) Show that a simple K-algebra A is separable if and only if its center is a separable extension field of K (cf. Ex. 5,7).

4.) Let K be either of characteristic zero, or a Galois-field. Show that a finite dimensional K-algebra is separable if and only if it is semi-simple.

5.) Let L be an inseparable extension field of K. Show that there exists a field $E \supset K$ such that $L \otimes_K E$ is not semi-simple. (Hint: If L is an inseparable extension of K, then $\mathrm{char}K = p > 0$, and $(K:1) = \infty$. There exists an element $\alpha \in L$ such that

$$\min_{L/K}(\alpha, X) = (X^p)^n + k_{n-1}(X^p)^{n-1} + \ldots + k_o,\ k_i \in K.$$

Let $E = K(k_o^{1/p}, \ldots, k_{n-1}^{1/p}) \neq K$. Then $\beta = \alpha^n + k_{n-1}^{1/p}\alpha^{n-1} + \ldots + k_o^{1/p}$ belongs to $L \otimes_K E$ and has the property that $\beta \neq 0$, but $\beta^p = 0$.)

6.) Prove 6.15. (Hint: Use the techniques from the proofs of 6.16 and 6.14.)

7.) Show that a direct sum of two algebras is separable if and only if each summand is separable. (Give two proofs: (i) use Ex. 6,6; (ii) use the definition (5.1).)

8.) Let S be a left artinian and left noetherian ring. Show that $P \in {}_S\underline{\underline{P}}^f$ is a progenerator if and only if P is a faithful left S-module (i.e., if $\mathrm{ann}_S(P) = 0$).

9.) Let $\underline{\underline{H}}$ be the quaternion algebra $\underline{\underline{H}} = \underline{\underline{Q}}(i,j,k)$ (cf. Ex. 5,9). For $a \in \underline{\underline{H}}$ compute $\mathrm{Tr}_{\underline{\underline{H}}/\underline{\underline{Q}}}(a)$ and $\mathrm{Trd}_{\underline{\underline{H}}/\underline{\underline{Q}}}(a)$; (If

$a = \alpha_0 + \alpha_1 i + \alpha_2 j + \alpha_3 k$ and if we put $\bar{a} = \alpha_0 + \alpha_1 i + \alpha_2 j + \alpha_3 k$, then $Pcrd_{\underline{H}/\underline{Q}} = (X-a)(X-\bar{a})$.

10.) Show that the trace and the reduced trace are symmetric.

§7. Projective covers

Essential epimorphisms are used to restate Nakayama's lemma. The projective modules over left semi-perfect rings are described. Left noetherian and left artinian rings are semi-perfect.

Although the classical approach to the method of lifting idempotents (cf. e.g. Curtis - Reiner [1], §77) is perhaps shorter, we shall present here the more general concept of projective covers (cf. Bass [7], Shu a [1]).

7.1 Definitions: Let S be a ring.

(i) An epimorphism $\varphi \in \mathrm{Hom}_S(M,N)$, $M,N \in {}_S\underline{\underline{M}}$ is called an essential epimorphism if for every $M' \in {}_S\underline{\underline{M}}$ and $\psi \in \mathrm{Hom}_S(M'M)$:

$\psi\varphi : M' \longrightarrow N$ is epic implies $\psi : M' \longrightarrow M$ is epic.

(ii) $P \in {}_S\underline{\underline{P}}^f$ is a projective cover of $M \in {}_S\underline{\underline{M}}^f$, if there exists an essential epimorphism $\varphi : P \longrightarrow M$.

It seems worthwhile to rephrase Nakayama's lemma (I,(4.18)) in terms of essential epimorphisms.

7.2 Lemma (Nakayama's lemma): Let S be a left noetherian ring, $M,N \in {}_S\underline{\underline{M}}^f$, and let $\varphi \in \mathrm{Hom}_S(M,N)$ be an epimorphism. If $\mathrm{Ker}\,\varphi \subset \mathrm{rad}\, S \cdot M$, then φ is an essential epimorphism.

Proof: (i) (I,(4.18)) implies (7.2): Let $\psi \in \mathrm{Hom}_S(M',M)$, $M' \in {}_S\underline{\underline{M}}$ be given such that $\psi\varphi : M' \xrightarrow{\psi} M \xrightarrow{\varphi} N$ is an epimorphism. Then $M = \mathrm{Im}\,\psi + \mathrm{Ker}\,\varphi = \mathrm{Im}\,\psi + \mathrm{rad}\, S \cdot M$. With (I,(4.18)) we conclude that ψ is epic.

(ii) (7.2) implies (I,(4.18)): Let M' be a submodule of $M \in {}_S\underline{\underline{M}}^f$ such that $M' + \mathrm{rad}\, S \cdot M = M$. If $\varphi : M \longrightarrow M/\mathrm{rad}\, S \cdot M$ is the canonical epimorphism and $\psi : M' \longrightarrow M$ the canonical injection, then, since φ is an essential epimorphism, and since $\psi\varphi$ is epic, ψ is epic; i.e., $M' = M$. #

7.3 <u>Lemma</u> (Uniqueness of projective covers): Let S be a ring and $M \in {}_S\underline{\underline{M}}^f$. If M has a projective cover P, then P - up to isomorphism - is uniquely determined by M.

<u>Proof</u>: Let $P, P' \in {}_S\underline{\underline{P}}^f$ be projective covers for $M \in {}_S\underline{\underline{M}}^f$. Then we can complete the diagram

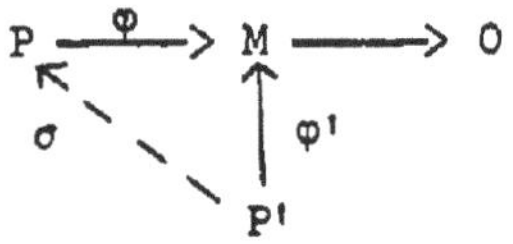

commutatively (cf. I, (2.9)). Since φ and φ' are essential epimorphisms, σ is an epimorphism, and thus, the sequence $0 \longrightarrow \operatorname{Ker}\sigma \longrightarrow P' \xrightarrow{\sigma} P \longrightarrow 0$ splits; i.e., there exists a monomorphism $\rho: P \longrightarrow P'$, such that $\rho\sigma = 1_P$. The commutative diagram

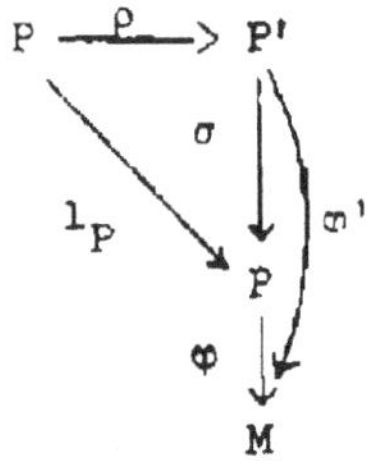

shows that ρ is an epimorphism, since φ' was essential. Whence $P \cong P'$. #

7.4 <u>Lemma</u>: Let S be a left noetherian ring and I a left ideal of S contained in $\operatorname{rad} S$ and let $M \in {}_S\underline{\underline{M}}^f$. Then either both M and M/IM have the same projective cover or neither of them has a projective cover.

<u>Proof</u>: If $P \in {}_S\underline{\underline{P}}^f$ is a projective cover for M, and if

$\varphi : P \longrightarrow M$ is the essential epimorphism, then (7.2) shows that $\varphi\sigma : P \xrightarrow{\varphi} M \xrightarrow{\sigma} M/IM$ - where σ is the canonical epimorphism - is an essential epimorphism; i.e., P is also a projective cover for M/IM. Conversely, if $P' \in {}_{\underline{\underline{S}}}\underline{P}^f$ is a projective cover for M/IM with essential epimorphism φ', then we define φ via the commutative diagram

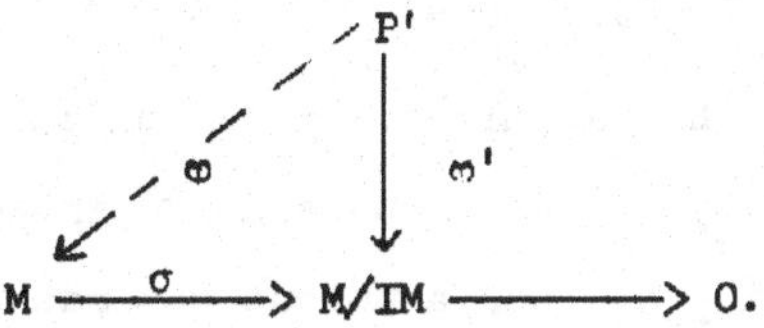

From (7.2) we conclude that σ is an essential epimorphism, hence φ is an essential epimorphism, because $\varphi' = \varphi\sigma$ is one. #

7.5 **Definitions**:

(i) A ring S is called left semi-primary, if $S/\text{rad}\, S$ is a left noetherian and left artinian ring (cf. I,(4.11).

(ii) We say that a ring S is left semi-perfect, if S is left semi-primary and for every idempotent $\bar{e} \in S/\text{rad}\, S$, there exists an idempotent $e \in S$ such that $e \longmapsto \bar{e}$ under the canonical homomorphism $S \longrightarrow S/\text{rad}\, S$.

7.6 **Theorem**: If S is a left semi-perfect ring, then every $M \in {}_{\underline{\underline{S}}}\underline{M}^f$ has a projective cover.

Proof: We denote by "$\bar{\ }$" the reduction modulo rad S. Given M; since $\bar{M} \in {}_{\bar{\underline{\underline{S}}}}\underline{M}^f$, it follows from (Ex. 7,3) that

$$\bar{M} \cong \oplus_{i=1}^{n} \bar{S}\,\bar{e}^{(\alpha_i)} ,$$

where $\{\bar{e}_i\}_{1 \le i \le n}$ is a complete set of non-equivalent primitive idempotents in $\bar{S}$ and the α_i are non-negative integers (cf. (5.6)). Since S is semi-perfect, there are idempotents $\{e_i\}_{1 \le i \le n}$ of S

such that $\overline{Se}_i = S\bar{e}_i$, $1 \leq i \leq n$. We put $P = \oplus_{i=1}^n Se_i^{(\alpha_i)} \in {}_S\underline{\underline{P}}^f$ and $\varphi : P \longrightarrow \oplus_{i=1}^n \bar{S}\bar{e}_i^{(\alpha_i)}$, the canonical homomorphism. Then it follows from (7.2) that P is a projective cover for $\bar{M}$, and (7.4) shows that P is a projective cover for M. #

7.6 Theorem: Let S be a left semi-perfect ring. If ${}_SS = \oplus_{i=1}^n P_i^{(\alpha_i)}$ is a decomposition of ${}_SS$ into indecomposable submodules P_i, $P_i \not\cong P_j$ for $i \neq j$, then:

(i) Every $P \in {}_S\underline{\underline{P}}^f$ has a unique expression as $P \cong \oplus_{i=1}^n P_i^{(\beta_i)}$, where the $\{\beta_i\}_{1 \leq i \leq n}$ are non-negative integers.

(ii) Each P_i, $1 \leq i \leq n$, has a unique maximal submodule $\operatorname{rad} S \cdot P_i$.

(iii) $\{P_i/\operatorname{rad} S \cdot P_i\}_{1 \leq i \leq n}$ are the non-isomorphic simple left $S/\operatorname{rad} S$-modules.

The proof is straightforward. #

Exercises §7:

1.) Let S be a left noetherian ring and I a left ideal in S. Show that the canonical epimorphism $S \longrightarrow S/I$ is an essential epimorphism if and only if $I \subset \operatorname{rad} S$.

Moreover, if $M \in {}_SM^f$, then, for a submodule $N \subset M$, the canonical epimorphism $M \longrightarrow M/N$ is essential if and only if $N \subset \operatorname{rad}_S M$.

2.) Let S be a ring. Show that

(i) if S is left semi-primary then $(S)_n$ is left semi-primary.

(ii) if S is left semi-perfect then $(S)_n$ is left semi-perfect.

3.) Let S be a left noetherian and left artinian ring. Show that

S is semi-perfect. (Hint: rad S is nilpotent (cf. I, Ex. 4,6), say $(\mathrm{rad}\ S)^n = 0$. Then we have the following chain of canonical homomorphisms $S = S/(\mathrm{rad}\ S)^n \xrightarrow{\varphi_{n-1}} S/(\mathrm{rad}\ S)^{n-1} \xrightarrow{\varphi_{n-2}} \cdots \xrightarrow{\varphi_1} S/\mathrm{rad}\ S$. Given an idempotent $\overline{e} \in S/\mathrm{rad}\ S$, we pick $a \in S/(\mathrm{rad}\ S)^2$ such that $a\varphi_1 = \overline{e}$. Then $a^2 - a = z \in \ker \varphi_1 = \mathrm{rad}\ S/(\mathrm{rad}\ S)^2$. Now, $e_1 = (a-z)^2$ is an idempotent in $S/(\mathrm{rad}\ S)^2$ such that $e_1\varphi_1 = \overline{e}$. Continuing this way, one produces after (n-1) steps an idempotent e in S with the desired properties.)

Let $\varepsilon_1, \varepsilon_2$ be orthogonal idempotents in $\overline{S}$, and let e be an idempotent in S such that $\overline{e} = \varepsilon_1 + \varepsilon_2$. Show that there exist orthogonal idempotents e_1, e_2 of S such that $\overline{e}_1 = \varepsilon_1$, $\overline{e}_2 = \varepsilon_2$.

CHAPTER IV

MAXIMAL ORDERS

§1. Lattices and orders

The basic definitions for orders and lattices over orders are given, and some elementary properties are derived. The connections between the global, the local and the complete case are developed.

Let R be a Dedekind domain (cf. I, (7.1)) with quotient field K and A a finite dimensional K-algebra.

1.1 Definitions: (i) An R-order Λ in A is a subring of A with the same identity as A such that

α) $\Lambda \in {}_R\underline{M}^f$,

β) $K\Lambda = A$; i.e., Λ contains a K-basis of A.

(ii) A left Λ-lattice is a left Λ-module, which is at the same time an R-lattice in some finite dimensional K-vectorspace V (cf. I, (7.1)); i.e., a torsionfree R-module of finite type.

1.2 Remark: From now on we assume that A is a finite dimensional separable K-algebra (cf. III, (5.1)), and for an R-order Λ in A we use the following notation:

${}_\Lambda\underline{M}^o$ = the category of left Λ-lattices. (If no confusion can arise, we omit the word "left".)

$\underline{S}$ = set of all prime ideals in R.

1.3 Lemma: Let M be an R-lattice in A; i.e., $KM = A$. Then

$$\Lambda_\ell(M) = \{a \in A : aM \subset M\} \text{ and}$$

$$\Lambda_r(M) = \{a \in A : Ma \subset M\}$$

are R-orders in A, called the <u>left</u> and <u>right ring of multipliers of M</u> resp. (or shortly: <u>left</u> and <u>right order of M</u> resp.).

<u>Proof</u>: To show that $\Lambda_\ell(M)$ is an order, we observe that $\Lambda_\ell(M)$ is obviously a ring. Since $KM = A$, M contains a K-basis for A say $\{w_i\}_{1\le i\le n}$. Let M be generated over R by $\{m_i\}_{1\le i\le t}$, and choose $0 \ne r \in R$ such that $(rw_i)m_j \in M$, $1 \le i \le n$, $1 \le j \le t$. This is indeed possible:

$$m_j = \sum_{\ell=1}^{n} k_{j\ell} w_\ell, \quad 1 \le j \le t, \; k_{j\ell} \in K$$

$$w_i w_\ell = \sum_{s=1}^{n} k_{i\ell}^s w_s, \quad 1 \le i, \ell \le n, \; k_{i\ell}^s \in K.$$

Since K is the quotient field of R (cf. I, (6.6)), there exists $0 \ne r' \in R$ such that $r'k_{j\ell} \in R$, $1 \le j \le t$, $1 \le \ell \le n$, and $r'k_{i\ell}^s \in R$, $1 \le i, \ell, s \le n$. Then $r = r'^2$ has the desired property, and we have $rw_i M \subset M$, $1 \le i \le n$; i.e., $rw_i \in \Lambda_\ell(M)$, $1 \le i \le n$, and $\Lambda_\ell(M)$ contains a K-basis for A. Since $KM = A$, $M \cap R\cdot 1 \ne 0$, and we choose $0 \ne r_0 \in M \cap R\cdot 1$. Then $\Lambda_\ell(M)\cdot r_0 \subset M$, and $\Lambda_\ell(M)r_0 \in {}_R\underline{M}^f$, R being noetherian. Since $\Lambda_\ell(M)r_0 \cong_R \Lambda_\ell(M)$, $\Lambda_\ell(M) \in {}_R\underline{M}^f$, and $\Lambda_\ell(M)$ is an R-order in A. The proof for $\Lambda_r(M)$ is done similarly. #

In the proof of the next theorem we use essentially the fact, that A is a separable K-algebra. (1.4) in turn is used to guarantee the existence of maximal orders (cf. (4.6)).

1.4 <u>Theorem</u>: Let S be a subring of A with the same identity as A. Then S is an R-order in A if and only if

(i) $S \in {}_R\underline{M}$,

(ii) $KS = A$,

(iii) every $s \in S$ is integral over R.

Proof: If S is an R-order in A, then (i) and (ii) are satisfied by definition (cf. (1.1)), and (iii) is satisfied since $S \in {}_{\underline{R}}\underline{M}^f$ (cf. I, (5.4)).

Conversely: We have to show that S is finitely generated over R (cf. (1.1, α)). Let $\{w_i\}_{1\leq i\leq n}$ be a K-basis of A contained in S. For $s \in S$, we have $s = \sum_{i=1}^{n} k_i w_i$, $k_i \in K$, $1 \leq i \leq n$, and $sw_j = \sum_{i=1}^{n} k_i w_i w_j$, $1 \leq j \leq n$. We shall show that $\mathrm{Trd}_{A/K}(sw_j) \in R$ (cf. III, (6.11)).

1.4' Claim: If $a \in A$ is integral over R, then

$$\mathrm{Trd}_{A/K}(a), \mathrm{Nrd}_{A/K}(a) \in R.$$

Proof: It suffices to show that $\mathrm{Pcrd}_{A/K}(a,X) \in R[X]$ (cf. III, (6.11)). Since a is integral over R, it satisfies a monic polynomial $f(X) \in R[X]$ (cf. I, (5.2)). We shall first show that $g(X) = \min_{A/K}(a,X) \in R[X]$ (cf. III, (3.1)). We intend to apply Gauß' lemma (cf. I, Ex. 7.6). However, since Gauß' lemma is only formulated for principal ideal domains we use the technique of localization: $g(X) \in R[X] \iff g(X) \in R_{\underline{p}}[X]$, $\forall \underline{p} \in \underline{S}$ (cf. I, (8.6)), and since $R_{\underline{p}}$ is a principal ideal domain (cf. I, (8.3)), we may apply Gauß' lemma. We write $f(X) = g(X)h(X)$, where $h(X) \in K[X]$. Putting $g(X) = \alpha g_o(X)$, $h(X) = \beta/\gamma h_o(X)$, where $g_o(X), h_o(X) \in R_{\underline{p}}[X]$ are primitive polynomials (cf. I, Ex. 7.6) and $\alpha,\beta,\gamma \in R_{\underline{p}}$, we obtain $\gamma f(X) = \alpha\beta g_o(X)h_o(X)$. By Gauß' lemma (I, Ex. 7.6), $g_o(X)h_o(X)$ is a primitive polynomial, and since $f(X)$ is monic, $\gamma = \alpha\beta$; hence $f(X) = g_o(X)h_o(X)$, and $g_o(X)$ has leading coefficient 1. Since $g(X)$ was monic to start with, $g_o(X) = g(X) \in R_{\underline{p}}[X]$. Thus,

$g(X) \in R_{\underline{p}}[X]$, $\forall \underline{p} \in \underline{S}$ and therefore $g(X) \in R[X]$; i.e., $\min_{A/K}(a,X) \in R[X]$. From (III, (3.5) and (6.15)) follows that $\mathrm{Prd}_{A/K}(a,X)^n$ divides $\min_{A/K}(a,X)^n$. Now a similar argument as above shows that $\mathrm{Prd}_{A/K}(a,X) \in R[X]$. #

Returning to the proof of (1.4), we find

$$1.4'' \qquad \sum_{i=1}^{n} k_i \,\mathrm{Trd}_{A/K}(w_i w_j) = \mathrm{Trd}_{A/K}(s w_j) \in R, \quad 1 \leq j \leq n.$$

But since A is separable, the discriminant of the basis $\{w_i\}_{1\leq i\leq n}$ is different from zero (cf. III, (6.18)); i.e., $\det(\mathrm{Trd}_{A/K}(w_i w_j)) \neq 0$, and we may solve the system (1.4") with respect to the $\{k_i\}_{1\leq i\leq n}$; i.e., $k_i \in R\cdot\det(\mathrm{Trd}_{A/K}(w_i w_j))^{-1}$. Consequently $s \in \sum_{\ell=1}^{n} R\cdot\det(\mathrm{Trd}_{A/K}(w_i w_j))^{-1} w_\ell$. Since this holds for all $s \in S$, and since R is noetherian, we conclude $S \in {}_{R}\underline{M}^{f}$; i.e., S in an R-order in A. #

1.5 Lemma: Let L be a finite dimensional separable extension field of K, and let R' be the integral closure of R in L. Then R' is a Dedekind domain. Moreover, if $\underline{p} \in \underline{S}$, then $R'_{\underline{p}} = R_{\underline{p}} \otimes_R R'$ is a semi-local Dedekind domain; i.e., it has only finitely many prime ideals. In particular, $R'_{\underline{p}}$ is a principal ideal domain.

Proof: By (III, Ex. 6.3), L is a separable K-algebra, and by (1.4), R' is a finitely generated R-module. Thus, R' is noetherian since R is noetherian (cf. I, (4.1)). Obviously R' is integrally closed in L, and L is the quotient field of R'. Thus, it remains to show that every prime ideal in R' is maximal. Let $\underline{p}'$ be a prime ideal in R', and put $\underline{p} = R \cap \underline{p}'$; then $\underline{p}$ is a prime ideal in R, whence it is maximal, R being a Dedekind domain. Since $R/\underline{p}$ is a field, $R'/\underline{p}'$ is a finite dimensional $R/\underline{p}$-algebra which is an integral domain; i.e., a field by (Ex.1.1). Thus, $\underline{p}'$ is a maximal ideal. Now if $\underline{p}$ is a prime ideal in R, then $S = R \setminus \{\underline{p}\}$ is a multiplicative system in R', and R'_S is a Dedekind domain. The only maximal ideals in R'_S are the ones containing $\underline{p}R'_S$; i.e., R'_S has only a finite number of prime ideals (cf. I, (7.2)) and thus it is a principal ideal domain (cf. I, (7.8)). #

1.6 Notation: For $\underline{p} \in \underline{S}$, we let $R_{\underline{p}}$ be the localization of R at $\underline{p}$ (cf. I, (6.6)) and $\hat{R}_{\underline{p}}$ with quotient field $\hat{K}_{\underline{p}}$ the completion of R at $\underline{p}$ (cf. I, (9.13). For $X \in {}_R\underline{M}^f$, we identify $R_{\underline{p}} \otimes_R X = X_{\underline{p}}$ (cf. I, (6.4)), $\hat{R}_{\underline{p}} \otimes_R X = \hat{X}_{\underline{p}}$ (cf. I, (9.8), (9.13)), $K \otimes_R X = KX$ (cf. I, (6.4)) and, for $Y \in {}_K\underline{M}^f$, $\hat{K}_{\underline{p}} \otimes_K Y = \hat{K}_{\underline{p}} Y$. We have the natural inclusions: $X \subset X_{\underline{p}} \subset KX$, $X \subset \hat{X}_{\underline{p}} \subset \hat{K}_{\underline{p}} X_{\underline{p}} = \hat{K}_{\underline{p}} X$.

1.7 Lemma: If Λ is an R-order in A, then $\Lambda_{\underline{p}}$ is an $R_{\underline{p}}$-order in A and $\hat{\Lambda}_{\underline{p}}$ is an $\hat{R}_{\underline{p}}$-order in $\hat{A}_{\underline{p}}$. Moreover, for $M \in {}_{\Lambda}\underline{M}^o$, we have $M_{\underline{p}} \in {}_{\Lambda_{\underline{p}}}\underline{M}^o$ and $\hat{M}_{\underline{p}} \in {}_{\hat{\Lambda}_{\underline{p}}}\underline{M}^o$.

Proof: $\Lambda_{\underline{p}}$ is a ring with the same identity as A (cf. I, Ex. 5.1) and $\Lambda_{\underline{p}} \in {}_{R_{\underline{p}}}\underline{M}^f$. Moreover, $K\Lambda_{\underline{p}} = K \otimes_{R_{\underline{p}}} R_{\underline{p}} \otimes_R \Lambda = A$; thus, $\Lambda_{\underline{p}}$ is an $R_{\underline{p}}$-order in A. The same argument shows that $\hat{\Lambda}_{\underline{p}}$ is an $\hat{R}_{\underline{p}}$-order in $\hat{A}_{\underline{p}}$. Similarly, the statements for lattices are proved. Observe that $X \in {}_R\underline{M}^f$ is an R-lattice if and only if X is R-torsion free, and that $R_{\underline{p}} \otimes_R -$ and $\hat{R}_{\underline{p}} \otimes_R -$ are exact functors on ${}_R\underline{M}^o$ (cf. I, (6.5) and I, (9.17)). It should be observed that $\hat{A}_{\underline{p}}$ is a separable $\hat{K}_{\underline{p}}$-algebra. #

1.8 <u>Theorem</u>: Let Λ be an R-order in A.

(i) For $M \in {}_{\Lambda}\underline{\underline{M}}^{o}$, we have $M = \bigcap_{\underline{p} \in \underline{S}} M_{\underline{p}}$.

(ii) If $M, N \in {}_{\Lambda}\underline{\underline{M}}^{o}$ are such that $KM = KN$, then $M_{\underline{p}} = N_{\underline{p}}$ for almost all $\underline{p} \in \underline{S}$.

(iii) Let $\{M(\underline{p})\}_{\underline{p} \in \underline{S}}$ be a family of $R_{\underline{p}}$-lattices, such that $KM(\underline{p}) = V \in {}_{A}\underline{\underline{M}}^{f}$ is the same for every $\underline{p} \in \underline{S}$. If there exists $N \in {}_{\Lambda}\underline{\underline{M}}^{o}$ such that $N_{\underline{p}} = M(\underline{p})$ for almost all $\underline{p} \in \underline{S}$, then there exists $M \in {}_{\Lambda}\underline{\underline{M}}^{o}$ such that $M_{\underline{p}} = M(\underline{p})$ for every $\underline{p} \in \underline{S}$.

(iv) If $M, N \in {}_{\Lambda}\underline{\underline{M}}^{o}$ are such that $KM = KN$, and $M \supset N$, then

$$M/N \cong \bigoplus_{\underline{p} \in \underline{S}} M_{\underline{p}}/N_{\underline{p}} \quad \text{as} \quad \Lambda\text{-modules.}$$

The <u>proof</u> follows easily from (I, (8.6) - I, (8.9)) and is left as an exercise. #

1.9 <u>Theorem</u>: Let $\Lambda_{\underline{p}}$ be an $R_{\underline{p}}$-order in A and let $L \in {}_{A}\underline{\underline{M}}^{f}$ be fixed. Then there exists a one-to-one, inclusion preserving correpondence between $\underline{\underline{M}}_{p}(L) = \{M \in {}_{\Lambda_{\underline{p}}}\underline{\underline{M}}^{o} : KM \cong L\}$ and $\underline{\underline{M}}_{p}(\hat{L}_{\underline{p}}) = \{\hat{M} \in {}_{\hat{\Lambda}_{\underline{p}}}\underline{\underline{M}}^{o} : \hat{K}_{\underline{p}}\hat{M} \cong \hat{L}_{\underline{p}}\}$. The correspndence is given by

$$\begin{array}{ccc} \underline{\underline{M}}_{p}(L) & & \underline{\underline{M}}_{p}(\hat{L}_{\underline{p}}) \\ M & \longrightarrow & \hat{M}_{\underline{p}} \\ M = L \cap \hat{M} & \longleftarrow & \hat{M} \end{array}$$

The <u>proof</u> follows easily from (I, (9.14)) and is left as an exercise. #

1.10 <u>Corollary</u>: (1.8) remains valid if the localizations are replaced by the completions; in particular, (iv) implies $M_{\underline{p}}/N_{\underline{p}} \cong \hat{M}_{\underline{p}}/\hat{N}_{\underline{p}}$.

1.11 Definitions: Let Λ be an R-order in A.

(i) $M \in {}_{\Lambda}\underline{\underline{M}}^{o}$ is called reducible if there exists $N \subset M$ such that $\mathrm{rank}_R(N) < \mathrm{rank}_R(M)$, where $\mathrm{rank}_R(X) = \dim_K(KX)$, for $X \in {}_{R}\underline{\underline{M}}^{o}$. Observe, that $N \in {}_{\Lambda}\underline{\underline{M}}^{o}$.

(ii) $M \in {}_{\Lambda}\underline{\underline{M}}^{o}$ is said to decompose if there exist $M_1, M_2 \in {}_{\Lambda}\underline{\underline{M}}^{o}$, $M_1, M_2 \neq 0$ such that $M = M_1 \oplus M_2$.

Remark: It should be observed that irreducibility and simplicity are two different concepts for Λ-modules: A Λ-lattice M can never be a simple Λ-module since Λ is not artinian (if $0 \neq r \in R$ is a non-unit, then $M \supsetneqq Mr$).

1.12 Lemma: Let $L \in {}_{\Lambda}\underline{\underline{M}}^{f}$. Then there exists $M \in {}_{\Lambda}\underline{\underline{M}}^{o}$ such that $KM = L$.

Proof: Let $\{v_i\}_{1 \le i \le n}$ be a K-basis for L, and put

$$M = \sum_{i=1}^{n} \Lambda v_i = \{\sum_{i=1}^{n} \lambda_i v_i : \lambda_i \in \Lambda\}.$$

Then $M \in {}_{\Lambda}\underline{\underline{M}}^{f}$ and since $M \subset L$ and $\Lambda \in {}_{R}\underline{\underline{M}}^{f}$, $M \in {}_{\Lambda}\underline{\underline{M}}^{o}$. Moreover, $KM = L$. #

1.13 Lemma: Let $M \in {}_{\Lambda}\underline{\underline{M}}^{o}$. Then M is irreducible if and only if KM is a simple A-module.

Proof: If M is reducible then KM cannot be simple. Conversely, assume that KM is not simple, say $0 \neq L \subsetneqq KM$. Then

$$N = L \cap M \in {}_{\Lambda}\underline{\underline{M}}^{o}$$

is an R-pure submodule of M (cf. I, (7.4) and the proof of I, (7.3)). Moreover, $\mathrm{rank}_R(N) < \mathrm{rank}_R(M)$, and M is reducible.#

1.14 **Lemma**: Let $\Lambda_1 \subset \Lambda_2$ be two R-orders in A. For $M \in {}_{\Lambda_2}\underline{M}^f$, $N \in {}_{\Lambda_2}\underline{M}^o$ we have $\mathrm{Hom}_{\Lambda_1}(M,N) = \mathrm{Hom}_{\Lambda_2}(M,N)$.

Proof: Trivially, $\mathrm{Hom}_{\Lambda_1}(M,N) \supset \mathrm{Hom}_{\Lambda_2}(M,N)$. For the other inclusion let $\varphi \in \mathrm{Hom}_{\Lambda_1}(M,N)$. For $\lambda \in \Lambda_2$ pick $0 \neq r \in R$ such that $r\lambda \in \Lambda_1$ (cf. Ex. 1.4). Then $r(\lambda(m\varphi)) = (r\lambda)(m\varphi) = (r\lambda m)\varphi = r((\lambda m)\varphi)$, $\forall m \in M$; i.e., $r(\lambda(m\varphi)) = r((\lambda m)\varphi)$. Since N is R-torsion-free, this implies $\lambda(m\varphi) = (\lambda m)\varphi$; i.e., $\varphi \in \mathrm{Hom}_{\Lambda_2}(M,N)$. #

1.15 **Lemma**: Let Λ be an R-order in A and $M \in {}_{\Lambda}\underline{M}^o$. Then $\mathrm{End}_A(KM)$ is a separable K-algebra, and $\mathrm{End}_\Lambda(M)$ is an R-order in $\mathrm{End}_A(KM)$.

Proof: Let $KM \cong \bigoplus_{i=1}^{s} L_i^{(\alpha_i)}$, where $\{L_i\}_{1\leq i\leq s}$ are non-isomorphic simple left A-modules. Then $\mathrm{End}_A(KM) \underset{\text{ring}}{\cong} \bigoplus_{i=1}^{s} (D_i)_{\alpha_i}$, where $D_i = \mathrm{End}_A(L_i)$, $1 \leq i \leq s$, are skewfields over K. By Wedderburn's structure theorem (III, (5.5)), $A \underset{\text{ring}}{\simeq} \bigoplus_{i=1}^{s} (D_i)_{n_i} \oplus A_o$. Since the separability of A implies that the center of D_i is separable (cf. III, field extension of K, $1 \leq i \leq s$, $\mathrm{End}_A(KM)$ is separable (cf. III, Ex. 6,3; Ex. 6,7). To show that $\mathrm{End}_\Lambda(M)$ is an R-order in $\mathrm{End}_A(KM)$, we observe that $\mathrm{End}_\Lambda(M) \in {}_{R}\underline{M}^f$ (cf. III, Ex. 1,3) and $K \otimes_R \mathrm{End}_\Lambda(M) \cong \mathrm{End}_A(KM)$ (cf. III, (1.3)). But $K \otimes_R \mathrm{End}_\Lambda(M) = K\cdot\mathrm{End}_\Lambda(M)$ (cf. I, (6.4)). #

Exercises §1:

In these exercises, R is a Dedekind domain with quotient field K, A a separable finite dimensional K-algebra and Λ an R-order in A.

1. Let B be a finite dimensional commutative K-algebra, which

is at the same time an integral domain. Show that B is a field. (Hint: Use I, Ex. 4,6 and III, (5.5)).

2.) Let Λ_1 and Λ_2 be R-orders in A. Show that there exists $0 \neq r \in R$, such that $r\Lambda_1 \subset \Lambda_2$.

§2. The method of lifting idempotents for orders over a complete Dedekind domain

It is proved, that over a complete Dedekind domain $\hat{R}$, every $\hat{R}$-order is semi-perfect.

Let R be a Dedekind domain with quotient field K, $\underline{p}$ a fixed prime ideal in R and $\hat{R}$, with quotient field $\hat{K}$ and radical $\hat{\pi}\hat{R}$, the $\underline{p}$-adic completion of R under the $\underline{p}$-adic topology (cf. I, §9). Let $\hat{A}$ be a finite dimensional $\hat{K}$-algebra and $\hat{\Lambda}$ an $\hat{R}$-order in $\hat{A}$.

2.1 Theorem: $\hat{\Lambda}$ is semi-perfect.

We shall prove the more general:

2.2 Theorem: Let $0 \neq \hat{I}$ be a two-sided $\hat{\Lambda}$-ideal in $\hat{A}$, contained in $\operatorname{rad}\hat{\Lambda}$. Then $\hat{\Lambda}/\hat{I}$ is noetherian and artinian, and the idempotents in $\hat{\Lambda}/\hat{I}$ can be lifted to idempotents in $\hat{\Lambda}$.

Proof: $\hat{\Lambda}/\hat{I}$ is noetherian as homomorphic image of a noetherian ring (cf. I,(4.3)). To show that it is also artinian, we observe, that $\hat{R} \underset{\neq}{\supset} \hat{I} \cap \hat{R} \cdot 1 \neq 0$ is a proper ideal in $\hat{R}$. Since $\hat{R}$ is a local principal ideal domain with radical $\hat{\pi}\hat{R}$ (cf. I,(9.13)), $\hat{R} \cap \hat{I} = \hat{\pi}^n\hat{R}$ for some $n \in \underline{\underline{N}}$. But $\hat{R}/\hat{\pi}^n\hat{R}$ is an artinian ring, and since $\hat{\Lambda}/\hat{\pi}^n\hat{\Lambda}$ is a finite $\hat{R}/\hat{\pi}^n\hat{R}$-algebra, it is also artinian (cf. Ex. 2,3). Consequently, $\hat{\Lambda}/\hat{I}$, as a homomorphic image of $\hat{\Lambda}/\hat{\pi}^n\hat{\Lambda}$, is artinian (cf. I,(4.3)). For this we did not need that $\hat{I} \subset \operatorname{rad}\hat{\Lambda}$, only $\hat{K}\hat{I} = \hat{A}$. Since $\hat{\pi}\hat{R}$ is hausdorff (cf. I,(9.6)) and, since $\hat{\Lambda}$ is $(\hat{\pi}\hat{R})$-hausdorff, $\hat{\Lambda} = \varprojlim \hat{\Lambda}/(\hat{\pi}\hat{\Lambda})^n$ (cf. I,(9.8)). Above we have seen that $\hat{I} \supset \hat{\pi}^n\hat{\Lambda}$; conversely, since $\hat{I} \subset \operatorname{rad}\hat{\Lambda}$, we have $(\hat{I} + \hat{\pi}\hat{\Lambda})/\hat{\pi}\hat{\Lambda} \subset \operatorname{rad}(\hat{\Lambda}/\hat{\pi}\hat{\Lambda})$ (cf. I, Ex. 4,5); but $\hat{\Lambda}/\hat{\pi}\hat{\Lambda}$ is artinian and noetherian. Thus $\operatorname{rad}(\hat{\Lambda}/\hat{\pi}\hat{\Lambda})$ is nilpotent, (cf. I, Ex. 4,6), and hence there exists $m \in \underline{\underline{N}}$ with $I^m \subset \hat{\pi}\hat{\Lambda}$. We now apply (I,(9.11)), to conclude, that $\hat{\Lambda} = \varprojlim \hat{\Lambda}/\hat{I}^n$. Finally we show that the idempotents from $\hat{\Lambda}/\hat{I}$ can be lifted. We recall that we have a chain of natural epimorphisms

$$\hat\Lambda = \varprojlim \hat\Lambda/\hat I^n \ldots \hat\Lambda/\hat I^n \longrightarrow \hat\Lambda/\hat I^{n-1} \longrightarrow \ldots \longrightarrow \hat\Lambda/\hat I^2 \xrightarrow{\varphi_{21}} \hat\Lambda/\hat I.$$

If $\bar e$ is an idempotent in $\hat\Lambda/\hat I$, we choose $x \in \hat\Lambda/\hat I^2$ with $\varphi_{2,1} : x \longmapsto \bar e$. Then $z = x^2 - x \in \ker \varphi_{2,1}$, and $(x-z)^2 = e_2$ is an idempotent in $\hat\Lambda/\hat I^2$ such that $\varphi_{2,1} : e_2 \longmapsto \bar e$ (cf. III, Ex. 7,3). Continuing this way, we construct a family of idempotents e_i in $\hat\Lambda/\hat I^i$ such that $\varphi_{i,1} : e_i \longmapsto \bar e$. Then $\varphi_i \sigma_i : \hat\Lambda \longrightarrow \hat\Lambda/\hat I^i$, where σ_i is right multiplication by e_i, satisfies (I,(9.2)), and hence there exists a unique endomorphism $\sigma : \hat\Lambda \longrightarrow \hat\Lambda$ completing the diagram

$$\begin{array}{ccc} \hat\Lambda & \overset{\sigma}{\dashrightarrow} & \hat\Lambda \\ {\scriptstyle\varphi_i}\downarrow & & \downarrow{\scriptstyle\varphi_i} \\ \hat\Lambda/\hat I^i & \xrightarrow{\ \sigma_i\ } & \hat\Lambda/\hat I^i \end{array}$$

and it is easily checked, that $\sigma^2 = \sigma$; i.e., there exists an idempotent e in $\hat\Lambda$ such that $\varphi_1 : e \longmapsto \bar e$.

2.3 Corollary: Under the hypotheses of (2.2), if $e \in \hat\Lambda$ is an idempotent such that $\varphi_1 : e \longmapsto \bar e_1 + \bar e_2$, where $\bar e_1$ and $\bar e_2$ are orthogonal idempotents in $\hat\Lambda/\hat I$, then there exist orthogonal idempotents $e_1, e_2 \in \hat\Lambda$ such that $\varphi_1 : e_i \longmapsto \bar e_i$, $i=1,2$.

Proof: We choose $x' \in \hat\Lambda/\hat I^2$ such that $\varphi_{21} : x' \longmapsto \bar e_1$. Now we put $x = (e\varphi_2)x'(e\varphi_2)$. Then $e_1^{(2)} = (x-z)^2$ with $z = x^2 - x$ is an idempotent in $\hat\Lambda/\hat I^2$ such that $(e\varphi_2)e_1^{(2)} = e_1^{(2)}(e\varphi_2)$. If we put now $e_2^{(2)} = (e\varphi_2) - e_1^{(2)}$, then $e\varphi_2 = e_1^{(2)} + e_2^{(2)}$ and $e_1^{(2)}$ and $e_2^{(2)}$ are idempotent. Now we continue as in the proof of (2.2) to obtain the desired result. #

2.4 Lemma: If $\hat I$ is a two-sided $\hat\Lambda$-ideal in $\hat\Lambda$ and $\hat I \subset \operatorname{rad}\hat\Lambda$; then

$$\hat\Lambda/\operatorname{rad}\hat\Lambda \cong \hat\Lambda/\hat I \big/ \operatorname{rad}(\hat\Lambda/\hat I).$$

Proof: By I,(4,17) we have $\operatorname{rad}\hat\Lambda/\hat I = \operatorname{rad}(\hat\Lambda/\hat I)$, from which the lemma follows immediately. #

2.5 **Remark**: (2.4) does not depend on the fact, that $\hat{\Lambda}$ is complete; i.e., it is valid also for $\Lambda_{\underline{p}}$ etc.

2.6 **Lemma**: Let $\underline{p} \in \underline{S}$ and denote by $\Lambda_{\underline{p}}$ the localization of the R-order Λ in A at $\underline{p}$. Then $\underline{p}\Lambda_{\underline{p}} \subset \text{rad } \Lambda_{\underline{p}}$, and hence $\psi: \Lambda_{\underline{p}} \longrightarrow \Lambda_{\underline{p}}/\underline{p}\Lambda_{\underline{p}}$ is an essential epimorphism (cf. III,(7.2)).

Proof: It suffices to show, that $\underline{p}\Lambda_{\underline{p}}$ is contained in all maximal left $\Lambda_{\underline{p}}$-ideals (cf. I,(4.15)). If not, there would exist a maximal left $\Lambda_{\underline{p}}$-ideal $I_{\underline{p}}$ such that $I_{\underline{p}} + \underline{p}\Lambda_{\underline{p}} = \Lambda_{\underline{p}}$. But $\Lambda_{\underline{p}} \in {}_{R_{\underline{p}}}\underline{M}^f$, so this contradicts Nakayama's lemma (cf. I,(4.18)). #

Exercises §2:

1.) Let $A = \oplus_{i=1}^{n} (K)_{n_i}$ and let $\Lambda_{\underline{p}}$ be an $R_{\underline{p}}$-order in A. Show, that $\Lambda_{\underline{p}}$ is semi-perfect. (Hint: Show that for every idempotent $\hat{e}$ in $\hat{\Lambda}_{\underline{p}}$, there exists an idempotent e in $\Lambda_{\underline{p}}$ such that $\hat{\Lambda}_{\underline{p}}e = \hat{\Lambda}_{\underline{p}}\hat{e}$. Now, use (2.1).)

2.) If A is central simple, $A = (D)_n$, and if $\hat{K} \otimes_K D$ is a skewfield, show that every $R_{\underline{p}}$-order $\Lambda_{\underline{p}}$ in A is semi-perfect.

3.) Let S be an artinian and noetherian commutative ring and B a finite S-algebra. Show that B is left artinian.

§3. Projective lattices and progenerators over orders

It is shown, that projective lattices and progenerators are preserved under localization, completion and reduction modulo a prime ideal $\underline{p}$ in R. For projective lattices the equivalence of local isomorphism, isomorphism over the completion and isomorphism modulo $\underline{p}$ is proved. Morita equivalences preserve irreducible, indecomposable and projective lattices.

Let R be a Dedekind domain with quotient field K and A a finite dimensional separable K-algebra. We use the notation of (1.6).

3.1 Theorem: $M \in {}_{\Lambda}\underline{M}^{o}$ is projective (resp. a generator) if and only if $M_{\underline{p}} \in {}_{\Lambda_{\underline{p}}}\underline{M}^{o}$ is projective (resp. a generator) for every $\underline{p} \in \underline{S}$.

Proof: Since $R_{\underline{p}} \otimes -$ preserves direct sums, localization preserves projective modules as well as generators (cf. I,(2.9), III,(1.10)). Conversely, assume that $M_{\underline{p}} \in {}_{\Lambda_{\underline{p}}}\underline{P}^{f}, \forall \underline{p} \in \underline{S}$. By (I, Ex. 8,3 and III, Ex. 1,3)

3.1'
$$\mathrm{Ext}^1_{\Lambda}(M,X) = E_o \oplus (\oplus_{\underline{p} \in \underline{S}} E_{\underline{p}}),$$

where E_o is an R-lattice and $\{E_{\underline{p}}\}_{\underline{p} \in \underline{S}}$ are the $\underline{p}$-primary components of the torsion part of $\mathrm{Ext}^1_{\Lambda}(M,X)$ (cf. I,(8.9)). Since $R_{\underline{q}} \otimes_R E_{\underline{p}} = E_{\underline{p}}$ or 0, depending on whether $\underline{p} = \underline{q}$ or not, and by (III,(1.2), I,(6.4)), we obtain from our assumption $\mathrm{Ext}^1_{\Lambda_{\underline{p}}}(M_{\underline{p}},X_{\underline{p}}) = E_{o\underline{p}} \oplus E_{\underline{p}}$ $= 0$ for every $\underline{p} \in \underline{S}$ (cf. II, (4.2)). Hence $(E_o)_{\underline{p}} = 0$ as well as $E_{\underline{p}} = 0$, for all $\underline{p} \in \underline{S}$. Now since $E_{o\underline{p}} \neq 0$, $\forall \underline{p} \in \underline{S}$, E_o being an R-lattice, unless $E_o = 0$, we conclude from (3.1') that $\mathrm{Ext}^1_{\Lambda}(M,X) = 0$, $\forall X \in {}_{\Lambda}\underline{M}^{f}$, i.e., $\underline{M} \in {}_{\Lambda}\underline{P}^{f}$.

Assume now, that all localizations $M_{\underline{p}}$ of $M \in {}_{\Lambda}\underline{M}^{o}$ are generators. For $X \in {}_{\Lambda}\underline{M}^{f}$, we write $X = X_o \oplus (\bigoplus_{i=1}^{n} X_{\underline{p}_i})$, where $X_o \in {}_{\Lambda}\underline{M}^{o}$ and $X_{\underline{p}_i}$ are the primary components of X, (observe that the above decomposition is also a decomposition as Λ-modules). Given now $0 \neq \varphi \in \mathrm{Hom}_\Lambda(X,X')$, we write $\varphi = \varphi_o \oplus (\oplus_{i=1}^n \varphi_i)$, $\varphi_i \in \mathrm{Hom}_\Lambda(X_{\underline{p}_i},X')$, $\varphi_o \in \mathrm{Hom}_\Lambda(X_o,X')$. If $\varphi_o \neq 0$, then the commutative diagram

$$\begin{array}{ccc} \mathrm{Hom}_\Lambda(M,X_o) & \hookrightarrow & \mathrm{Hom}_{\Lambda_{\underline{p}}}(M_{\underline{p}},X_{o_{\underline{p}}}) \\ \downarrow \mathrm{hom}(1_M,\varphi_o) & & \downarrow \mathrm{hom}(1_{M_{\underline{p}}},\varphi_{o_{\underline{p}}}) \\ \mathrm{Hom}_\Lambda(M,X') & \longrightarrow & \mathrm{Hom}_{\Lambda_{\underline{p}}}(M_{\underline{p}},X'_{\underline{p}}) \end{array}$$

shows $\mathrm{hom}(1_M,\varphi_o) \neq 0$ since $\varphi_{o_{\underline{p}}} \neq 0$ for all $\underline{p}$; i.e., $\mathrm{hom}(1_M,\varphi) \neq 0$. If $\varphi_o = 0$, then for some i, $\varphi_i \neq 0$, and

$$\mathrm{Hom}_\Lambda(M,X_{\underline{p}_i}) \hookrightarrow \mathrm{Hom}_{\Lambda_{\underline{p}_i}}(M_{\underline{p}_i},X_{\underline{p}_i}).$$

Thus a similar diagram as above shows that $\mathrm{hom}(1_M,\varphi_i) \neq 0$; i.e., $\mathrm{hom}(1_M,\varphi) \neq 0$, and M is a generator (cf. III,(1.10)). #

3.2 <u>Theorem</u>: $M_{\underline{p}} \in {}_{\Lambda_{\underline{p}}}\underline{M}^o$ is projective (resp. a generator) if and only if $\hat{M}_{\underline{p}} \in {}_{\hat{\Lambda}_{\underline{p}}}\underline{M}^o$ is projective (resp. a generator).

<u>Proof</u>: By (I,(9.13)), $\hat{R}_{\underline{p}} \otimes_{R_{\underline{p}}} -$ is an exact functor on ${}_{\Lambda_{\underline{p}}}\underline{M}^f$; moreover, $\hat{R}_{\underline{p}} \otimes_{R_{\underline{p}}} -$ is also a faithful functor on ${}_{\Lambda_{\underline{p}}}\underline{M}^f$ (cf. I,(9.12) and I,(9.1)). Now the result follows from (III,(1.11)). (It should be observed, that in the proof of (III,(1.11)) only the fact, that $B \otimes_S -$ is a faithful exact functor on ${}_{\underline{C}}\underline{M}^f$ was used.) #

3.3 <u>Remark</u>: For $\underline{p} \in \underline{S}$, $R/\underline{p}$ is a field, and thus $\bar{\Lambda}_{\underline{p}} = \Lambda_{\underline{p}}/\underline{p}\Lambda_{\underline{p}} = \Lambda/\underline{p}\Lambda$ is a finite dimensional $\bar{R}_{\underline{p}} = R/\underline{p}R = R_{\underline{p}}/\underline{p}R_{\underline{p}}$-algebra. Moreover, $\bar{M}_{\underline{p}} = M/\underline{p}M \overset{nat}{\cong} \bar{R}_{\underline{p}} \otimes_R M, \forall M \in {}_\Lambda\underline{M}^o$ (cf. I,(3.18)).

3.4 <u>Theorem</u>: $M_{\underline{p}} \in {}_{\Lambda_{\underline{p}}}\underline{M}^o$ is projective (resp. faithfully projective), if and only if $\bar{M}_{\underline{p}} = M_{\underline{p}}/\underline{p}M_{\underline{p}} \in {}_{\bar{\Lambda}_{\underline{p}}}\underline{M}^f$ is projective (resp. faithfully projective.)

<u>Proof</u>: We have $\bar{M}_{\underline{\underline{p}}} \overset{nat}{\cong} \bar{R}_{\underline{\underline{p}}} \otimes_{R_{\underline{\underline{p}}}} M_{\underline{\underline{p}}}$ (cf. I,(3.18)). If $M_{\underline{\underline{p}}}$ is projective (resp. faithfully projective), then $\bar{M}_{\underline{\underline{p}}}$ is projective (resp. faithfully projective), since $\bar{R}_{\underline{\underline{p}}} \otimes_{R_{\underline{\underline{p}}}} -$ is an additive functor mapping free modules into free modules.

<u>Conversely</u>, let $\bar{M}_{\underline{\underline{p}}} \in {}_{\bar{\Lambda}_{\underline{\underline{p}}}}\underline{\underline{P}}^f$. Since we have a unitary ring homomorphism $\Psi : \Lambda_{\underline{\underline{p}}} \longrightarrow \bar{\Lambda}_{\underline{\underline{p}}}$, we obtain from the change of ring theorem (II,(4.6)) $hd_{\Lambda_{\underline{\underline{p}}}}(\bar{M}_{\underline{\underline{p}}}) \leq hd_{\bar{\Lambda}_{\underline{\underline{p}}}}(\bar{M}_{\underline{\underline{p}}}) + hd_{\Lambda_{\underline{\underline{p}}}}(\bar{\Lambda}_{\underline{\underline{p}}})$; i.e., $hd_{\Lambda_{\underline{\underline{p}}}}(\bar{M}_{\underline{\underline{p}}}) \leq hd_{\Lambda_{\underline{\underline{p}}}}(\bar{\Lambda}_{\underline{\underline{p}}})$. However, the exact sequence $0 \longrightarrow \Lambda_{\underline{\underline{p}}} \xrightarrow{\varphi} \Lambda_{\underline{\underline{p}}} \xrightarrow{\Psi} \bar{\Lambda}_{\underline{\underline{p}}} \longrightarrow 0$, where φ is multiplication by π, $\pi R_{\underline{\underline{p}}} = \text{rad } R_{\underline{\underline{p}}}$, (cf. I,(8.3)), implies $hd_{\Lambda_{\underline{\underline{p}}}}(\bar{\Lambda}_{\underline{\underline{p}}}) = 1 + hd_{\Lambda_{\underline{\underline{p}}}}(\Lambda_{p})$ (cf. II,(4.5)); i.e., $hd_{\Lambda_{\underline{\underline{p}}}}(\bar{\Lambda}_{\underline{\underline{p}}}) = 1$. Thus $hd_{\Lambda_{\underline{\underline{p}}}}(\bar{M}_{\underline{\underline{p}}}) \leq 1$. The exact sequence of $\Lambda_{\underline{\underline{p}}}$-modules

$$0 \longrightarrow M_{\underline{\underline{p}}} \xrightarrow{\sigma} M_{\underline{\underline{p}}} \xrightarrow{\tau} \bar{M}_{\underline{\underline{p}}} \longrightarrow 0,$$

where σ is multiplication by π and τ is the canonical homomorphism, gives rise to the exact sequence (cf. II,(3.10)), for $N_{\underline{\underline{p}}} \in {}_{\Lambda_{\underline{\underline{p}}}}\underline{\underline{M}}^f$,

$$\text{Ext}^1_{\Lambda_{\underline{\underline{p}}}}(\bar{M}_{\underline{\underline{p}}}, N_{\underline{\underline{p}}}) \longrightarrow \text{Ext}^1_{\Lambda_{\underline{\underline{p}}}}(M_{\underline{\underline{p}}}, N_{\underline{\underline{p}}}) \xrightarrow{\sigma^*} \text{Ext}^1_{\Lambda_{\underline{\underline{p}}}}(M_{\underline{\underline{p}}}, N_{\underline{\underline{p}}}) \longrightarrow 0,$$

since $hd_{\Lambda_{\underline{\underline{p}}}}(\bar{M}_{\underline{\underline{p}}}) \leq 1$ (cf. II,(4.3)). However, σ^* is induced from the multiplication by π, and thus, σ^* itself is multiplication by π (cf. II,(2.2) and II,(3.4)). Thus, $\pi \text{Ext}^1_{\Lambda_{\underline{\underline{p}}}}(M_{\underline{\underline{p}}}, N_{\underline{\underline{p}}}) = \text{Ext}^1_{\Lambda_{\underline{\underline{p}}}}(M_{\underline{\underline{p}}}, N_{\underline{\underline{p}}})$. But by Nakayama's lemma (I,(4.18)), this implies $\text{Ext}^1_{\Lambda_{\underline{\underline{p}}}}(M_{\underline{\underline{p}}}, N_{\underline{\underline{p}}}) = 0$ (cf. III, Ex. 1,3); i.e., $M_{\underline{\underline{p}}} \in {}_{\Lambda_{\underline{\underline{p}}}}\underline{\underline{P}}^f$ (cf. II, Ex. 4,2).

Finally let $\bar{M}_{\underline{\underline{p}}}$ be a $\bar{\Lambda}_{\underline{\underline{p}}}$-progenerator. Then, by (III,(1.10)), there exists a $\bar{\Lambda}_{\underline{p}}$-epimorphism $\bar{\varphi} : \bar{M}_{\underline{p}}^{(n)} \longrightarrow \bar{\Lambda}_{\underline{p}}$, for some $n \in \underline{N}$. However $\bar{\varphi}$ is also a $\Lambda_{\underline{p}}$-epimorphism, because of the ring epimorphism $\Lambda_{\underline{p}} \longrightarrow \bar{\Lambda}_{\underline{p}}$. Since with $M_{\underline{p}}$ also $M_{\underline{p}}^{(n)}$ has already been proven $\Lambda_{\underline{p}}$-projective, we obtain the following commutative diagram of $\Lambda_{\underline{p}}$-maps:

$$\begin{array}{ccccc} & & M_{\underline{p}}^{(n)} & & \\ & \swarrow^{\alpha} & \downarrow \tau & & \\ & & M_{\underline{p}}^{(n)} & & \\ & & \downarrow \varphi & & \\ \Lambda_{\underline{p}} & \xrightarrow{\ \psi\ } & \bar{\Lambda}_{\underline{p}} & \longrightarrow & 0 , \end{array}$$

and it remains to show that α is epic, using the fact that $\alpha\psi = \tau\varphi$ is epic. But this follows from III,(7.1),(7.2) since ψ is an essential epimorphism (cf. (2.6)). #

3.5 <u>Theorem</u>: Let $P_1, P_2 \in {}_{\Lambda_{\underline{p}}}\underline{P}^f$. Then $P_1 \cong_{\Lambda_{\underline{p}}} P_2$ if and only if $\bar{P}_{1_{\underline{p}}} \cong_{\bar{\Lambda}_{\underline{p}}} \bar{P}_{2_{\underline{p}}}$.

<u>Proof</u>: Let $\varphi : \bar{P}_{1_{\underline{p}}} \xrightarrow{\ \sim\ } \bar{P}_{2_{\underline{p}}}$ be a $\bar{\Lambda}_{\underline{p}}$-isomorphism; then it is also a $\Lambda_{\underline{p}}$-isomorphism. If $\sigma_1 : P_1 \longrightarrow \bar{P}_{1_{\underline{p}}}$ and $\sigma_2 : P_2 \longrightarrow \bar{P}_{2_{\underline{p}}}$ are the canonical homomorphisms, then $P_1 \xrightarrow{\sigma_1\varphi} \bar{P}_{2_{\underline{p}}}$ and $P_2 \xrightarrow{\sigma_2} \bar{P}_{2_{\underline{p}}}$ are essential epimorphisms (cf. (2.5) and III,(7.2)); i.e., P_1 and P_2 are projective covers for $\bar{P}_{2_{\underline{p}}}$ (cf. III,(7.1)). From (III,(7.3)) we conclude $P_1 \cong_{\Lambda_{\underline{p}}} P_2$. The other direction is trivial. #

3.6 <u>Corollary</u>: Let $P_1, P_2 \in {}_{\Lambda_{\underline{p}}}\underline{P}^f$. Then $P_1 \cong_{\Lambda_{\underline{p}}} P_2$ if and only if $\hat{P}_{1_{\underline{p}}} \cong_{\hat{\Lambda}_{\underline{p}}} \hat{P}_{2_{\underline{p}}}$.

Proof: One sees at once that (3.5) remains valid if the localization is replaced by the completion. Let $\hat{P}_{1_{\underline{p}}} \cong_{\hat{\Lambda}_{\underline{p}}} \hat{P}_{2_{\underline{p}}}$; then, by (3.5), $\overline{\hat{P}}_{1_{\underline{p}}} \cong \overline{\hat{P}}_{2_{\underline{p}}}$. But from (1.10) we conclude that $\overline{P}_{1_{\underline{p}}} \cong \overline{P}_{2_{\underline{p}}}$, and another application of (3.5) shows that $P_1 \cong_{\Lambda_{\underline{p}}} P_2$. #

3.7 Theorem: Let Λ be an R-order in A and $E \in {}_{\Lambda}\underline{\underline{M}}^{o}$ a progenerator (cf. III,(1.9)). If we put $\Omega = \mathrm{End}_{\Lambda}(E)$, then the Morita equivalence between ${}_{\Lambda}\underline{\underline{M}}^{f}$ and ${}_{\Omega}\underline{\underline{M}}^{f}$ (cf. III,(2.1))

(i) preserves lattices,

(ii) preserves inclusions,

(iii) preserves irreducible and indecomposable lattices,

(iv) preserves projective lattices and progenerators.

(v) $KE \in {}_{A}\underline{\underline{M}}^{f}$ is a progenerator.

(vi) For a fixed $\underline{p} \in \underline{\underline{S}}$ we donote by "–" reduction modulo $\underline{p}$. Then $\overline{E} \in {}_{\overline{\Lambda}}\underline{\underline{M}}^{o}$ is a progenerator, and for $M \in {}_{\Lambda}\underline{\underline{M}}^{o}$ we have a natural isomorphism

$$\overline{\mathrm{Hom}_{\Lambda}(E,M)} \cong \mathrm{Hom}_{\overline{\Lambda}}(\overline{E},\overline{M}),$$

and $\overline{\Omega} \cong \mathrm{End}_{\overline{\Lambda}}(\overline{E})$ as rings.

Proof: From (1.15) we know that Ω is an R-order in $\mathrm{End}_{A}(KM)$. Thus, to show (i) it suffices to prove that $\mathrm{Hom}_{\Lambda}(M_1,M_2)$ is R-torsion-free if $M_1,M_2 \in {}_{\Lambda}\underline{\underline{M}}^{o}$ (cf. Ex. 1,3). But if $\varphi \in \mathrm{Hom}_{\Lambda}(M_1,M_2)$ is such that $r\varphi = 0$ for some $0 \neq r \in R$, then $r\mathrm{Im}\,\varphi = 0$; i.e., $\mathrm{Im}\,\varphi = 0$, since M_2 is an R-lattice. Hence $\mathrm{Hom}_{\Lambda}(M_1,M_2)$ is an R-lattice. (ii) follows from (III,(2.1)) and (iv) from (III,(1.10) and the proof of III,(1.11), since $h^{E} \sim E^{*} \otimes_{\Lambda} -$ is a faithful exact functor. (v) is clear because KE is a faithful projective A-module (cf. III, Ex. 6,8). But then (iii) follows from (III,(2.1) and IV,(1.13)). (3.1) and (3.4) imply that $\overline{E}$ is a progenerator. Finally to prove the remaining part of (vi) we establish the more general result:

3.8 Theorem: Let S be a noetherian ring and I a two-sided ideal in S such that S/I is artinian. If $E \in {}_S\underline{\underline{M}}^f$ is a progenerator then $E/IE \in {}_{S/I}\underline{\underline{M}}^f$ is a progenerator, and for $M \in {}_S\underline{\underline{M}}^f$ we have a natural isomorphism

$$\mathrm{Hom}_{S/I}(E/IE, M/IM) \cong \mathrm{Hom}_S(E,M)/\mathrm{Hom}_S(E,IM).$$

Moreover, if I = rad S, then $\mathrm{Hom}_S(E,IE) \cong \mathrm{rad}\, T$, where $T = \mathrm{End}_S(E)$.

Proof: Since $M/IM \overset{\text{nat}}{\cong} S/I \otimes_R M$ and since $S/I \otimes_R -$ is an additive functor, $E/IE \in {}_S\underline{\underline{M}}^f$ is a progenerator. For $M \in {}_S\underline{\underline{M}}^f$ we have the exact sequence

$$0 \longrightarrow IM \longrightarrow M \longrightarrow M/IM \longrightarrow 0.$$

But E is projective, and we obtain the exact sequence

$$0 \longrightarrow \mathrm{Hom}_S(E,IM) \longrightarrow \mathrm{Hom}_S(E,M) \longrightarrow \mathrm{Hom}_S(E,M/IM) \longrightarrow 0;$$

i.e., $\mathrm{Hom}_S(E,M/IM) \cong \mathrm{Hom}_S(E,M)/\mathrm{Hom}_S(E,IM)$. However, for every $\varphi \in \mathrm{Hom}_S(E,M/IM)$, $\mathrm{Ker}\,\varphi \supset IE = \mathrm{Ker}\,\psi$, where ψ is the canonical epimorphism $E \longrightarrow E/IE$. Hence there exists (cf. I, Ex. 2,3) a unique $\tau \in \mathrm{Hom}_S(E/IE,M/IM)$ such that $\varphi = \psi\tau$. Consequently

$$\mathrm{Hom}_S(E,M/IM) = \mathrm{Hom}_S(E/IE,M/IM) = \mathrm{Hom}_{S/I}(E/IE,M/IM),$$

and for every M we have a natural isomorphism

$$\mathrm{Hom}_{S/I}(E/IE,M/IM) \cong \mathrm{Hom}_S(E,M)/\mathrm{Hom}_S(E,IM).$$

In particular, for M = E, this is a ring isomorphism.

If now I = rad S, then we have a Morita equivalence between

$$S/I \text{ and } \mathrm{End}_{S/I}(E/IE,E/IE) = T_1.$$

Since S/I is artinian and noetherian every finitely generated left module is projective. Thus, the same must hold for T_1; i.e., T_1 is semi-simple. The ring isomorphism

$$\mathrm{End}_{S/I}(E/IE) \cong \mathrm{End}_S(E)/\mathrm{Hom}_S(E,IE)$$

shows $\mathrm{Hom}_S(E,IE) \supset \mathrm{rad}\, T$. Conversely, the canonical epimorphism $\varphi: E \longrightarrow E/IE$ is an essential epimorphism (cf. III,(7.2)). But Morita equivalences preserve essential epimorphisms (cf. Ex. 3,4); hence

$$\mathrm{hom}(1_E, \varphi) : \mathrm{End}_S(E) \longrightarrow \mathrm{Hom}_S(E, E/IE)$$

is an essential epimorphism. Whence

$$\mathrm{Hom}_S(E, IE) = \mathrm{rad}\ T \quad (\text{cf. III, Ex. } 7{,}1). \quad \#$$

We now return to the proof of (3.7,vi). With (3.8) we get for $I = \underline{\underline{p}}\Lambda$

$$\mathrm{Hom}_{\bar{\Lambda}}(\bar{E},\bar{M}) = \mathrm{Hom}_{\Lambda}(E,M)/\mathrm{Hom}_{\Lambda}(E,\underline{\underline{p}}M).$$

But it is easily seen - using e.g. localizations - that

$\mathrm{Hom}_{\Lambda}(E,\underline{\underline{p}}M) = \underline{\underline{p}}\mathrm{Hom}_{\Lambda}(E,M)$. Thus $\mathrm{Hom}_{\bar{\Lambda}}(\bar{E},\bar{M}) \cong \overline{\mathrm{Hom}_{\Lambda}(E,M)}$. #

3.9 Corollary: Let $\Lambda^\#$ be an $R^\#$-order in A and $J^\# = \mathrm{rad}\,\Lambda^\#$, where $-^\#$ denotes the localization at some fixed prime ideal in R. If $E^\# \in {}_{\Lambda^\#}M^o$ is a progenerator then so is $E^\#/J^\#E^\#$. Moreover, we have a natural isomorphism of modules

$$\mathrm{Hom}_{\Lambda^\#/J^\#}(E^\#/J^\#E^\#, M^\#/J^\#M^\#) \cong \mathrm{Hom}_{\Lambda^\#}(E^\#,M^\#)/\mathrm{Hom}_{\Lambda^\#}(E^\#,J^\#M^\#),$$

and a ring isomorphism

$$\mathrm{End}_{\Lambda^\#/J^\#}(E^\#/J^\#E^\#) \cong \Omega^\#/\mathrm{rad}\,\Omega^\#,$$

where $\Omega^\# = \mathrm{End}_{\Lambda^\#}(M^\#)$. Moreover, $\mathrm{Hom}_{\Lambda^\#}(E^\#,J^\#M^\#) \supset (\mathrm{rad}\,\Omega^\#)\mathrm{Hom}_{\Lambda^\#}(E^\#,M^\#)$.

The proof is an immediate consequence of (3.8). #

Exercises §3:

1.) In $(\underline{\underline{Q}})_2$ - this is a separable $\underline{\underline{Q}}$-algebra by (III,(5.16) - we consider the $\underline{\underline{Z}}$-orders

$$\Lambda_1 = \left\{ \begin{pmatrix} z_1 & z_3 \\ z_2 & z_4 \end{pmatrix} : z_i \in \underline{\underline{Z}} \right\},$$

$$\Lambda_2 = \left\{ \begin{pmatrix} z_1 & p^{-1}z_3 \\ pz_2 & z_4 \end{pmatrix} : z_i \in \underline{\underline{Z}} \right\}, \text{ p a fixed rational prime,}$$

$$\Lambda_3 = \left\{ \begin{pmatrix} z_1 & z_3 \\ pz_2 & z_4 \end{pmatrix} : z_i \in \mathbb{Z} \right\}, \quad p \text{ a fixed rational prime.}$$

Show, that Λ_i, $1 \leq i \leq 3$, are $\mathbb{Z}$-orders in $(\mathbb{Q})_2$ and show, that $\Lambda_1 \cong \Lambda_2$ as rings.

2.) Show that in the examples above ${}_{\Lambda_i}\underline{P}^f \supset {}_{\Lambda_i}\underline{M}^o$, $1 \leq i \leq 3$.

(Hint: It suffices to show that every irreducible Λ_i-lattice is projective. To show this, one should observe that with every Λ_i-lattice one can assosiate a family of matrices via the following construction: Let $w_1, \ldots, w_n$ be a $\mathbb{Z}$-basis for $M \in {}_{\Lambda_i}\underline{M}^o$, then

$\lambda w_k = \sum_{j=1}^n z_{kj}(\lambda) w_j$, for every $\lambda \in \Lambda_i$.

Now, one associates with M the family of matrices $(z_{kj}(\lambda))$, $\lambda \in \Lambda_i$, and proves that $M' \cong M$ if and only if $(z_{kj}(\lambda)) \sim (z_{kj}(\lambda)')$, where $\underline{A} \sim \underline{B}$ means $\exists$ an invertible matrix $\underline{U}$ with entries in $\mathbb{Z}$ such that $(z_{kj}(\lambda))\underline{U} = \underline{U}(z_{kj}(\lambda)')$, $\forall \lambda \in \Lambda_i$ (cf. Introduction). Show that

(i) $$\Lambda_1 \begin{pmatrix} 1 & 0 \\ 0 & 0 \end{pmatrix} \cong \Lambda_1 \begin{pmatrix} 0 & 1 \\ 0 & 0 \end{pmatrix}$$

and that both modules are progenerators;

(ii) $$\Lambda_2 \begin{pmatrix} 1 & 0 \\ 0 & 0 \end{pmatrix} \cong \Lambda_2 \begin{pmatrix} 0 & 1 \\ 0 & 0 \end{pmatrix}$$

and that both modules are progenerators. Use (III,(2.1)) to conclude, that there exists - up to isomorphism - only one irreducible Λ_i-lattice, i=1,2.

(iii) $$\Lambda_3 \begin{pmatrix} 1 & 0 \\ 0 & 0 \end{pmatrix} \not\cong \Lambda_3 \begin{pmatrix} 0 & 1 \\ 0 & 0 \end{pmatrix}$$

and that both modules are projective Λ_3-modules. If now M is an irreducible Λ_3-lattice, show that either $\mathrm{End}_{\mathbb{Z}}(M) = \Lambda_1$ or $\mathrm{End}_{\mathbb{Z}}(M) = \Lambda_2$.

Use this to show that there are exactly two classes of non-isomorphic irreducible Λ_3-lattices.

3.) Show that

$$\Lambda = \left\{ \begin{pmatrix} z_1 & z_3 \\ z_2 & z_1 + pz_4 \end{pmatrix}, z_i \in \mathbb{Z} \right\}, \text{ p a rational prime number,}$$

has no projective irreducible Λ-lattices.

4.) Show that a Morita equivalence between two rings preserves essential epimorphisms (cf. III,(2.1), III,(7.1)).

§4. Maximal orders

The goal of this section is to prove that maximal orders exist and are hereditary; their two-sided ideals form a group under multiplication.

Let R be a Dedekind domain with quotient field K, and A a finite dimensional separable K-algebra. $\underline{\underline{S}}$ denotes the set of all prime ideals in R.

4.1 Definitions:

(i) An R-order Λ in A is called hereditary, if every $M \in {}_{\Lambda}\underline{\underline{M}}^{o}$ is projective.

(ii) An R-order Λ in A is called maximal, if it is not properly contained in any other R-order in A.

4.2 Lemma: The following conditions are equivalent for an R-order Λ in A:

(i) Λ is hereditary

(ii) $\Lambda_{\underline{\underline{p}}}$ is hereditary, for every $\underline{\underline{p}} \in \underline{\underline{S}}$

(iii) $\hat{\Lambda}_{\underline{\underline{p}}}$ is hereditary, for every $\underline{\underline{p}} \in \underline{\underline{S}}$.

Proof: (iii) $\Longrightarrow$ (i) by (3.1), and (iii) $\Longrightarrow$ (ii) by (3.2).

(i) $\Longrightarrow$ (ii) For a given $M(\underline{\underline{p}}) \in {}_{\Lambda_{\underline{\underline{p}}}}\underline{\underline{M}}^{o}$, $\underline{\underline{p}}$ fixed, we put $L=KM(\underline{\underline{p}}) \in {}_{\Lambda}\underline{\underline{M}}^{f}$.

Let $N \in {}_{\Lambda}\underline{\underline{M}}^{o}$ be a Λ-lattice such that $KN = L$ (cf. (1.12)).

By (1.8) there exists $M \in {}_{\Lambda}\underline{\underline{M}}^{o}$ such that $M_{\underline{\underline{p}}} = M(\underline{\underline{p}})$ and $M_{\underline{\underline{q}}} = N_{\underline{\underline{q}}}$, $\underline{\underline{p}} \neq \underline{\underline{q}} \in \underline{\underline{S}}$. Now, the statement follows from (3.1).

(ii) $\Longrightarrow$ (iii) Let $\hat{M} \in {}_{\hat{\Lambda}_{\underline{\underline{p}}}}\underline{\underline{M}}^{o}$, $\underline{\underline{p}}$ fixed. Then $\hat{K}_{\underline{\underline{p}}}\hat{M} \in {}_{\hat{A}_{\underline{\underline{p}}}}\underline{\underline{P}}^{f}$, since $\hat{A}_{\underline{\underline{p}}}$ is semi-simple, and there exists $\hat{L} \in {}_{\hat{A}_{\underline{\underline{p}}}}\underline{\underline{M}}^{f}$ such that $\hat{K}_{\underline{\underline{p}}}\hat{M} \oplus \hat{L} \cong_{\hat{A}_{\underline{\underline{p}}}} \hat{A}_{\underline{\underline{p}}}^{(n)}$.

Let $\hat{X} \in {}_{\hat{\Lambda}_{\underline{\underline{p}}}}\underline{\underline{M}}^{o}$ be such that $\hat{K}_{\underline{\underline{p}}}\hat{X} = \hat{L}$ (cf. (1.12)). Then

${}_A A^{(n)} \cap (\hat{M} \oplus \hat{X}) \in {}_{\Lambda_{\underline{\underline{p}}}}\underline{\underline{M}}^o$, is such that $\hat{R}_{\underline{\underline{p}}} \otimes_{R_{\underline{\underline{p}}}} ({}_A A^{(n)} \cap (\hat{M} \oplus \hat{X})) \cong \hat{M} \oplus \hat{X}$ (cf. (1.9) and (1.10)). Since $\Lambda_{\underline{\underline{p}}}$ is hereditary, ${}_A A^{(n)} \cap (\hat{M} \oplus \hat{X}) \in {}_{\Lambda_{\underline{\underline{p}}}}\underline{\underline{P}}^f$, and $\hat{M} \oplus \hat{X} \in {}_{\hat{\Lambda}_{\underline{\underline{p}}}}\underline{\underline{P}}^f$ (cf. (3.2)); i.e., $\hat{M} \in {}_{\hat{\Lambda}_{\underline{\underline{p}}}}\underline{\underline{P}}^f$ (cf. I,(2.9)). #

4.3 Lemma: Let Λ be a hereditary R-order in A, and let $A = \oplus_{i=1}^n Ae_i$ be the decomposition of A into simple K-algebras, where the $\{e_i\}_{1\le i\le n}$ are the central idempotents in A (cf. III,(5.5)). Then $\Lambda = \oplus_{i=1}^n \Lambda e_i$, where Λe_i is a hereditary R-order in Ae_i, $1\le i\le n$. Moreover, Λ contains a complete set of primitive orthogonal idempotents of A.

Proof: Since Λ is hereditary, every indecomposable Λ-lattice is also irreducible. For, in the proof of (1.13) it has been shown, that every reducible Λ-lattice M contains an R-pure submodule (cf.I,(7.4)) of smaller R-rank. Since Λ is hereditary, M decomposes. In particular, ${}_\Lambda\Lambda$ is a direct sum (not necessarily unique) of irreducible left Λ-lattices; i.e., ${}_\Lambda\Lambda = \oplus_{i=1}^m \Lambda e_i'$. Obviously the $\{e_i'\}_{1\le i\le m}$ form a complete set of orthogonal primitive idempotents (cf. (1.13)) of A. Adding up the A-equivalent primitive idempotents to yield the central idempotents $\{e_i\}_{1\le i\le n}$ of A (cf. III,(5.7), (5.8)), we obtain $\Lambda = \oplus_{i=1}^n \Lambda e_i$; where Λe_i is an R-order in Ae_i, $1\le i\le n$. Since ${}_{\Lambda e_i}\underline{\underline{M}}^f \subset {}_{\Lambda}\underline{\underline{P}}^f$, Λe_i is a hereditary R-order in Ae_i, $1\le i\le n$. #

4.4 Remark: Let Λ be an R-order in A and let $\{e_i\}_{1\le i\le n}$ be the central primitive idempotents of A (cf. III,(5.7)). Then Λe_i is an R-order in Ae_i, $1\le i\le n$, and $\Lambda \subset \oplus_{i=1}^n \Lambda e_i$.

4.5 Remark: To treat maximal orders, it suffices to assume that A is a central simple K-algebra. Let Γ be a maximal R-order in the separable K-algebra A, and let $\{e_i\}_{1\le i\le n}$ be the central primitive idempotents of A. Because of the maximality of Γ, we must have $\Gamma = \oplus_{i=1}^{n} \Gamma e_i$ (cf. (4.4)), where Γe_i is a maximal R-order in Ae_i, $1 \le i \le n$. If R_i is the integral closure of R in the center K_i of Ae_i, $1 \le i \le n$, then R_i is a Dedekind domain with quotient field K_i (cf. (1.5) and III, Ex. 6,3), and, since R_i is an R-order in K_i (cf. proof of (1.5) and (1.4)), $R_i \in {}_{R}\underline{\underline{M}}^f$. Because Γe_i is maximal, $R_i \subset \Gamma e_i$, and we may view Γe_i as maximal R_i-order in the central simple K_i-algebra Ae_i.

4.6 __Theorem:__ Every R-order in A is contained in a maximal one; in particular, there exist maximal R-orders in A.

__Proof:__ It is clear, that there exist R-orders in A (cf. (1.3)). Let Λ be an R-order in A, and let

$$\Lambda = \Lambda_0 \subset \Lambda_1 \subset \dots \subset \Lambda_i \subset \dots$$

be an ascending chain of R-orders in A, containing Λ. Then $\tilde{\Lambda} = \bigcup_i \Lambda_i$ is a subring of A with the same identity as A, containing a K-basis for A. Moreover, since every element in $\tilde{\Lambda}$ is integral over R, $\tilde{\Lambda}$ is an R-order in A (cf. (1.4)). Since $\tilde{\Lambda} \in {}_{R}\underline{\underline{M}}^f$, $\tilde{\Lambda} = \Lambda_k$ for some k; i.e., Λ_k is maximal. This shows at the same time that any ascending chain of orders starting with Λ terminates. #

4.7 __Lemma:__ If B is a finite dimensional K-algebra, having non-zero radical, there do not exist maximal R-orders in A.

__Proof:__ There exist R-orders in B (observe, that in the proof of (1.3) we have not used the fact, that A was separable). Let Λ be an R-order in B and put $N_i = \Lambda \cap (\text{rad } B)^i$. If $0 \ne r \in R$ is a non-unit, then

$$\Lambda_k = \Lambda + r^{-k}N_1 + r^{-2k}N_2 + \dots, \quad k = 0,1,2,\dots$$

are R-orders in B such that

4.7' $$\Lambda_0 \subsetneq \Lambda_1 \subsetneq \Lambda_2 \subsetneq \dots$$

is an infinite ascending chain of R-orders in B. Indeed, it should be observed, that the sum in the definition of Λ_k is finite, since rad B is nilpotent (cf. I, Ex. 4,6), say $(\text{rad } B)^{s+1} = 0$, but $(\text{rad } B)^s \neq 0$. Therefore Λ_k is an R-order in B. To show that the chain (4.7') is strictly increasing, let us assume that $\Lambda_k = \Lambda_{k+1}$. Then $r^{s(k+1)}(\Lambda + r^{-(k+1)}N_1 + \ldots + r^{-s(k+1)}N_s) = r^{s(k+1)}(\Lambda + r^{-k}N_1 + \ldots + r^{-sk}N_s)$, i.e., $N_s \subset r^s\Lambda$. But N_s is an R-pure submodule of Λ (cf. I,(7.4)). Thus $N_s \subset r^s\Lambda$ implies $r^sN_s = N_s$. But this is impossible, as one sees by localizing and applying Nakayama's lemma (cf. I,(4.18)), unless $N_s = 0$, and that we had excluded; i.e., $\Lambda_k \neq \Lambda_{k+1}$, $k=0,1,2,\ldots$. #

4.8 Lemma: Let Λ be an R-order in A. The following statements are equivalent:

(i) Λ is maximal.

(ii) $\Lambda_{\underline{\underline{p}}}$ is maximal for every $\underline{\underline{p}} \in \underline{\underline{S}}$.

(iii) $\hat{\Lambda}_{\underline{\underline{p}}}$ is maximal for every $\underline{\underline{p}} \in \underline{\underline{S}}$.

Proof: (i) $\Longrightarrow$ (ii) Let Λ be maximal, and assume that for some $\underline{\underline{p}} \in \underline{\underline{S}}$, $\Gamma_{\underline{\underline{p}}}$ is a maximal $R_{\underline{\underline{p}}}$-order in A containing $\Lambda_{\underline{\underline{p}}}$. By (1.8), there exists $M \in {}_{\Lambda}\underline{\underline{M}}^o$, such that $M_{\underline{\underline{p}}} = \Gamma_{\underline{\underline{p}}}$ and $M_{\underline{\underline{q}}} = \Lambda_{\underline{\underline{q}}}$, $\underline{\underline{p}} \neq \underline{\underline{q}} \in \underline{\underline{S}}$. From (1.8) it follows, that $M = (\bigcap_{\substack{\underline{\underline{q}} \in \underline{\underline{S}} \\ \underline{\underline{q}} \neq \underline{\underline{p}}}} \Lambda_{\underline{\underline{q}}}) \cap \Gamma_{\underline{\underline{p}}}$. This shows, that M is an R-order in A containing Λ. The maximality of Λ implies $M = \Lambda$; i.e., $\Lambda_{\underline{\underline{p}}} = \Gamma_{\underline{\underline{p}}}$ is maximal.

(ii) $\Longrightarrow$ (i) If conversely, $\Lambda_{\underline{\underline{p}}}$ is maximal, $\forall\, \underline{\underline{p}} \in \underline{\underline{S}}$, and if Γ is a maximal R-order in A containing Λ, then $\Gamma_{\underline{\underline{p}}} = \Lambda_{\underline{\underline{p}}}$, $\forall\, \underline{\underline{p}} \in \underline{\underline{S}}$; i.e., $\Lambda = \bigcap_{\underline{\underline{p}} \in \underline{\underline{S}}} \Lambda_{\underline{\underline{p}}} = \bigcap_{\underline{\underline{p}} \in \underline{\underline{S}}} \Gamma_{\underline{\underline{p}}} = \Gamma$ (cf. (1.8)), and Λ is maximal.

(ii) $\Longleftrightarrow$ (iii) It follows easily from (1.9), that $\Lambda_{\underline{\underline{p}}}$ is maximal if and only if $\hat{\Lambda}_{\underline{\underline{p}}}$ is maximal. #

Remark: We are going to show next, that maximal R-orders in A are hereditary.

4.9 __Definition:__ Let Λ be an R-order in A. A two-sided Λ-ideal P in A (i.e., $P \in {}_{\Lambda}\underline{\underline{M}}^{o}$, $P \subset \Lambda$, such that KP = A) is called a __prime ideal in Λ__, if for any two-sided Λ-ideals $0 \neq I_1, I_2 \subset \Lambda$ in A $I_1 I_2 \subset P$ implies $I_1 \subset P$ or $I_2 \subset P$.

__Note:__ A __Λ-ideal__ shall, in the sequel, always mean a two-sided Λ-lattice $I \subset \Lambda$, such that KI = A; and for a prime ideal P in Λ, we shall always assume $P \neq 0$ and $P \neq \Lambda$.

4.10 __Lemma:__ Let P be a prime ideal in Λ. Then $R \cdot 1 \cap P = \underline{\underline{p}}$ is a prime ideal in R, and Λ/P is a simple $R/\underline{\underline{p}}$-algebra; in particular, P is a maximal two-sided ideal in Λ.

__Proof:__ We put $\underline{\underline{a}} = R \cdot 1 \cap P$; then $\underline{\underline{a}}$ is an ideal in R. If $a, b \in R$ and $ab \in \underline{\underline{a}}$, then $P \supset (ab)\Lambda = (a\Lambda)(b\Lambda)$. Since P is prime, $a\Lambda \subset P$ or $b\Lambda \subset P$; i.e., $a \in \underline{\underline{a}}$ or $b \in \underline{\underline{a}}$, and $\underline{\underline{a}} = \underline{\underline{p}}$ is a prime ideal in R. Moreover, Λ/P is a finite dimensional $R/\underline{\underline{p}}$-algebra.

Let $\varphi : \Lambda \longmapsto \Lambda/P$ be the canonical homomorphism. If Λ/P were not simple, then it is easily seen that it would contain two non-zero two-sided ideals $\bar{I}_1$ and $\bar{I}_2$ such that $\bar{I}_1 \bar{I}_2 = 0$ (cf. Ex. 4,6). If we put $I_i = \varphi^{-1}(\bar{I}_i)$, i = 1,2, then $I_1 I_2 \subset P$; i.e., $\bar{I}_1$ or $\bar{I}_2$ must be zero, a contradiction. #

4.11 __Lemma:__ Let I be a non-trivial Λ-ideal. Then I contains a product of prime ideals.

__Proof:__ If I is not prime, then there exist Λ-ideals J_1 and J_2, such that $J_1 J_2 \subset I$ but $J_1 \not\subset I$ and $J_2 \not\subset I$. We put $I_1 = J_1 + I$ and $I_2 = J_2 + I$; then $I \subsetneq I_1 \subsetneq \Lambda$, $I \subsetneq I_2 \subsetneq \Lambda$ and $I_1 I_2 \subset I$. If $I_1 = \Lambda$, then $I_1 I_2 = I_2 \not\subset I$. Now we repeat this process with I_1 and I_2. This construction has to stop after finitely many steps, since Λ is noetherian (cf. I,(4.1)); i.e., after, say n, steps all ideals are prime (cf. (4.10)), and $I \supset \prod_{i=1}^{m} P_i$, where $\{P_i\}_{1 \leq i \leq m}$

are prime ideals in Λ. #

4.12 <u>Definition</u>: Let I be a non-zero <u>fractional Λ-ideal</u>, (i.e., I is a full two-sided Λ-lattice in A). Then we define

$$I^{-1} = \{x \in A : I x I \subset I\} ;$$

I^{-1} is a Λ-lattice and $KI^{-1} = A$.

<u>Proof</u>: Since $I^{-1} \supset \Lambda s$ for some $0 \neq s \in R$ we only have to show, that I^{-1} is a finitely generated R-module. But KI = A implies $R \cdot 1 \cap I \neq 0$. Let $0 \neq r \in R \cdot 1 \cap I$; then $rI^{-1}r \subset I$; i.e., $I^{-1}r^2 \subset I$. Since $I \in {}_R\underline{\underline{M}}^o$, so is $I^{-1}r^2$, and hence I^{-1} is a finitely generated R-module. #

4.13 <u>Lemma</u>: For any (fractional) Λ-ideal I we have

$$\Lambda \subset \Lambda_l(I), \ \Lambda \subset \Lambda_r(I),$$

$$I^{-1} = \{x \in A: Ix \subset \Lambda_l(I)\}$$

$$= \{x \in A: xI \subset \Lambda_r(I)\}, \text{ in particular}$$

$$II^{-1} \subset \Lambda_l(I) \text{ and } I^{-1}I \subset \Lambda_r(I).$$

If I is integral (i.e., $I \subset \Lambda$) then $I^{-1} \supset \Lambda$.

<u>Proof</u>: This is left as an exercise for the reader. #

4.14 <u>Lemma</u>: If Γ is a maximal R-order in A and if I is a two-sided Γ-ideal $0 \subsetneq I \subsetneq \Gamma$, then $I^{-1} \supsetneq \Gamma$.

<u>Proof</u>: Assume the contrary; i.e., $I^{-1} = \Gamma$ and let P be a maximal two-sided Γ-ideal containing I. Then $P^{-1} \subset I^{-1}$; whence $P^{-1} = \Gamma$ (cf (4,13)). (Since Γ is maximal, $\Lambda_r(J) = \Lambda_l(J) = \Gamma$, for all Γ-ideals.) Choose $0 \neq r \in P \cap R \cdot 1$; then by (4.11) $P \supset r\Gamma \supset \prod_{i=1}^{n} P_i$ for some set of prime ideals $\{P_i\}_{1 \leq i \leq n}$ in Γ which is choosen such that n is minimal. Since P is prime, $P = P_j$ for some $1 \leq j \leq n$. We put $I_1 = \prod_{i=1}^{j-1} P_i$, $I_2 = \prod_{i=j+1}^{n} P_i$. Then $\Gamma \supset r^{-1}I_1PI_2$ implies $PI_2 \supset PI_2r^{-1}I_1PI_2$; i.e., $PI_2r^{-1}I_1 \subset \Gamma$. Thus $I_2r^{-1}I_1 \subset P^{-1} = \Gamma$ and $I_1I_2 \subset r\Gamma \subset P$, a contradiction to the minimality of n. #

4.15 Theorem: Let Γ be a maximal R-order in A. If $0 \neq I \neq \Gamma$ is a Γ-ideal, then

$$I^{-1}I = II^{-1} = \Gamma.$$

In other words, the Γ-ideals and their inverses generate a group with Γ as identity. If Λ is an R-order in A, and if $0 \neq I \neq \Lambda$ is a Λ-ideal, I is called an invertible Λ-ideal, if there exists a Λ-lattice I' in A with $I'I = II' = \Lambda$. In that case I' is uniquely determined and $I' = I^{-1}$.

Proof: I^{-1} is a two-sided Γ-lattice, and thus $0 \neq J = II^{-1} \subset \Gamma$ is a two-sided Γ-ideal. Since Γ is maximal, we may apply (4.14): If $J \neq \Gamma$, then $J^{-1} \supsetneq \Gamma$. But $JJ^{-1} \subset \Gamma$ implies $I^{-1}J^{-1} \subset I^{-1}$(cf. (4.13)); since I^{-1} is a Γ-lattice (cf. (4.12)), $\Lambda_r(I^{-1}) = \Gamma$ (cf. (1.3)); i.e., $J^{-1} \subset \Gamma$, a contradiction; i.e., $J = II^{-1} = \Gamma$. Similarly one shows, that $I^{-1}I = \Gamma$. #

4.16 Theorem: Let Γ be a maximal R-order in A. Then every Γ-ideal I, $0 \neq I \neq \Gamma$ is a unique product of prime ideals.

Proof: (i) If P_1 and P_2 are prime ideals in Γ, then $P_1P_2 = P_2P_1$. In fact, this is obviously true if $P_1 = P_2$. If $P_1 \neq P_2$, then $P_1(P_1^{-1}P_2P_1) = P_2P_1 \subset P_2$ and since $P_1 \not\subset P_2$ (cf. (4.10)), $P_1^{-1}P_2P_1 \subset P_2$ (observe $P_1^{-1}P_2P_1 \subset \Gamma$) because P_2 is prime (cf. (4.9)); i.e., $P_2P_1 \subset P_1P_2$. Similarly one shows $P_1P_2 \subset P_2P_1$; i.e., $P_1P_2 = P_2P_1$.

(ii) We assume that not every proper Γ-ideal can be written as a product of prime ideals, and we choose a largest proper Γ-ideal I which is not a product of prime ideals. Since Γ is noetherian, we may choose a maximal ideal P -which is necessarily a prime ideal-, such that $P \supset I$. But then

$$\Gamma = PP^{-1} \supsetneq IP^{-1} \supsetneq I \quad ,$$

where the inequalities follow from (4.15) and (4.14). Hence IP^{-1} is a product of prime ideals. It follows that $I = IP^{-1}P$ is a product of prime ideals, a contradiction.

(iii) The uniqueness of this factorization follows easily by induction on the length of the product, using (i) (cf. Ex. 4,1). #

4.17 Remark: One should observe the similarity between the proofs of the pr ceeding theorems and the proofs of the corresponding theorems for Dedekind domains (cf. Ex. 4,1).

4.18 Theorem: Let Λ be an R-order in A and I an invertible Λ-ideal (cf. (4.15)). Then ${}_\Lambda I \in {}_{\Lambda}\underline{\underline{M}}^o$ and $I_\Lambda \in \underline{\underline{M}}^o_\Lambda$ are progenerators (cf. III,(1.9)).

Proof: Since $I^{-1}I = \Lambda$, there exist $x_i \in I^{-1}$, $y_i \in I$, $1 \leq i \leq n$, such that $\sum_{i=1}^n x_i y_i = 1$. But with $x_i \in I^{-1}$ we may associate $\varphi_i : {}_\Lambda I \longrightarrow {}_\Lambda\Lambda$ (cf. (4.12)), $\varphi_i : \alpha \longmapsto \alpha x_i$, $\alpha \in I$. Thus, $1_{{}_\Lambda I} = (\sum_{i=1}^n \varphi_i \otimes y_i)^{\mu}{}_\Lambda I$ (cf. III,(1.4)), and ${}_\Lambda I \in {}_{\Lambda}\underline{\underline{P}}^f$ by (III,(1.5)). But with $x_i \in I^{-1}$ we can also associate $\psi_i \in \mathrm{Hom}_\Lambda(I_\Lambda, \Lambda_\Lambda)$, $1 \leq i \leq n$, $\psi_i : I_\Lambda \longrightarrow \Lambda_\Lambda$ (cf. (4.12)), $\psi_i : \alpha \longmapsto x_i\alpha$. Thus, $1 = (\sum_{i=1}^n \psi_i \otimes y_i)^{\tau}I_\Lambda$, and $I_\Lambda \in \underline{\underline{M}}^o_\Lambda$ is a generator in $\underline{\underline{M}}^o_\Lambda$ (cf. III,(1.9)). Now, applying a similar argument, using $II^{-1} = \Lambda$, one shows, that $I_\Lambda \in \underline{\underline{P}}^f_\Lambda$ and ${}_\Lambda I \in {}_{\Lambda}\underline{\underline{M}}^o$ is a generator; i.e., ${}_\Lambda I \in {}_{\Lambda}\underline{\underline{M}}^o$ and $I_\Lambda \in \underline{\underline{M}}^o_\Lambda$ are progenerators. #

4.19 Theorem (Auslander - Goldman [1]): Let Γ be a maximal R-order in the separable K-algebra A. Then Γ is hereditary (cf. (4.1)).

Proof: Because of (4.2) and (4.8) it suffices to show, that $\Gamma_{\underline{\underline{p}}}$ is hereditary for every $\underline{\underline{p}} \in \underline{\underline{S}}$. We shall use a technique similar to that of the proof of (3.4). Let $N = \mathrm{rad}\, \Gamma_{\underline{\underline{p}}}$; then $N \in {}_{\Gamma_{\underline{\underline{p}}}}\underline{\underline{P}}^f$ (cf. (4.15) and (4.18)). For $M \in {}_{\underline{\underline{\Gamma_p}}}\underline{\underline{M}}^o$, we have the exact sequence

$$0 \longrightarrow M \xrightarrow{\varphi} M \longrightarrow M/\pi M \longrightarrow 0,$$

where $\pi R_{\underline{\underline{p}}} = \text{rad } R_{\underline{\underline{p}}}$ and φ is multiplication by π. We shall show below, that $\text{hd}_{\underline{\underline{\Gamma_p}}}(M/\pi M) \leq 1$ (cf. II,(4.1)). Taking this for granted for the moment, we obtain from the above sequence the exact sequence, for every $X \in {}_{\underline{\underline{\Gamma_p}}}\underline{\underline{M}}^f$,

$$\text{Ext}^1_{\underline{\underline{\Gamma_p}}}(M/\pi M, X) \longrightarrow \text{Ext}^1_{\underline{\underline{\Gamma_p}}}(M, X) \xrightarrow{\varphi^*} \text{Ext}^1_{\underline{\underline{\Gamma_p}}}(M, X) \longrightarrow 0$$

(cf. II,(4.3)). Here φ^* is still multiplication by π; we have $\text{Ext}^1_{\underline{\underline{\Gamma_p}}}(M,X) = \pi \text{Ext}^1_{\underline{\underline{\Gamma_p}}}(M,X) = 0$, by Nakayama's lemma, and so, $M \in {}_{\underline{\underline{\Gamma_p}}}\underline{\underline{P}}^f$.

It remains to show, that $\text{hd}_{\underline{\underline{\Gamma_p}}}(M/\pi M) \leq 1$. $\Gamma_{\underline{\underline{p}}}/\pi \Gamma_{\underline{\underline{p}}}$ is a finite dimensional $R_{\underline{\underline{p}}}/\pi R_{\underline{\underline{p}}}$-algebra, and $M/\pi M \in {}_{\underline{\underline{\Gamma_p}}/\pi \underline{\underline{\Gamma_p}}}\underline{\underline{M}}^f$. Let $M/\pi M = M_o \underset{\neq}{\supset} M_1 \underset{\neq}{\supset} \cdots \underset{\neq}{\supset} M_s \underset{\neq}{\supset} 0$ be a composition series of $M/\pi M$ (cf. I,(4.7)). Then the composition factors M_i/M_{i+1}, $0 \leq i \leq s$, are simple left $\Gamma_{\underline{\underline{p}}}/\pi \Gamma_{\underline{\underline{p}}}$-modules. Hence $N(M_i/M_{i+1}) = 0$, $0 \leq i \leq s$, (cf. I,(4.15) and (2.4)) and the M_i/M_{i+1} are $\Gamma_{\underline{\underline{p}}}/N$-modules. The change of ring theorem (cf. II,(4.6)) implies

$$\text{hd}_{\underline{\underline{\Gamma_p}}}(M_i/M_{i+1}) \leq \text{hd}_{\underline{\underline{\Gamma_p}}}(\Gamma_{\underline{\underline{p}}}/N) + \text{hd}_{\underline{\underline{\Gamma_p}}/N}(M_i/M_{i+1}).$$

Since $\Gamma_{\underline{\underline{p}}}/N$ is a finite dimensional semi-simple $R_{\underline{\underline{p}}}/\pi R_{\underline{\underline{p}}}$-algebra (cf. I,(4.17)), $\text{hd}_{\underline{\underline{\Gamma_p}}/N}(M_i/M_{i+1}) = 0$ (cf. II,(4.2)). However, the exact sequence

$$0 \longrightarrow N \longrightarrow \Gamma_{\underline{\underline{p}}} \longrightarrow \Gamma_{\underline{\underline{p}}}/N \longrightarrow 0$$

implies $hd_{\Gamma_{\underline{\underline{p}}}}(\Gamma_{\underline{\underline{p}}}/N) = 1 + hd_{\Gamma_{\underline{\underline{p}}}}(N)$ (cf. II,(4.5)). But $N \in {}_{\Gamma_{\underline{\underline{p}}}}\underline{\underline{P}}^f$ implies $hd_{\Gamma_{\underline{\underline{p}}}}(\Gamma_{\underline{\underline{p}}}/N) = 1$ (cf. II,(4.2)). Thus, we conclude $hd_{\Gamma_{\underline{\underline{p}}}}(M_i/M_{i+1}) \leq 1$, $0 \leq i \leq s$. The exact sequence of $\Gamma_{\underline{\underline{p}}}$-modules

$$0 \longrightarrow M_{i+1} \longrightarrow M_i \longrightarrow M_i/M_{i+1} \longrightarrow 0$$

implies

$$hd_{\Gamma_{\underline{\underline{p}}}}(M_i) \leq \max(hd_{\Gamma_{\underline{\underline{p}}}}(M_{i+1}),\ hd_{\Gamma_{\underline{\underline{p}}}}(M_i/M_{i+1}))$$
$$= \max(hd_{\Gamma_{\underline{\underline{p}}}}(M_{i+1}),\ 1) \text{ (cf. II,(4.4)).}$$

Using induction, we obtain $hd_{\Gamma_{\underline{\underline{p}}}}(M/\pi M) \leq 1$. #

<u>Exercises §4</u>:

1.) Let R be a Dedekind domain. Show that every proper integral ideal $\underline{\underline{a}}$ in R has a unique representation:

$$\underline{\underline{a}} = \prod_{i=1}^{n} \underline{\underline{p}}_i^{(\alpha_i)}, \text{ where } \underline{\underline{p}}_i,\ 1 \leq i \leq n,$$

are different prime ideals in R and $\alpha_i \in \underline{\underline{N}}$, $1 \leq i \leq n$. (Hint: Show

(i) $\underline{\underline{a}}$ contains a product of prime ideals (cf. (4.11)), using the fact that prime ideals are maximal;

(ii) $\underline{\underline{a}}^{-1} \supsetneq R$, using the fact that R is integrally closed in its quotient field (cf. (4.14));

(iii) $\underline{\underline{a}}^{-1}\underline{\underline{a}} = R$ (cf. (4.15));

(iv) $\underline{\underline{a}} = \prod_{i=1}^{n} \underline{\underline{p}}_i^{(\alpha_i)}$ is a unique factorization (cf. (4.16).)

2.) Let R be a Dedekind domain. Show that every R-lattice is projective. (Hint: Let $M \in {}_R\underline{\underline{M}}^o$ be an R-lattice. We use induction on rank (M). For rank (M) = 1, the statement follows from Ex. 1. For the step n to n + 1, use an argument similar to that in the proof of (I,(7.3)).)

3.) A finite dimensional extension field K of $\underline{\underline{Q}}$ is called an

<u>algebraic number field</u>. If R is the integral closure of $\underline{\underline{Z}}$ in K, R is called the <u>ring of algebraic integers in K</u>. Show that R is a Dedekind domain. (Hint: R is the unique maximal $\underline{\underline{Z}}$-order in the separable $\underline{\underline{Q}}$-algebra K (cf. III, Ex. 5,8).)

4.) Let A be a separable finite dimensional K-algebra, where K is the quotient field of the Dedekind domain R. If A is commutative, show that there is exactly one maximal R-order in A. If A is not commutative, give an example where there are more than one maximal orders (cf. Ex. 2,1).

5.) Let Λ be an R-order in the separable finite dimensional K-algebra. Show that the following conditions are equivalent:

(i) Λ is hereditary;

(ii) every left Λ-ideal is projective;

(iii) every irreducible Λ-lattice is projective.

6.) Let A be a finite dimensional K-algebra. If A is not simple, show that there exist two-sided non-zero A-ideals whose product is zero.

7.) Let Γ be a maximal order in the separable K-algebra A. Let $0 \neq I$ be a two-sided fractional Γ-ideal. Show that I is a unique product of prime ideals and their inverses. Use this to to show that the non-zero fractional Γ-ideals from a group under multiplication.

§ 5. Maximal orders and progenerators

Maximal orders are characterized by the property that every faithful lattice is a progenerator. If $\Gamma_{\underline{\underline{p}}}$ is a maximal $R_{\underline{\underline{p}}}$-order in A, then $\Gamma_{\underline{\underline{p}}}$ is a principal ideal ring and the Krull-Schmidt theorem is valid for $\Gamma_{\underline{\underline{p}}}$-lattices. Moreover for $\Gamma_{\underline{\underline{p}}}$-lattices $M_{\underline{\underline{p}}}$, $N_{\underline{\underline{p}}}$ we have $M_{\underline{\underline{p}}} \cong N_{\underline{\underline{p}}}$ if and only if $KM_{\underline{\underline{p}}} \cong KN_{\underline{\underline{p}}}$.

We keep the notation of the previous sections.

5.1 Lemma: Let Λ be an R-order in A, such that every faithful Λ-lattice is a generator. Then Λ is maximal.

Proof: Let Γ be a maximal R-order in A containing Λ. Then ${}_\Lambda\Gamma$ is a faithful Λ-lattice, hence a generator; i.e., per definition (cf. III,(1.9)) $\operatorname{Im}\tau_{{}_\Lambda\Gamma} = \Lambda$, where $\tau_{{}_\Lambda\Gamma} : \Gamma \otimes_{\operatorname{Hom}_\Lambda(\Gamma,\Gamma)} \operatorname{Hom}_\Lambda({}_\Lambda\Gamma, {}_\Lambda\Lambda) \longrightarrow \Lambda$, $\gamma \otimes \varphi \longrightarrow \gamma\varphi$. By (1.14), $\operatorname{Hom}_\Lambda(\Gamma,\Gamma) = \operatorname{Hom}_\Gamma(\Gamma,\Gamma) \cong \Gamma$. Under this isomorphism $\operatorname{Hom}_\Lambda({}_\Lambda\Gamma, {}_\Lambda\Lambda) \cong F_1(\Lambda : \Gamma) = \{x \in A : \Gamma x \subset \Lambda\}$ is the left conductor of Γ in Λ. Hence $\operatorname{Im}\tau = \Gamma \cdot F_1(\Lambda : \Gamma) = F_1(\Lambda : \Gamma)$, since $F_1(\Lambda : \Gamma) \in {}_{\Gamma}\underline{\underline{M}}^f$. Thus $F_1(\Lambda : \Gamma) = \Lambda = \Gamma\Lambda$; i.e., $\Gamma = \Lambda$, and Λ is maximal. #

5.2 Theorem: Let $\hat{D}$ be a separable skewfield over $\hat{K}_{\underline{\underline{p}}}$. Then

(i) there exists exactly one maximal $\hat{R}_{\underline{\underline{p}}}$-order $\hat{\Omega}$ in $\hat{D}$,

(ii) rad $\hat{\Omega}$ is the unique maximal two-sided ideal in $\hat{\Omega}$,

(iii) every left ideal in $\hat{\Omega}$ is two-sided, and it is a power of rad $\hat{\Omega}$,

(iv) every $\hat{\Omega}$-lattice is a progenerator,

(v) all irreducible $\hat{\Omega}$-lattices are isomorphic.

Proof: Let $\hat{\Omega}$ be a maximal $\hat{R}_{\underline{\underline{p}}}$-order in $\hat{D}$.

We show first, that rad $\hat{\Omega}$ is the unique maximal left $\hat{\Omega}$-ideal. Let $\hat{I}$ be a left $\hat{\Omega}$-ideal with $\hat{\Omega} \neq \hat{I} \supset$ rad $\hat{\Omega}$. Then $\hat{I}/$rad $\hat{\Omega}$ is a left ($\hat{\Omega}/$rad $\hat{\Omega}$)-ideal. Now, from the method of lifting idempotents (cf. (2.1)), it follows that $\hat{\Omega}/$rad $\hat{\Omega}$ is a skewfield; i.e., $\hat{I}$ = rad $\hat{\Omega}$, and rad $\hat{\Omega}$ is the unique maximal left ideal in $\hat{\Omega}$ (cf. I,(4.16)), in particular, rad $\hat{\Omega}$ is the unique prime ideal in $\hat{\Omega}$. From (4.16) we conclude

$$\hat{\pi}\hat{\Omega} = (\text{rad } \hat{\Omega})^e$$

for some positive integer e (observe, that $\hat{\pi}\hat{\Omega}$ is a two-sided ideal). Let now $0 \neq \hat{I} \neq \hat{\Omega}$ be a left ideal in $\hat{\Omega}$ and let n be the largest integer such that $(\text{rad } \hat{\Omega})^n \supset \hat{I}$ but $(\text{rad } \hat{\Omega})^{n+1} \not\supset \hat{I}$; observe that $\bigcap_{m \in \underline{\underline{N}}} (\text{rad } \hat{\Omega})^m = 0$ (cf. proof of I,(9.11)). Then $\hat{\Omega} \supset (\text{rad } \hat{\Omega})^{-n}\hat{I}$, and, if rad $\hat{\Omega} \supset (\text{rad } \hat{\Omega})^{-n}\hat{I}$, then $(\text{rad } \hat{\Omega})^{n+1} \supset \hat{I}$; i.e., $\hat{\Omega} \supset (\text{rad } \hat{\Omega})^{-n}\hat{I} \underset{\neq}{\supset}$ rad $\hat{\Omega}$. Since rad $\hat{\Omega}$ is the unique maximal left ideal, $\hat{\Omega} = (\text{rad } \hat{\Omega})^{-n}\hat{I}$, hence $(\text{rad } \hat{\Omega})^n = \hat{I}$, and $\hat{I}$ is a two-sided $\hat{\Omega}$-ideal.

If $\hat{\Omega}_1$ is another $\hat{R}_{\underline{\underline{p}}}$-order in $\hat{D}$, then $\hat{\Omega}\hat{\Omega}_1$ is a two-sided $\hat{\Omega}$-ideal in $\hat{D}$; whence $\hat{\Omega} = \Lambda_r(\hat{\Omega}\hat{\Omega}_1) \supset \hat{\Omega}_1$; (it should be observed, that there exists $0 \neq r \in \hat{R}_{\underline{\underline{p}}}$, such that $r\hat{\Omega}\hat{\Omega}_1 \subset \hat{\Omega}$, and $r\hat{\Omega}\hat{\Omega}_1$ is isomorphic as two-sided $\hat{\Omega}$-ideal to $\hat{\Omega}\hat{\Omega}_1$). It remains to show that every $\hat{\Omega}$-lattice is a progenerator. By (4.19), every $\hat{\Omega}$-lattice $\hat{M}$ is projective, and it remains to show that for every right ideal $\hat{I} \neq \hat{\Omega}$ in $\hat{\Omega}$, $\hat{I}\hat{M} \neq \hat{M}$ (cf. III,(1.7) and III,(1.10)). But $\hat{I} \subset$ rad $\hat{\Omega}$, and by Nakayama's lemma (I,(4.18)), rad $\hat{\Omega} \cdot \hat{M} \underset{\neq}{\subset} \hat{M}$. If $\hat{M}$ is an irreducible $\hat{\Omega}$-lattice, then $\hat{M} \cong {}_{\hat{\Omega}}\hat{\Omega}$ because of the validity of the Krull-Schmidt theorem (cf. (4.19), (2.1) and III,(7.7)), thus all irreducible $\hat{\Omega}$-lattices are isomorphic. #

From the proof of (5.2) follows immediately:

5.3 <u>Corollary</u>: Let $\hat{A}$ be a separable $\hat{K}_{\underline{p}}$-algebra and $\hat{\Lambda}$ an $\hat{R}_{\underline{p}}$-order in $\hat{A}$, such that ${}_{\hat{\Lambda}}\hat{\Lambda}$ is indecomposable as module. Then

(i) rad $\hat{\Lambda}$ is the unique maximal left ideal in $\hat{\Lambda}$.

(ii) every projective $\hat{\Lambda}$-lattice is a progenerator.

5.4 <u>Theorem</u>: Let $\hat{A}$ be a simple separable $\hat{K}_{\underline{p}}$-algebra and $\hat{\Gamma}$ a maximal $\hat{R}_{\underline{p}}$-order in $\hat{A}$. Then

(i) rad $\hat{\Gamma}$ is the unique prime ideal in $\hat{\Gamma}$,

(ii) every $\hat{\Gamma}$-lattice is a progenerator,

(iii) all irreducible $\hat{\Gamma}$-lattices are isomorphic.

(iv) If $\hat{\Gamma}_1$ is another maximal $\hat{R}_{\underline{p}}$-order in $\hat{A}$, then there is a Morita-equivalence between ${}_{\hat{\Gamma}}\underline{M}^o$ and ${}_{\hat{\Gamma}_1}\underline{M}^o$.

<u>Proof</u>: Let $\hat{A} = (\hat{D})_n$, where $\hat{D}$ is a separable skewfield over $\hat{K}_{\underline{p}}$, and let $\hat{\Omega}$ be the unique maximal $\hat{R}_{\underline{p}}$-order in $\hat{D}$ (cf. (5.2)). Then

$$\hat{\Gamma} = \mathrm{End}_{\hat{\Omega}}({}_{\hat{\Omega}}\hat{\Omega}^{(n)}) = (\hat{\Omega})_n$$

is an $\hat{R}_{\underline{p}}$-order in $\hat{A}$ (cf. (1.15)). Since ${}_{\hat{\Omega}}\hat{\Omega}^{(n)}$ is a progenerator for ${}_{\hat{\Omega}}\underline{M}^o$, we have a Morita-equivalence between ${}_{\hat{\Omega}}\underline{M}^o$ and ${}_{\hat{\Gamma}}\underline{M}^o$. In particular, every $\hat{\Gamma}$-lattice is a progenerator, and consequently, $\hat{\Gamma}$ is a maximal $\hat{R}_{\underline{p}}$-order in $\hat{A}$ (cf. (5.1)). From (III, Ex. 5,6) it follows that rad $\hat{\Gamma}$ $=(\mathrm{rad}\ \hat{\Omega})_n$. Thus $\hat{\Gamma}/\mathrm{rad}\ \hat{\Gamma} = (\hat{\Omega}/\mathrm{rad}\ \hat{\Omega})_n$ is a simple algebra and hence has no two-sided ideals; i.e., rad $\hat{\Gamma}$ is the unique maximal two-sided ideal in $\hat{\Gamma}$.

If now $\hat{\Gamma}_1$ is any $\hat{R}_{\underline{p}}$-order in $\hat{A}$, then $\hat{\Gamma}\,\hat{\Gamma}_1$ is a faithfully projective $\hat{\Gamma}$-lattice, since every left $\hat{\Gamma}$-lattice is a progenerator, and we have a Morita-equivalence between ${}_{\hat{\Gamma}}\underline{M}^o$ and ${}_{\hat{\Gamma}_1}\underline{M}^o$; in fact, $\mathrm{End}_{\hat{\Gamma}}(\hat{\Gamma}\,\hat{\Gamma}_1)$ is an $\hat{R}_{\underline{p}}$-order in $\hat{A}$ containing $\hat{\Gamma}_1$; i.e., $\mathrm{End}_{\hat{\Gamma}}(\hat{\Gamma}\,\hat{\Gamma}_1) = \hat{\Gamma}_1$. Thus, every $\hat{\Gamma}_1$-lattice is a progenerator (cf. (3.7)), and it remains to show, that rad $\hat{\Gamma}_1$ is the unique prime ideal in $\hat{\Gamma}_1$. For this it suffices to show, that $\hat{\Gamma}_1/\mathrm{rad}\ \hat{\Gamma}_1$ is a simple $\hat{R}_{\underline{p}}/\pi\hat{R}_{\underline{p}}$-algebra. From (3.9) it follows that we have a Morita-equivalence between ${}_{\hat{\Gamma}/\mathrm{rad}\hat{\Gamma}}\underline{M}^f$ and

$\mathrm{End}_{\hat{\Gamma}/\mathrm{rad}\hat{\Gamma}}(\hat{\Gamma}\hat{\Gamma}_1/(\mathrm{rad}\hat{\Gamma})\hat{\Gamma}\hat{\Gamma}_1)^{\underline{\underline{M}}^o}$. But in (3.9) it was shown that we have a natural ring isomorphism

$$\mathrm{End}_{\hat{\Gamma}/\mathrm{rad}\hat{\Gamma}}(\hat{\Gamma}\hat{\Gamma}_1/(\mathrm{rad}\ \hat{\Gamma})\hat{\Gamma}\hat{\Gamma}_1) \cong \hat{\Gamma}_1/\mathrm{rad}\ \hat{\Gamma}_1 \ .$$

Thus, we have a Morita equivalence between ${}_{\hat{\Gamma}/\mathrm{rad}\hat{\Gamma}}\underline{\underline{M}}^o$ and ${}_{\hat{\Gamma}_1/\mathrm{rad}\hat{\Gamma}_1}\underline{\underline{M}}^o$. Since $\hat{\Gamma}/\mathrm{rad}\ \hat{\Gamma}$ has only one isomorphism class of simple modules, the same is true for $\hat{\Gamma}_1/\mathrm{rad}\ \hat{\Gamma}_1$ (cf. III,(2.1)); i.e., $\hat{\Gamma}_1/\mathrm{rad}\ \hat{\Gamma}_1$ is simple (cf. III,(5.3)). (iii) follows readily from (5.2,v). #

5.5 __Theorem__: Let Λ be an R-order in A. Then Λ is maximal if and only if every faithful Λ-lattice is a progenerator. If Γ_1 and Γ_2 are two maximal R-orders in A hen we have a Morita equivalence between ${}_{\hat{\Gamma}_1}\underline{\underline{M}}^o$ and ${}_{\hat{\Gamma}_2}\underline{\underline{M}}^o$. In addition, being maximal is invariant under Morita equivalence.*)

__Proof__: For the first part, it suffices to show that for a maximal R-order Γ, every faithful Γ-lattice is a progenerator (cf. (5.1)).

(i) $M \in {}_{\Gamma}\underline{\underline{M}}^o$ is __faithful__ if and only if KM is faithful, if and only if KM contains every simple left A-module with multiplicity > 0 (cf. III, Ex. 6,8). Now, let $\{e_i\}_{1 \le i \le n}$ be a complete system of non-equivalent primitive idempotents of A, and let $\{e_i'\}_{1 \le i \le n}$ be the corresponding central idempotents (cf. III,(5.5) and III,(5.6)). If $KM = \oplus_{i=1}^{n} Ae_i^{(\alpha_i)}$, then $\mathrm{ann}_A(KM) = \oplus_{\{i:\alpha_i=0\}} Ae_i'$. This shows that KM is faithful if and only if $\alpha_i > 0$, $1 \le i \le n$.

(ii) Since KM is a faithful A-module if and only if $\hat{K}_{\underline{\underline{p}}}M$ is a faithful $\hat{A}_{\underline{\underline{p}}}$-module for every $\underline{\underline{p}} \in \underline{\underline{S}}$ (cf. III, Ex. 6,8), if suffices to show that every faithful $\hat{\Gamma}_{\underline{\underline{p}}}$-module is a progenerator in ${}_{\hat{\Gamma}_{\underline{\underline{p}}}}\underline{\underline{M}}$ for every $\underline{\underline{p}} \in \underline{\underline{S}}$ (cf. (3.1),(3.2)).

---

*) More precisely: If $E \in {}_{\Gamma}\underline{\underline{M}}^o$ is a progenerator, then Γ is a maximal R-order in A if and only if $\mathrm{End}_{\Gamma}(E)$ is one.

(iii) Let $\hat{\Gamma}_{\underline{\underline{p}}} = \oplus_{i=1}^{n} \hat{\Gamma}_i$ be the decomposition of $\hat{\Gamma}_{\underline{\underline{p}}}$ into maximal $\hat{R}_{\underline{\underline{p}}}$-orders in simple $\hat{K}_{\underline{\underline{p}}}$-algebras $\hat{A}_i$ (cf. (4.5)). Since $\hat{K}_{\underline{\underline{p}}}M$ is a faithful $\hat{A}_{\underline{\underline{p}}}$-module, $\hat{\Gamma}_i\hat{M}_{\underline{\underline{p}}} \neq 0$ for every $1 \le i \le n$. Thus each $\hat{\Gamma}_i\hat{M}_{\underline{\underline{p}}}$ is a progenerator for ${}_{\hat{\Gamma}_i}\underline{\underline{M}}^o$, $1 \le i \le n$ (cf. (5.4)). Thus, $\hat{M}_{\underline{\underline{p}}}$ is a progenerator in ${}_{\hat{\Gamma}_{\underline{\underline{p}}}}\underline{\underline{M}}^o$.

(iv) If Γ_1 and Γ_2 are maximal R-orders in A, then $\Gamma_1 \Gamma_2 \in {}_{\Gamma_1}\underline{\underline{M}}^o$ is a progenerator, $\Gamma_1 \Gamma_2$ being a faithful Γ_1-lattice, and $\mathrm{End}_{\Gamma_1}(\Gamma_1\Gamma_2)=\Gamma_2$, since Γ_2 is maximal. Thus, we have a Morita equivalence between ${}_{\Gamma_1}\underline{\underline{M}}^o$ and ${}_{\Gamma_2}\underline{\underline{M}}^o$. The first part of the proof shows that being maximal is invariant under Morita equivalence (cf. (5.1) and III,(2.6)). #

5.6 Theorem: Let $\hat{\Gamma}$ be a maximal $\hat{R}_{\underline{\underline{p}}}$-order in $\hat{A}$. Then $\hat{\Gamma}$ is a principal ideal ring; i.e., every left $\hat{\Gamma}$-ideal in $\hat{A}$ can be generated by one element.

Proof: Since $\hat{\Gamma}$ is hereditary (cf. (4.19)), the Krull-Schmidt theorem is valid for ${}_{\hat{\Gamma}}\underline{\underline{M}}^o$ (cf. (2.1) and III,(7.7)). If $\hat{P}_1,\dots,\hat{P}_n$ are the non-isomorphic indecomposable direct summands of $\hat{\Gamma}$, then $\{\hat{P}_i\}_{1 \le i \le n}$ are all the non-isomorphic irreducible $\hat{\Gamma}$-lattices, since for hereditary orders, a lattice is indecomposable if and only if it is irreducible (cf. proof of (4.3)) and since in ${}_{\hat{\Gamma}}\underline{\underline{M}}^o$ the Krull-Schmidt theorem is valid. If now $\hat{\Gamma} = \oplus_{j=1}^{m} \hat{\Gamma}_j$ is the decomposition of $\hat{\Gamma}$ into maximal $\hat{R}_{\underline{\underline{p}}}$-orders in simple $\hat{K}_{\underline{\underline{p}}}$-algebras (cf. (4.5)), then each $\hat{P}_i$ is a $\hat{\Gamma}_j$-lattice for some $1 \le j \le m$. Moreover, if $\hat{P}_i, \hat{P}_k \in {}_{\hat{\Gamma}_j}\underline{\underline{M}}^o$ for some j, then $\hat{P}_i \cong \hat{P}_k$, since $\hat{P}_i$ and $\hat{P}_j$ are progenerators for $\hat{\Gamma}_j$ (cf. (5.4)). Thus, $\hat{K}_p\hat{P}_i \not\cong \hat{K}_p\hat{P}_k$ for $i \neq k$. Hence m = n is the number of simple components into which A decomposes. Thus, $K_{\underline{\underline{p}}}\hat{M} \cong K\hat{N}_{\underline{\underline{p}}}$ if and only if

$\hat{M} \cong \hat{N}$ for $\hat{M},\hat{N} \in {}_{\hat{\Gamma}}\underline{\underline{M}}^{o}$. If now $\hat{I}$ is left $\hat{\Gamma}$-ideal in $\hat{A}$, then $\hat{I} \cong {}_{\hat{\Gamma}}\hat{\Gamma}$, i.e., there exists $\varphi \in \mathrm{Hom}_{\hat{\Gamma}}(\hat{I},\hat{\Gamma}) \subset \mathrm{Hom}_{\hat{A}}(\hat{A},\hat{A}) = \hat{A}$, such that $\hat{I}\varphi = \hat{\Gamma}$; but φ is given by right multiplication with a regular element $a \in A$; i.e., $\hat{\Gamma} = \hat{I}\, a^{-1}$, and $\hat{I}$ is principal. (It should be observed, that an ideal $\hat{I}$ is always such that $\hat{K}_{\underline{\underline{p}}}\hat{I} = \hat{A}$.) #

5.7 __Corollary:__ Let $\Gamma^{\#}$ be a maximal $R_{\underline{\underline{p}}}$-order in A. Then for $M,N \in {}_{\Gamma^{\#}}\underline{\underline{M}}^{o}$, $M \cong N \Leftrightarrow KM \cong KN$. Moreover, the __Krull-Schmidt theorem__ is valid for ${}_{\Gamma^{\#}}\underline{\underline{M}}^{o}$, and $\Gamma^{\#}$ is a principal ideal ring.

__Proof:__ The first statement follows from ((3.6),(4.19) and (5.6)). If now, for $M \in {}_{\Gamma^{\#}}\underline{\underline{M}}^{o}$, $M \cong \oplus_{i=1}^{n} M_i \cong \oplus_{j=1}^{t} N_j$, are two decompositions into indecomposable $\Gamma^{\#}$-lattices, then KM_i and KN_j are simple A-modules (cf. proof of (4.3)). From the Krull-Schmidt theorem for A-modules (cf. I,(4.10)) and from the first part of the corollary, it follows, that $n = t$ and $N_i \cong M_i$, if necessary after renumbering. Then $\Gamma^{\#}$ is necessarily a principal ideal ring (cf. proof of (5.6)). #

5.8 __Corollary:__ All maximal $R_{\underline{\underline{p}}}$-orders in A are conjugate; i.e., if $\hat{\Gamma}_1^{\#}$ and $\hat{\Gamma}_2^{\#}$ are two maximal $R_{\underline{\underline{p}}}$-orders in A, then there exists a regular element a in A, such that $\Gamma_1^{\#} = a\,\Gamma_2^{\#}\,a^{-1}$; and for every unit a in A, $a\,\Gamma_1^{\#}\,a^{-1}$ is a maximal $R_{\underline{\underline{p}}}$-order in A.

__Proof:__ The first statement follows immediately from the proof of (5.5) and from (5.7). For the rest it should be observed, that $a^{-1}\Gamma_1^{\#}\,a = \mathrm{End}_{\Gamma_1^{\#}}(\Gamma_1^{\#}\,a)$ (cf. (5.5)). #

5.9 __Theorem:__ Let Γ be a maximal R-order in A, C = center of Γ and $\underline{\underline{p}} \in \underline{\underline{S}}$. Then the number of prime ideals in Γ containing $\underline{\underline{p}}\Gamma$ is finite. It is equal to the number of prime ideals in C containing $\underline{\underline{p}} \cdot C$, and also to the number of simple algebras into which $\hat{A}_{\underline{p}}$ splits. Moreover, rad $\Gamma_{\underline{p}} = \bigcap_{P=\text{prime ideal in } \Gamma_{\underline{p}}} P$.

Proof: The number of prime ideals in Γ containing $\underline{p}\Gamma$ is the same as the number of prime ideals in $\Gamma_{\underline{p}}$, as is easily seen. From (1.9) (observe, that (1.9) remains also valid for two-sided lattices) it follows, that the prime ideals in $\Gamma_{\underline{p}}$ and the prime ideals in $\hat{\Gamma}_{\underline{p}}$ are in one-to-one correspondence. Let $\hat{\Gamma}_{\underline{p}} = \oplus_{i=1}^{n} \hat{\Gamma}_i$, where $\hat{\Gamma}_i$, $1 \le i \le n$, are maximal $\hat{R}_{\underline{p}}$-orders in the simple components of $\hat{A}_{\underline{p}}$. In (5.3) we have shown that rad $\hat{\Gamma}_i = \hat{P}_i$ is the unique maximal ideal in $\hat{\Gamma}_i$. Then $\hat{P}'_i = \hat{P}_i \oplus (\oplus_{j \neq i} \hat{\Gamma}_j)$, $1 \le i \le n$, are the unique maximal ideals in $\hat{\Gamma}_{\underline{p}}$. This shows that the number of prime ideals in Γ containing $\underline{p}\,\Gamma$ is the same as the number of simple components of $\hat{A}_{\underline{p}}$. A similar argument applied to the center of Γ shows that this number is equal to the number of maximal ideals in C containing $\underline{p}C$. Since the maximal left ideals in $\Gamma_{\underline{p}}$ and the maximal left ideals in $\hat{\Gamma}_{\underline{p}}$ are in one-to-one correspondence (cf. (1.9)) rad $\hat{\Gamma}_p = \hat{R}_{\underline{p}}$ rad Γ_p. Since rad $\hat{\Gamma}_{\underline{p}} = \bigcap_{i=1}^{n} \hat{P}'_i$, rad $\Gamma_p = \bigcap_{i=1}^{n} P_i$, where the intersection is taken over all prime ideals in Γ_p. #

Exercises §5:

1.) Let R be a semi-local Dedekind domain (i.e., R has only finitely many maximal ideals) with quotient field K. If Γ is a maximal R-order in the separable K-algebra A, then (5.7) is valid for ${}_{\Gamma}\underline{M}^{o}$. Moreover, rad $\Gamma = \bigcap P$, where the intersection is taken over all prime ideals in Γ. (Hint: In view of the proofs of (5.6) and (5.7), it suffices to show that for $M, N \in {}_{\Gamma}\underline{M}^{o}$,

$$KM \cong KN \Longleftrightarrow M \cong N.$$

Let $\underline{p}_1,\dots,\underline{p}_n$ be the prime ideals in R. It follows from (5.7), that

$$KM \cong KN \Longleftrightarrow M_{\underline{p}_i} \cong N_{\underline{p}_i}\ ,\ 1 \le i \le n;$$

i.e., $\exists\ a_{\underline{p}_i} \in \mathrm{Hom}_{\Gamma_{\underline{p}}}(M_{\underline{p}_i}, N_{\underline{p}_i}) \subset \mathrm{Hom}_A(KM,KN)$, such that $M_{\underline{p}_i}\, a_{\underline{p}_i} = N_{\underline{p}_i}$. We may assume, that KM = KN and that $N \subset M$ and that $a_{\underline{p}_i} \in \mathrm{End}_{\Gamma}(M)$. Now, $\underline{p}_i \cdot \mathrm{End}_{\Gamma}(M)$, $1 \le i \le n$, satisfy the hypotheses of the Chinese remainder theorem (cf. I,(7.7)). Hence $M \cong N$.)

2.) Let R be a Dedekind domain with quotient field K, L a finite separable extension of K and Γ the integral closure of R in L. For a prime ideal $\underline{p}$ in R, we have

$$\underline{p}\Gamma = \prod_{i=1}^{n} P_i^{e_i}\ ,$$

where P_i, $1 \le i \le n$, are prime ideals in Γ; e_i is called the <u>ramification index of P_i over $\underline{p}$</u>. Then Γ/P_i is an extension field of $R/\underline{p}$ of degree f_i, $1 \le i \le n$. f_i is called the <u>residue class degree</u>, and we have

$$[L : K] = \sum_{i=1}^{n} e_i f_i\ .$$

(Hint: Use (5.9): $\hat{L}_{\underline{p}}$ is the direct sum of n extension fields $\hat{L}_i$ of $\hat{K}_{\underline{p}}$, and $\hat{\Gamma}_{\underline{p}} = \oplus_{i=1}^{n} \hat{\Gamma}_i$; rad $\Gamma_i = \hat{P}_i'$. Then $\hat{P}_i'$ has residue class degree f_i and ramification index e_i, and $e_i f_i = [\hat{L}_i : \hat{K}_{\underline{p}}]$, whence the above formula.)

3.) Let $\Gamma_{\underline{p}}$ be a maximal $R_{\underline{p}}$-order in a finite dimensional K-algebra. Show that every left $\Gamma_{\underline{p}}$-ideal is principal.

§6. Maximal orders in skewfields over complete fields

The arithmetic structure of the maximal order in a complete central skewfield is clarified; all possible complete skewfields are constructed.

In this section, $\hat{R}$ with quotient field $\hat{K}$ is the $\underline{p}$-adic completion of a Dedekind domain R (cf. I,(§9)) with respect to a prime ideal $\underline{p}$ of R, with rad $\hat{R} = \hat{\pi}\hat{R}$, and $\overline{R} = \hat{R}/\underline{p}\hat{R}$.

6.1 Hensel's Lemma: Let $f(X) \in \hat{R}[X]$, and assume that there are polynomials $g_o(X)$, $h_o(X) \in \hat{R}[X]$ satisfying

(i) $f(X) - g_o(X)h_o(X) \in \hat{\pi}\hat{R}[X]$,

(ii) $g_o(X)\hat{R}[X] + h_o(X)\hat{R}[X] + \hat{\pi}\hat{R}[X] = \hat{R}[X]$,

(iii) $g_o(X)$ is monic.

Then there exist polynomials $g(X)$, $h(X) \in \hat{R}[X]$, such that

(i') $f(X) = g(X)h(X)$,

(ii') $g_o(X) - g(X) \in \hat{\pi}\hat{R}[X]$ and $h_o(X) - h(X) \in \hat{\pi}\hat{R}[X]$,

(iii') $g(X)$ is monic and degree $g(X)$ = degree $g_o(X)$.

In particular, a separable polynomial that has a root in $\overline{R}$ also has a root in $\hat{R}$.

Proof: We have an isomorphism

$\sigma : \varprojlim(\hat{R}[X]/\hat{\pi}^n\hat{R}[X]) \longrightarrow \hat{R}[X]$, since $\hat{R}[X]/\hat{\pi}^n\hat{R}[X] \cong (\hat{R}/\hat{\pi}^n\hat{R})[X]$.

We set $S_i = \hat{R}[X]/\hat{\pi}^i\hat{R}[X] \cong (\hat{R}/\hat{\pi}^i\hat{R})[X]$, $i = 1,2,\ldots$, and let φ_i, φ_{ij}, for $i \geq j$, be the canonical homomorphisms, (cf. I,(9.4)),

$$\varphi_i : \hat{R}[X] \longrightarrow S_i$$

$$\varphi_{ij} : S_i \longrightarrow S_j = S_i/\hat{\pi}^j S_i \ .$$

Now we construct sequences $\{g_i\}$, $\{h_i\} \in \varprojlim S_i$ so that their images $g(X)$ and $h(X)$ under σ have the desired properties: We put

$$g_1 = (g_o(X))\varphi_1 \text{ and } h_1 = (h_o(X))\varphi_1.$$

The conditions

(i") $(f(X))\varphi_1 = g_1 h_1$,

(ii") $g_1 S_1 + h_1 S_1 + \hat{\pi} S_1 = S_1$, $(g_o(X))\varphi_1 - g_1\varphi_{1,1} = 0$, $(h_o(X))\varphi_1 - h_1\varphi_{1,1} = 0$,

(iii") g_1 is monic and degree g_1 = degree $g_o(X)$,

are then obviously satisfied for $i = 1$. Assume now that they hold for $i = n$. Because $\varphi_{n+1,n}$ is epic we can choose $g'_{n+1}, h'_{n+1} \in S_{n+1}$, so that g'_{n+1} is a monic polynomial and of the same degree as g_n, and that $g'_{n+1}\varphi_{n+1,n} = g_n$ and $h'_{n+1}\varphi_{n+1,n} = h_n$. Since $\varphi_{n+1}\varphi_{n+1,n} = \varphi_n$ and by assumption (i"), there exists $s \in S_{n+1}$, so that

$$(f(X))\varphi_{n+1} - g'_{n+1}h'_{n+1} = \hat{\pi}^n s \in \operatorname{Ker}\varphi_{n+1,n} = \hat{\pi}^n S_{n+1}.$$

By induction assumption (ii") and since $\operatorname{Ker}\varphi_{n+1,n} \subset \hat{\pi} S_{n+1}$, we have $g'_{n+1}S_{n+1} + h'_{n+1}S_{n+1} + \hat{\pi} S_{n+1} = S_{n+1}$, and thus, since $\hat{\pi}^{n+1}S_{n+1} = 0$, there are $x, y \in S_{n+1}$ such that

$$\hat{\pi}^n s = g'_{n+1}\hat{\pi}^n x + h'_{n+1}\hat{\pi}^n y.$$

Moreover, y can be chosen so that its degree is strictly less than that of g'_{n+1}. For, if degree $y \geq$ degree g'_{n+1}, then there exist q, $y' \in S_{n+1}$ so that $y = q\, g'_{n+1} + y'$ and degree y' < degree g'_{n+1}. (This follows simply from the fact that g'_{n+1} is monic.) But then y can be replaced by y' and x by $x + q\, h'_{n+1}$. Now we set

$$g_{n+1} = g'_{n+1} + \hat{\pi}^n y, \quad h_{n+1} = h'_{n+1} + \hat{\pi}^n x.$$

Our induction assumtions are now easily verified for $i = n + 1$. Furthermore we have

$$g_{n+1}\varphi_{n+1,n} = g_n \text{ and } h_{n+1}\varphi_{n+1,n} = h_n .$$

It follows that our sequences do indeed belong to $\varprojlim S_1$ and have the desired properties (cf. I,(9.2)). #

6.2 Theorem: Let $\hat{A}$ be a finite dimensional separable skewfield over $\hat{K}$. Then $a \in \hat{A}$ is integral over $\hat{R}$ if and only if $N_{\hat{A}/\hat{K}}(a) \in \hat{R}$.

Proof: One shows as in (1.4') that $N_{\hat{A}/\hat{K}}(a) \in \hat{R}$ whenever a is integral over $\hat{R}$ (cf. III,(6.15)). Thus we may assume that $a \in \hat{A}$ with $N_{\hat{A}/\hat{K}}(a) \in \hat{R}$ and it remains to show that $\min_{\hat{A}/\hat{K}}(a,X) \in \hat{R}[X]$ (cf. III,(3.1)). Since $\hat{K}[X]/(\min_{\hat{A}/\hat{K}}(a,X))$ is isomorphic to a subring of the skewfield $\hat{A}$, which does not contain zero divisors, $\min_{\hat{A}/\hat{K}}(a,X)$ is irreducible. However, since $\min_{\hat{A}/\hat{K}}(a,X)$ divides $Pc_{\hat{A}/\hat{K}}(a,X)$, (cf. III,(3.4)), and $Pc_{\hat{A},\hat{K}}(a,X)$ divides $\min_{\hat{A}/\hat{K}}(a,X)^n$, (cf. III,(3.5)), this implies that, for some $m \geq n$,

$$\min_{\hat{A}/\hat{K}}(a,X)^m = Pc_{\hat{A}/\hat{K}}(a,X) \quad \text{(cf. Ex. 6,2)}.$$

Now, $Pc_{\hat{A}/\hat{K}}(a,X)$ has leading ceofficient 1 and constant term in $\hat{R}$, since $N_{\hat{A}/\hat{K}}(a,X) \in \hat{R}$ (cf. III,(3.2')). Since $\hat{R}$ is integrally closed in $\hat{K}$ the same is true of $\min_{\hat{A}/\hat{K}}(a,X)$. Thus it suffices to show that, whenever

$$f(X) = X^m + k_{m-1}X^{m-1} + \dots + r_o \in \hat{K}[X]$$

is an irreducible polynomial with $r_o \in \hat{R}$, then $f(X) \in \hat{R}[X]$. Multiplying $f(X)$ by some $0 \neq r \in \hat{R}$ we may assume that

$$f_1(X) = \alpha_m X^m + \alpha_{m-1}X^{m-1} + \dots + \alpha_o = rf(X) \in \hat{R}[X]$$

is a primitive irreducible polynomial (cf. I, Ex. 7,6). It remains to show that α_m is a unit in $\hat{R}$. Assume, to the contrary, that $\alpha_m \in \hat{\pi}\hat{R}$. But then $\alpha_o = \alpha_m r_o \in \hat{\pi}\hat{R}$, since $r_o \in \hat{R}$, and there exists a largest number m', $0 < m' < m$, for which $0 \neq \alpha_{m'} \notin \hat{\pi}\hat{R}$. It follows that

$$f_1(X) \equiv \alpha_{m'}(X^{m'} + \alpha_{m'}^{-1}\alpha_{m'-1}X^{m'-1} + \dots + \alpha_{m'}^{-1}\alpha_o) \bmod \hat{\pi}\hat{R}[X],$$

and $f(X)$ is reducible by Hensel's lemma (cf. 6.1), a contradiction.

Thus, $\min_{\hat{A}/\hat{K}}(a,X) \in \hat{R}[X]$. #

6.3 Corollary: Let $\hat{A}$ be a finite dimensional separable skewfield over $\hat{K}$. Then the unique maximal $\hat{R}$-order $\hat{\Gamma}$ in $\hat{A}$ is

$$\hat{\Gamma} = \{a \in \hat{A} : a \text{ is integral over } \hat{R}\} = \{a \in \hat{A} : N_{\hat{A}/\hat{K}}(a) \in \hat{R}\}.$$

Proof: Because of the uniqueness of $\hat{\Gamma}$ (cf. (5.2)) and by (6.2) it suffices to show that every integral element of $\hat{A}$ is contained in some $\hat{R}$-order. But this in fact, is true for any finite dimensional algebra A over the field of quotients K of a Dedekind domain R: Let $a \in A$ be integral over R and pick a K-basis $\{w_i\}_{1 \leq i \leq n}$ for A. R[a] is then an R-module of finite type and the module $M = \sum_{i=1}^{n} R[a]w_i$ is an R-lattice in A whose left order $\Lambda_1(M)$ contains a, (cf. (1.3)). #

6.4 Notation: As before, $\hat{R}$ with quotient field $\hat{K}$ and rad $\hat{R} = \hat{\pi}\hat{R}$ stands for the $\underline{p}$-adic completion of a Dedekind domain. In addition we assume now that $\hat{R}/\hat{\pi}\hat{R} = \bar{R}$ is a finite field, $[\bar{R} : 1] = q$. Let $\hat{A}$ be a finite dimensional separable skewfield over $\hat{K}$ with unique maximal $\hat{R}$-order $\hat{\Gamma}$ and set $\hat{P} = \text{rad } \hat{\Gamma} = \hat{\Gamma}\gamma$, (cf. (5.6)).

6.5 Lemma: Let K be an algebraic number field and R the integral closure of $\mathbb{Z}$ in K. Then $R/\underline{p}$ is a finite field for every prime ideal $\underline{p}$ of R.

Proof: We may view K as a separable $\mathbb{Q}$-algebra and R as the maximal $\mathbb{Z}$-order in K, (cf. (6.3)). For any prime ideal $\underline{p}$ in R, $\underline{p} \cap \mathbb{Z} = p\mathbb{Z}$ is a prime ideal in $\mathbb{Z}$, and $R/\underline{p}$ is a finite dimensional $\mathbb{Z}/p\mathbb{Z}$-algebra. Since $\mathbb{Z}/p\mathbb{Z}$ is a finite field, so is $R/\underline{p}$. #

6.6 Theorem (Hasse [2]): $\hat{\Gamma}/\hat{P}$ is a finite extension field of $\bar{R}$; its degree f over $\bar{R}$ is called the residue class degree of $\hat{A}$ over $\hat{K}$. For some positive integer e, called the ramification order of $\hat{A}$ over $\hat{K}$, we have $\hat{\pi}\hat{\Gamma} = \hat{P}^e$ and $[\hat{A} : \hat{K}] = e \cdot f$.

Proof: From (5.2) it follows that $\hat{\pi}\hat{\Gamma}$ is a power of $\hat{P}$ and that, since there exist no finite skewfields (cf. III,(6.7)). $\hat{\Gamma}/\hat{P}$ is an extension field of $\bar{R}$, (cf. (4.10)). (Observe that rad $\hat{R} = \hat{R} \cap$ rad $\hat{\Gamma}$, whence we may identify $(\hat{R}$ + rad $\hat{\Gamma})/$rad $\hat{\Gamma}$ with $\bar{R} = \hat{R}/$rad $\hat{R}$.) Finally, $[\hat{A} : \hat{K}] = [\hat{\Gamma} : \hat{R}] = [\hat{\Gamma}/\hat{\pi}\hat{\Gamma} : \bar{R}] = [\hat{\Gamma}/\hat{P}^e : \bar{R}]$, and because of the module isomorphisms

$$\hat{P}^i/\hat{P}^{i+1} = \hat{\Gamma}\gamma^i/\hat{\Gamma}\gamma^{i+1} \cong \hat{\Gamma}/\hat{\Gamma}\gamma = \hat{\Gamma}/\hat{P},$$

we obtain $[\hat{A} : \hat{K}] = [\hat{\Gamma}/\hat{P}^e : \bar{R}] = e[\hat{\Gamma}/\hat{P} : \bar{R}] = e \cdot f$. #

6.7 Corollary: If $\hat{K}$ is the center of $\hat{A}$, then $e = f = \sqrt{[\hat{A} : \hat{K}]}$.

Proof: According to (III,(6.5)), $[\hat{A} : \hat{K}] = m^2$, where m is the dimension of a maximal subfield of $\hat{A}$ over $\hat{K}$. $\hat{\Gamma}/\hat{P}$ being a finite separable extension field of $\bar{R}$, we have $\hat{\Gamma}/\hat{P} = \bar{R}(\bar{\omega})$, for some $\bar{\omega} \in \hat{\Gamma}/\hat{P}$, (cf. III, Ex. 5,8). If ω is a preimage of $\bar{\omega}$ in $\hat{\Gamma}$, then $\hat{K}(\omega)$ is a subfield of $\hat{A}$ and $m \geq [\hat{K}(\omega) : \hat{K}] \geq f$. On the other hand, $\hat{K}(\gamma)$ is also a subfield of $\hat{A}$. We claim that $m \geq [\hat{K}(\gamma) : \hat{K}] = e$. If not, we would have a relation $\sum_{i=0}^{e-1} r_i \gamma^i = 0$, $r_i \in \hat{R}$, $0 \leq i \leq e-1$, with at least one $r_i \notin \hat{\pi}\hat{R}$. This implies

$$\sum_{r_i \notin \hat{\pi}\hat{R}} r_i \gamma^i = - \sum_{r_i \in \hat{\pi}\hat{R}} r_i \gamma^i \in \hat{\pi}\hat{\Gamma} = \hat{\Gamma}\gamma^e.$$

Let j be the smallest integer i such that $r_i \notin \hat{\pi}\hat{R}$, then r_j is a unit in $\hat{R}$; we obtain $\gamma^j \in \hat{\Gamma}\gamma^{j+1}$, hence $\hat{\Gamma}\gamma^j = \hat{\Gamma}\gamma^{j+1}$, and thus a contradiction (cf. (4.16),(5.2)). Now it follows from (6.6) that $m = e = f$. #

6.8 Theorem (Hasse [2]): Let $\{\omega_i\}_{1 \leq i \leq f}$ be inverse images of an $\bar{R}$-basis for $\hat{\Gamma}/\hat{P}$ in $\hat{\Gamma}$. Then an $\hat{R}$-basis for $\hat{\Gamma}$ is given by $\{\omega_i \gamma^j\}$, $1 \leq i \leq f$, $0 \leq j \leq e-1$.

Proof: By construction the elements $\omega_i \gamma^j + \hat{\pi}\hat{\Gamma}$, with $1 \leq i \leq f$, $0 \leq j \leq e-1$, form an $\bar{R}$-basis for $\hat{\Gamma}/\hat{\pi}\hat{\Gamma}$. Therefore $\sum_{j=0}^{e-1}\sum_{i=1}^{f} \hat{R}\omega_i \gamma^j + \hat{\pi}\hat{\Gamma} = \hat{\Gamma}$, and from Nakayama's lemma (I,(4.18)) we conclude that

$\sum_{j=0}^{e-1}\sum_{i=1}^{f} \hat{R}\omega_i\gamma^j = \hat{\Gamma}$. The lemma now follows from the fact that $[\hat{\Gamma} : \hat{R}] = e \cdot f$. #

6.9 <u>Theorem</u> (Hasse [2]): Let $\hat{A}$ have residue class degree f and ramification order e over $\hat{K}$. Then $\hat{A}$ contains a primitive (q^f-1)-th root of 1, say ω, and $\bar{\omega} = \omega + \text{rad}\,\hat{\Gamma}$ is a primitive (q^f-1)-th root of 1 in $\hat{\Gamma}/\text{rad}\,\hat{\Gamma}$. The subfield $\hat{L} = \hat{K}(\omega)$ is called a <u>field of inertia for $\hat{A}$ over $\hat{K}$</u>. $\hat{L}$ has ramification order 1 over $\hat{K}$; i.e., it is <u>unramified</u> of residue class degree f over $\hat{K}$.

Conversely, if $\rho \in \hat{A}$ is a primitive n-th root of 1, and the characteristic p of the residue class field $\bar{R}$ does not divide n, then $n \mid (q^f-1)$ and the field $\hat{K}(\rho)$ is unramified and is a subfield of some field of inertia $\hat{L} = \hat{K}(\omega)$ of $\hat{A}$, where ω is a primitive (q^f-1)-th root of 1.

Moreover, $\hat{A} = \hat{K}(\omega, \gamma)$ for any $\gamma \in \hat{A}$ such that $\text{rad}\,\hat{\Gamma} = \hat{\Gamma}\gamma$, and $\hat{A}$ has ramification order e and residue class degree 1 over $\hat{L}$; i.e., $\hat{A}$ is <u>totally ramified</u> over $\hat{L}$.

<u>Proof</u>: Since $\hat{\Gamma}/\text{rad}\,\hat{\Gamma}$ is a finite field of degree f over $\bar{R}$, its multiplicative group is cyclic of order q^f-1 and is therefore generated by a primitive (q^f-1)-th root of 1, say $\bar{\omega}_0$. Let $\omega_0 \in \hat{\Gamma}$ be a preimage of $\bar{\omega}_0$. If $\hat{\Delta}$ is the maximal $\hat{R}$-order in the field $\hat{K}(\omega_0)$, then $\text{rad}\hat{\Delta} = \hat{\Delta} \cap \text{rad}\hat{\Gamma}$. Indeed, if $\text{rad}\,\hat{\Delta} = \hat{\Delta}\cdot\delta_0$, δ_0 is not a unit in $\hat{\Gamma}$. Since $\delta_0^{-1} \in \hat{\Gamma}$, implies that δ_0^{-1} is integral over $\hat{R}$, in $\hat{A}$, and consequently $\delta_0^{-1} \in \hat{K}(\omega_0)$ is integral over $\hat{R}$; i.e., $\delta_0^{-1} \in \hat{\Delta}$ (cf. (6.2)). Thus $\hat{\Delta} \cap \text{rad}\,\hat{\Gamma} \supset \text{rad}\,\hat{\Delta}$; on the other hand $\text{rad}\,\hat{\Delta}$ is the unique maximal ideal in $\hat{\Delta}$ (cf. (5.2)). Since surely $\hat{\Delta} \neq \text{rad}\,\hat{\Gamma} \cap \hat{\Delta}$, we have $\hat{\Delta} \cap \text{rad}\,\hat{\Gamma} = \text{rad}\,\hat{\Delta}$; and consequently $(\hat{\Delta} + \text{rad}\hat{\Gamma})/\hat{\Gamma} \cong \hat{\Delta}/\text{rad}\,\hat{\Delta}$, and we may view $\hat{\Gamma}/\text{rad}\,\hat{\Gamma}$ as an extension field of $\hat{\Delta}/\text{rad}\,\Delta$, and $\omega_0 \in \hat{\Delta}$, since $\omega_0 \in \hat{\Gamma} \cap \hat{K}(\omega_0)$. $\text{rad}\hat{\Delta} = \hat{\Delta} \cap \text{rad}\hat{\Gamma}$

implies that $\omega_0^s - 1 \in \operatorname{rad}\hat{\Delta}$ if and only if $\omega_0^s - 1 \in \operatorname{rad}\hat{\Gamma}$. Therefore, $\bar{\bar{\omega}}_0 = \omega_0 + \operatorname{rad}\hat{\Delta}$ is a primitive (q^f-1)-th root of 1 in $\hat{\Delta}/\operatorname{rad}\hat{\Delta}$, and consequently, $\hat{\Delta}/\operatorname{rad}\hat{\Delta} = \hat{\Gamma}/\operatorname{rad}\hat{\Gamma} = \bar{R}(\bar{\bar{\omega}}_0)$. Now since $\bar{R}(\bar{\bar{\omega}}_0)$ is of degree f over $\bar{R}$, there are polynomials $g_0(X)$, $h_0(X) \in \hat{R}[X]$ such that

$$X^{q^f-1} - 1 \equiv g_0(X)h_0(X) \bmod \hat{\pi}\hat{R}[X],$$

where $\bar{g}_0(X)$ is irreducible of degree f over $\bar{R}$ and $\bar{g}_0(\bar{\bar{\omega}}_0) = 0$, here $\bar{g}_0(X)$ stands for the image of $g_0(X)$ under the canonical homomorphism $\hat{R}[X] \longrightarrow \bar{R}[X]$. Moreover, we may assume $g_0(X)$ to be monic, and since the roots of $X^{q^f}-1$ are all distinct $\bar{g}_0(X)\bar{R}[X] + \bar{h}_0(X)\bar{R}[X] = \bar{R}[X]$, so that, by Hensel's lemma for some $g(X)$, $h(X) \in \hat{R}[X]$, $X^{q^f-1} - 1 = g(X)h(X)$, where $g(X)$ is monic of degree f over $\hat{R}$ and irreducible, since $\bar{g}(X) = \bar{g}_0(X)$ (cf. Gauss's lemma (I, Ex. 7.6)). Now, $\hat{\Delta}$ is a Dedekind domain, complete with respect to the $\operatorname{rad}\hat{\Delta}$-adic topology and for some $g_1(X) \in \hat{\Delta}[X]$ we have

$$g(X) \equiv (X-\omega_0)g_1(X) \bmod (\operatorname{rad}\hat{\Delta}[X]),$$

and

$$(X-\omega_0)\hat{\Delta}[X] + g_1(X)\hat{\Delta}[X] + \operatorname{rad}\hat{\Delta}[X] = \hat{\Delta}[X].$$

And Hensel's lemma yields the existence of $\omega \in \hat{\Delta}$, such that $g(\omega) = 0$ and $\omega - \omega_0 \in \operatorname{rad}\hat{\Delta}$. Moreover, since $\bar{\bar{\omega}}_0 = \omega + \operatorname{rad}\hat{\Delta}$ is a primitive (q^f-1)-th root of 1, so is ω. Now set $\hat{L} = \hat{K}(\omega)$, and let $\hat{\Omega}$ be the maximal $\hat{R}$-order in $\hat{L}$. Then $\omega \in \hat{\Omega}$, (cf. (6.3)), $\omega \notin \operatorname{rad}\hat{\Omega}$ and $[\hat{\Omega}/\operatorname{rad}\hat{\Omega} : \bar{R}] = f$, since $\bar{g}(X)$ is irreducible over $\bar{R}$. Thus $\hat{L}$ has residue class degree f, and therefore ramification order 1 over $\hat{R}$, (cf. (6.6)), hence $\operatorname{rad}\hat{\Omega} = \hat{\pi}\hat{\Omega}$, and $\hat{L}$ is unramified over $\hat{K}$. Now suppose that $\rho \in \hat{A}$ is a primitive n-th root of 1 and that $p \nmid n$. By Hensel's lemma $\hat{R}$ contains a primitive $(q-1)$-th root of 1, say δ. Since $(n,q) = 1$, there exists a smallest positive integer s such that $q^s \equiv 1 \bmod n$. $\omega_o = \delta\varrho$ is then a primitive (q^s-1)-th root of 1 in $\hat{A}$ and belongs to the maximal $\hat{R}$-order $\hat{\Delta}$ of the field $\hat{K}(\omega_o) = \hat{K}(\varrho)$. $\hat{\Delta}/\mathrm{rad}\,\hat{\Delta}$ is a subfield of $\hat{\Gamma}/\mathrm{rad}\,\hat{\Gamma}$ (cf. above) and contains the (q^s-1)-th root $\bar{\omega}_o = \omega_o + \mathrm{rad}\,\hat{\Delta}$ of 1. We show that $\bar{\omega}_o$ is a primitive (q^s-1)-th root of 1. Assume $\bar{\omega}_o^{q^t-1} = 1$, for $t \leq s$ (observe that the multiplicative group of $\hat{\Delta}/\mathrm{rad}\,\hat{\Delta}$ is cyclic of order $(q^\tau-1)$, for some $\tau \leq s$.) Since $\omega_o = \omega_o^{q^s}$, we have

$$\omega_o - \omega_o^{q^t} = \omega_o^{q^{sr}} - (\omega_o^{q^t})^{q^{sr}} \equiv (\omega_o - \omega_o^{q^t})^{q^{sr}} \bmod (\mathrm{rad}\,\hat{\Delta})^{sr}$$

for all positive integers r, using the fact that $x^{q^m} - y^{q^m} \equiv (x-y)^{q^m} \bmod (\mathrm{rad}\,\hat{\Delta})^m$, since $q \in \mathrm{rad}\,\hat{\Delta}$. But then, since by assumption $\omega_o - \omega_o^{q^t} \in \mathrm{rad}\,\hat{\Delta}$, it follows that $\omega_o - \omega_o^{q^t} \in \bigcap_n (\mathrm{rad}\,\hat{\Delta})^n = 0$ (cf. Herstein's lemma (I,(9.1)) and Nakayama's lemma (I,(4.18))); hence $s=t=\tau$. Thus we conclude, as in the first part of the proof, that $\hat{K}(\omega_o) = \hat{K}(\varrho)$ contains a primitive (q^s-1)-th root of 1 that is a root of an irreducible polynomial of degree s, $\hat{K}(\varrho)$ being a field, and ω_o is a power of this root; thus $[\hat{K}(\varrho) : \hat{K}] = s = [\hat{\Delta}/\mathrm{rad}\,\hat{\Delta} : \bar{R}]$, $\hat{K}(\varrho)$ is unramified over $\hat{K}$, and n divides q^f-1 since it divides q^s-1 and q^s-1 clearly divides q^f-1.

To show that $\hat{K}(\rho)$ is contained in some field of inertia of A over K we set $\hat{A}' = \{a \in \hat{A} : a\rho = \rho a\}$ and view $\hat{A}'$ over $\hat{K}(\rho)$. If $\hat{\Gamma}'$ is the maximal $\hat{R}(\rho)$-order in $\hat{A}'$, then $\hat{\Gamma}' \subset \hat{\Gamma}$ (cf.(6.3)), and $\mathrm{rad}\,\hat{\Gamma}' = \hat{\Gamma}' \cap \mathrm{rad}\,\hat{\Gamma}$, since $\mathrm{rad}\,\hat{\Gamma}'$ consists of all non-units of $\hat{\Gamma}'$; moreover, $\hat{\Gamma} = \hat{\Gamma}' + \mathrm{rad}\,\hat{\Gamma}$;*) hence $\hat{\Gamma}'/\mathrm{rad}\,\hat{\Gamma}' \cong \hat{\Gamma}/\mathrm{rad}\,\hat{\Gamma}$, and $\hat{A}'$ contains a primitive (q^f-1)-th root ω of 1. But then $\hat{K}(\varrho, \omega) = \hat{K}(\omega)$ is a field of inertia for $\hat{A}$ over $\hat{K}$.

We record the following consequence of the above discussion:

6.9' <u>Remark</u>: $\hat{A}$ is unramified over $\hat{K}$ if and only if it can be obtained from $\hat{K}$ by the adjunction of a root of 1 whose order is relatively prime to the characteristic p of the residue class field $\bar{R}$.

---

*) Observe that $x = 1/n(nx - \sum_{i=0}^{n-1} \varrho^{-i} x \varrho^i + \sum_{i=0}^{n-1} \varrho^{-i} x \varrho^i) = a + b$, $\forall x \in \hat{\Gamma}$, where $a = 1/n \sum_{i=0}^{n-1}(x - \varrho^{-i} x \varrho^i) \in \mathrm{rad}\,\hat{\Gamma}$, since n is a unit in $\hat{\Gamma}$ and $\hat{\Gamma}/\mathrm{rad}\,\hat{\Gamma}$ is commutative, and $b = 1/n \sum_{i=0}^{n-1} \varrho^{-i} x \varrho^i \in \hat{\Gamma}'$.

Finally, if rad $\hat{\Gamma} = \hat{\Gamma}\gamma$, then it follows from (6.8) that $\hat{A} = \hat{L}(\gamma)$, whenever $\hat{L}$ is a field of inertia of $\hat{A}$. Since the residue class fields $\hat{\Gamma}/\text{rad}\ \hat{\Gamma}$ and $\hat{\Omega}/\text{rad}\ \hat{\Omega}$, where $\hat{\Omega}$ is the maximal $\hat{R}$-order in $\hat{L}$, are isomorphic, $\hat{A}$ has residue class degree 1 over $\hat{L}$ and thus must be totally ramified of ramification order e over $\hat{L}$. #

6.10 <u>Theorem</u> (Hasse [2]): Under the conditions of Theorem (6.9) the two Galois groups $\text{Gal}(\hat{L}/\hat{K}) \cong \text{Gal}((\hat{\Gamma}/\text{rad}\ \hat{\Gamma})/\bar{R})$ are isomorphic. They are cyclic of order f generated by the so-called <u>Frobenius automorphisms</u>, $\sigma : \omega \mapsto \omega^q$, $\bar{\sigma} : \bar{\omega} \mapsto \bar{\omega}^q$, resp.

<u>Proof</u>: We recall: If K_1 is a finite extension field of K, K_1 is called <u>separable over K</u> if $\min_{K_1/K}(a,X)$ is a separable polynomial for every $a \in K_1$. In that case K_1 is a <u>simple extension</u> of K; i.e., there exists $\alpha \in K_1$, such that $K_1 = K(\alpha)$. K_1 is said to be a <u>normal extension</u> of K, if every irreducible polynomial in K[X], that has a root in K_1, decomposes into linear factors in K_1. K_1 is called a <u>Galois extension</u> of K if it is finite, separable and normal over K. For $K(\alpha)$ to be a Galois extension of K it suffices that $\min_{K_1/K}(\alpha, X)$ be a separable polynomial over K, that factors completely in K_1. The Galois group $\text{Gal}(K_1/K)$ of K_1 over K consists of all automorphisms of K_1 that leave K elementwise fixed. If K_1 is a Galois extension of K, $K_1 = K(\alpha)$ and $\min_{K_1/K}(\alpha, X) = f(X)$, then, in $K_1[X]$:

$$f(X) = \prod_{i=1}^{n}(\alpha_i - X),\ \alpha_1 = \alpha,\ \alpha_i \neq \alpha_j \text{ for } i \neq j,$$

and $\text{Gal}(K_1/K)$ consists of $\{\varphi_i\}_{1 \le i \le n}$, where φ_i is induced by $\alpha \longmapsto \alpha_i$, $1 \le i \le n$. In particular $|\text{Gal}(K_1/K)| = [K_1 : K]$ = degree of f(X). Obviously, every extension K_1 of K that is obtained by adjoining a primitive s-th root ω of 1 to K, is a Galois extension, provided s does not divide the characteristic of K, for, $\min_{K_1/K}(\omega, X)$ has no repeated linear factors in any extension field of K, and factors completely in K_1, since ω is primitive.

Now we come to the <u>proof of (6.10)</u>: If $L = K(\omega)$ is a Galois extension of degree f of the field K by a primitive (q^f-1)-th root of 1, then $\min_{L/K}(\omega, X)$ is of degree f and divides $X^{q^f}-1$, and so all its roots are of the form ω^r. But, if $\omega \longmapsto \omega^r$ induces an automorphism of L, then $\omega^{r^f} = \omega$, and hence $r^f \equiv 1 \bmod (q^f-1)$. Now, this congruence has at most f solutions, while the f integers q^i, $0 \leq i < f$, are solutions. Hence the roots of $\min_{L/K}(\omega, X)$ are exactly the elements ω^{q^i}, with $i = 0,1,\dots,f-1$, and $\omega \longmapsto \omega^q$ induces an automorphism φ of L of order f belonging to the Galois group of L over K. Since this group is of order f it is generated by σ; i.e.,

$$\mathrm{Gal}(L/K) = \langle \sigma \rangle .$$

Now the desired result follows for both Galois groups from (6.9).

Moreover, the Frobenius automorphism of $\overline{L} = \hat{\Omega}/\mathrm{rad}\ \hat{\Omega} = \hat{\Gamma}/\mathrm{rad}\ \hat{\Gamma}$ is given by $\overline{\sigma} : l \longmapsto l^q$, $l \in \overline{L}$, since, as is easily verified, this is an automorphism of $\overline{L}$ that leaves $\overline{R}$ elementwise fixed and that it is of order f. (Observe that the charactestic p of $\overline{L}$ divides q and that q is prime to the order q^f-1 of the multiplicative group of $\overline{L}$, while $\overline{\sigma}^t = 1$ implies $\overline{\omega}^{q^t} = \overline{\omega}$, whence the order of $\overline{\sigma}$ is a multiple of f and therefore = f.) #

6.11 <u>Theorem</u> (Hasse [2]): Let $\hat{L} = \hat{K}(\omega)$, where ω is a primitive (q^f-1)-th root of 1. Then an element $k \in \hat{K}$ is the norm of an element $l \in \hat{L}$; i.e., $k = N_{\hat{L}/\hat{K}}(l)$, for some $l \in \hat{L}$, if and only if $k = u\,\hat{\pi}^{tf}$, for some unit $u \in \hat{R}$ and $t \in \underline{\underline{Z}}$.

<u>Proof</u>: By (Ex. 6,1) any element $k \in \hat{K}$ can be written uniquely as $k = u\,\hat{\pi}^t$, with $t \in \underline{\underline{Z}}$ and u a unit in $\hat{R}$. We adhere to the notation of (6.9).

(i) If $k = N_{\hat{L}/\hat{K}}(l)$, with $l \in \hat{L}$, we can write $l = u' \hat{\pi}^t$ with a unit $u' \in \hat{\Omega}$ and $t \in \underline{\underline{Z}}$, since $\hat{L}$ is unramified over $\hat{K}$ (cf. (6.9'); i.e., rad $\hat{\Omega} = \hat{\pi}\hat{\Omega}$. But then, the norm being multiplicative (cf. III, Ex. 3,1),

$$N_{\hat{L}/\hat{K}}(u' \hat{\pi}^t) = u N_{\hat{L}/\hat{K}}(\hat{\pi}^t) = u \hat{\pi}^{tf},$$

where $u = N_{\hat{L}/\hat{K}}(u')$ is a unit in $\hat{R}$.

(ii) Conversely, let $k = u \hat{\pi}^{tf}$ for some $t \in \underline{\underline{Z}}$. Since $N_{\hat{L}/\hat{K}}(\hat{\pi}^t) = \hat{\pi}^{tf}$, it suffices to show that every unit $u \in \hat{R}$ is the norm of an element $l \in \hat{L}$. In fact, this has to be shown only for units $u \in \hat{R}$, for which $u \equiv 1 \mod \hat{\pi}\hat{R}$. For, from the proof of (6.10) it follows that

$$Pc_{\hat{L}/\hat{K}}(\omega, X) = \min_{\hat{L}/\hat{K}}(\omega, X) = \prod_{i=0}^{f-1} (\omega^{q^i} - X),$$

since both polynomials have the same degree (cf. III, (3.4) and (3.5)). Thus

$$N_{\hat{L}/\hat{K}}(\omega) = \prod_{i=0}^{f-1} \omega^{q^i} = \omega^{\sum_{i=0}^{f-1} q^i} = \omega^{\frac{q^f-1}{q-1}}.$$

But $\omega_1 = \omega^{\frac{q^f-1}{q-1}}$ is a primitive $(q-1)$-th root of 1, and thus $\overline{N_{\hat{L}/\hat{K}}(\omega)}$ generates the multiplicative group $\bar{R}^* = \bar{R} \setminus \{0\}$ of $\bar{R}$. Consequently, given a unit $u \in \hat{R}$, we can determine $s \in \underline{\underline{N}}$ such that $u \equiv N_{\hat{L}/\hat{K}}(\omega^s) \mod (\hat{\pi}\hat{R})$; i.e., $u = \omega_1^s + \hat{\pi} r = (1 + \hat{\pi} r \omega_1^{-s})\omega_1^s =$ $= u_1 \cdot N_{L/K}(\omega^s)$, for some $r \in \hat{R}$, where $u_1 = 1 + \hat{\pi} r \omega_1^{-s} \equiv 1 \mod (\hat{\pi}\hat{R})$ and $u_1 \in \hat{R}$, since ω_1 is a unit in $\hat{R}$. Thus it suffices to show that u_1 is a norm.

Now, let $u \equiv 1 \mod (\hat{\pi}\hat{R})$ be given. We shall construct a sequence $\{l_i\}_{i \in \underline{\underline{N}}}$, $l_i \in \hat{\Omega}$ such that

$$N_{\hat{L}/\hat{K}}(l) = u, \text{ for } l = \varprojlim l_i \varphi_i \in \hat{\Omega} = \varprojlim \hat{\Omega}/\hat{\pi}^i\hat{\Omega}.$$

For this purpose we observe that

$$N_{\hat{L}/\hat{K}}(1 + \hat{\pi}^s l) \equiv (1 + \hat{\pi}^s Tr_{\hat{L}/\hat{K}}(l)) \mod (\hat{\pi}^{s+1}\hat{R}),$$

for all $s \in \underline{\underline{N}}$, $l \in \hat{L}$. Indeed, from (Ex. 6.2) it follows that

$$Pc_{\hat{L}/\hat{K}}(l,X) = \prod_{i=1}^{f}(X-\sigma_i(l))$$

where $\sigma_i \in Gal(\hat{L}/\hat{K})$. Thus $N_{\hat{L}/\hat{K}}(l) = \prod_{i=1}^{f}\sigma_i(l)$, $Tr_{\hat{L}/\hat{K}}(l) = \sum_{i=1}^{f}\sigma_i(l)$ and hence, for any $s \in \underline{\underline{N}}$,

$$N_{\hat{L}/\hat{K}}(1+\hat{\pi}^s l) = \prod_{i=1}^{f}(1+\hat{\pi}^s\sigma_i(l)) \equiv (1+\hat{\pi}^s Tr_{\hat{L}/\hat{K}}(l)) \bmod (\hat{\pi}^{s+1}\hat{R}).$$

Moreover, every $\bar{r} \in \bar{R}$ is the trace of some $\bar{l} \in \bar{L} = \hat{\Omega}/\mathrm{rad}\,\hat{\Omega}$. Since $\bar{L}$ is separable over $\bar{R}$, the discriminant of any basis of $\bar{L}$ over $\bar{R}$ is non-zero and thus there exists $\bar{l}_o \in \bar{L}$ such that $Tr_{\bar{L}/\bar{R}}(\bar{l}_o) = \bar{k} \neq 0$, (cf. III,(3.1), (6.18) and Ex. 6,3). But then

$$Tr_{\bar{L}/\bar{R}}(\bar{r}/\bar{k}\cdot\bar{l}_o) = \bar{r}, \text{ for all } \bar{r} \in \bar{R}.$$

Finally, to construct our sequence $\{l_i\}_{i\in\underline{\underline{N}}}$ we set $l_1 = 1$. Assume that $l_1,\ldots,l_i$ have already been constructed so that

$$l_j \in \hat{\Omega},\ 1\le j\le i,\ l_{j+1}-l_j \in \hat{\pi}^j\hat{\Omega},\ 1\le j\le i-1,\ N_{\hat{L}/\hat{K}}(l_j) \equiv u \bmod (\hat{\pi}^j\hat{R}),\quad 1\le j\le i.$$

We can choose $\zeta_i \in \hat{\Omega}$ such that

$$Tr_{\hat{L}/\hat{K}}(\zeta_i) \equiv (u/N_{\hat{L}/\hat{K}}(l_i)-1)\hat{\pi}^{-i} \bmod (\hat{\pi}\hat{R}).$$

This can be done since $N_{\hat{L}/\hat{K}}(l_i)$ is a unit in $\hat{R}$ and $u-N_{\hat{L}/\hat{K}}(l_i) \in \hat{\pi}^i\hat{R}$, and thus $\alpha = (u/N_{\hat{L}/\hat{K}}(l_i)-1)\hat{\pi}^{-i} \in \hat{R}$, and there exists $\bar{l} \in \bar{L}$ such that $Tr_{\bar{L}/\bar{R}}(\bar{l}) = \bar{\alpha}$, and if l is a preimage of $\bar{l}$ in $\hat{\Omega}$, then $Tr_{\hat{L}/\hat{K}}(l) \equiv \alpha \bmod (\hat{\pi}\hat{R})$. (Observe that $\overline{Pc_{\hat{L}/\hat{K}}(l,X)} = Pc_{\bar{L}/\bar{R}}(\bar{l},X)$, since $Gal(\hat{L}/\hat{K}) \cong Gal(\bar{L}/\bar{R})$ via reduction modulo $\hat{\pi}$.) Now we set $l_{i+1} = l_i(1+\zeta_i\hat{\pi}^i)$, and it follows from our induction hypotheses that $l_{i+1} \in \hat{\Omega}$, $l_{i+1}-l_i = l_i\zeta_i\hat{\pi}^i \in \hat{\pi}^i\hat{\Omega}$,

$$N_{\hat{L}/\hat{K}}(l_{i+1}) = N_{\hat{L}/\hat{K}}(l_i)N_{\hat{L}/\hat{K}}(1+\zeta_i\hat{\pi}^i) \equiv N_{\hat{L}/\hat{K}}(l_i)(1+\hat{\pi}^i Tr_{\hat{L}/\hat{K}}(\zeta_i)) \bmod \hat{\pi}^{i+1}\hat{R};$$

i.e., $N_{\hat{L}/\hat{K}}(l_{i+1}) \equiv N_{\hat{L}/\hat{K}}(l_i)(u/N_{\hat{L}/\hat{K}}(l_i)) \bmod \hat{\pi}^{i+1}\hat{R} \equiv u \bmod \hat{\pi}^{i+1}\hat{R}$.

Now we put $l = \varprojlim l_i\varphi_i \in \hat{\Omega}$, where $\varphi_i : \hat{\Omega} \longrightarrow \hat{\Omega}/\hat{\pi}^i\hat{\Omega}$ are the canonical homomorphisms. This limit exists, since, by construction $l_{i+j}-l_i \in \hat{\pi}^i\hat{\Omega}$, for all $i, j \in \underline{\underline{N}}$.

Moreover, the norm function is continuous with respect to the $\hat{\pi}$-adic topology, since all $\sigma \in \mathrm{Gal}(\hat{L}/\hat{K})$ are continuous (cf. Ex. 6,3). Thus we have found an $1 \in \hat{\Omega}$ such that $N_{\hat{L}/\hat{K}}(1) = u$, and it follows that every element in $\hat{K}$ of the form $u\,\hat{\pi}^{\,t\cdot f}$, with u a unit in $\hat{R}$ and $t \in \underline{\underline{Z}}$, is the norm of an element of $\hat{L}$. #

6.12 <u>Remark</u>: Let $\hat{A}$ be a separable skewfield over $\hat{K}$. Then $\hat{A}$ is a central skewfield over its center $\hat{C}$ and $\hat{C}$ is an extension field of $\hat{K}$. If e' and f' are the ramification order and the residue class degree resp. of $\hat{C}$ over $\hat{K}$, then $[\hat{C} : \hat{K}] = e'\cdot f'$ (cf. (6.6)). If e and f are the ramification order and the residue class degree resp. of $\hat{A}$ over $\hat{K}$, then it follows from the preceding theorems that $e = e'm$ and $f = f'm$, where $[\hat{A} : \hat{C}] = m^2$. Moreover,

$$\hat{A} = \hat{K}(\omega, \gamma) = \hat{C}(\omega, \gamma), \hat{C} = K(\omega^{(q^f-1)/(q^{f'}-1)}, \alpha),$$

where ω is a primitive (q^f-1)-th root of 1 and $\alpha\hat{\Sigma} = \mathrm{rad}\,\hat{\Sigma}$, $\gamma\hat{\Gamma} = \mathrm{rad}\,\hat{\Gamma}$, where $\hat{\Sigma}$ and $\hat{\Gamma}$ are the respective maximal $\hat{R}$-orders in $\hat{C}$ and $\hat{A}$. Moreover, $\hat{\Gamma}$ is also the maximal $\hat{\Sigma}$-order in $\hat{A}$ (cf. (6.3), and III, Ex. 3,4). We shall show below that actually γ can be chosen so that $\alpha = \gamma^m$. It follows that a complete description of the separable skewfields over $\hat{K}$ is obtained by investigating the two extreme cases:

(i) the <u>commutative case</u>: $\hat{A} = \hat{C}$, and

(ii) the <u>central</u> case: $\hat{C} = \hat{K}$.

(i) If $\hat{A}$ is <u>commutative</u> of residue class degree f and ramification order e over $\hat{K}$, then it is obtained from $\hat{K}$ by two field extensions. The first one is achieved by adjoining a primitive (q^f-1)-th root of 1, say ω, and leads to $\hat{K}(\omega)$. This field contains q^f-1 roots of 1, and, since in a field a polynomial of degree n has at most n roots, $\hat{K}(\omega)$ contains all roots of 1 of order prime to p that belong to $\hat{A}$

(cf. (6.9)), and is therefore characterized as the smallest subfield of $\hat{A}$ with this property. Thus $\hat{L} = \hat{K}(\omega)$ may properly be called <u>the field of inertia of $\hat{A}$</u>. Observe that $\hat{L}$ is also the largest unramified subfield of $\hat{A}$. Now let $\hat{\Gamma}$ and $\hat{\Omega}$ stand for the maximal $\hat{R}$-orders in $\hat{A}$ and $\hat{L}$ resp. Then rad $\hat{\Omega} = \hat{\pi}\hat{\Omega}$, $\hat{A}$ is totally ramified over $\hat{L}$ with maximal $\hat{\Omega}$-order $\hat{\Gamma}$, and $\hat{A}$ is obtained from $\hat{L}$ by adjoining a root of a polynomial (cf. (6.9) and the proof of (6.2)),

$$g(X) = X^e + c_{e-1}X^{e-1} + \dots + c_o \text{ with}$$

$$c_{e-1}, \dots, c_o \in \hat{\pi}\hat{\Omega},\ c_o \notin \hat{\pi}^2\hat{\Omega}.$$

Conversely, every polynomial of this form (a so-called <u>Eisenstein polynomial</u>) is irreducible over $\hat{L}$ (cf. Ex. 6,5), and thus leads to a totally ramified extension of $\hat{L}$. Moreover, if the characteristic p of $\bar{R}$ does not divide e, then $\hat{A} = \hat{L}(\alpha)$, with $\alpha^e = \hat{\pi}$. For, if rad $\hat{\Gamma} = \gamma\hat{\Gamma}$, $\gamma^e\hat{\Gamma} = \hat{\pi}\hat{\Gamma}$, then $\gamma^e = \hat{\pi}u_o$ for some unit $u_o \in \hat{\Gamma}$. If $p \nmid e$, then $\bar{x} \longmapsto \bar{x}^e$ is an automorphism of the finite residue class field $\hat{\Gamma}$/rad $\hat{\Gamma}$, thus the polynomial $X^e - \bar{u}_o$, with $\bar{u}_o = u_o$ + rad $\hat{\Gamma}$, has a root in $\hat{\Gamma}$/rad $\hat{\Gamma}$, and it follows from Hensel's lemma that there is a unit $u \in \hat{\Gamma}$ such that $u^e = u_o$, and $\alpha = \gamma u^{-1}$ has all the desired properties. We observe, that in this case $\hat{A}$ has a maximal totally ramified subfield, namely $\hat{K}(\alpha)$, and $\hat{A} = \hat{K}(\omega) \otimes_K \hat{K}(\alpha)$, where ω is a primitive (q^f-1)-th root of 1 and $\alpha^e = \hat{\pi}$. $\hat{A}$ is completely characterized by the two integers e and f. In case $e \mid p$, the situation is more complicated.

(ii) Let now $\hat{A}$ be a <u>central skewfield</u> of dimension m^2 over $\hat{K}$, with maximal $\hat{R}$-order $\hat{\Gamma}$. $\hat{A}$ then contains a primitive (q^m-1)-th root of 1 (cf. (6.9)), say ω, and $\hat{\Gamma}$/rad $\hat{\Gamma} = \bar{R}(\bar{\omega})$, where $\bar{\omega} = \omega$ + rad $\hat{\Gamma}$. $\hat{K}(\omega)$ is then an unramified subfield of $\hat{A}$, and since its dimension over $\hat{K}$ is m, it is a maximal subfield of $\hat{A}$. Every maximal unramified subfield of $\hat{A}$ is called a <u>field of inertia</u> of $\hat{A}$. In contrast to the commutative case, there are infinitely many such subfields here. In fact, the fields of inertia constitute exactly one conjugacy class of maximal subfields of $\hat{A}$. For, clearly, whenever $\hat{L} = \hat{K}(\omega)$ is a field of inertia, so is $a\hat{L}a^{-1} = \hat{K}(a\,\omega\, a^{-1})$ for any $0 \neq a \in \hat{A}$; and conversely any two fields of inertia, being extensions of $\hat{K}$ by some primitive (q^m-1)-th roots of $1, \omega$ and ω' resp. are isomorphic and thus conjugate (cf. III,(6.6)), i.e., $\omega' = a\omega a^{-1}$ for some $0 \neq a \in \hat{A}$. Moreover, as we shall see below, $\hat{A}$ contains a totally ramified subfield $\hat{K}(\gamma)$ of degree m over $\hat{K}$ with $\gamma^m = \hat{\pi}$. Moreover, conjugation by γ induces the Frobenius automorphism $\tau_r : \omega \longmapsto \omega^{q^r}$, for some r, $(r,m) = 1$, on one of the fields of inertia $\hat{K}(\omega)$, and on the conjugate fields $a\hat{K}(\omega)a^{-1}$ the Frobenius automorphism is induced by conjugation with $a\gamma a^{-1}$. r is an invariant of $\hat{A}$ and Hasse has constructed to every pair $\{m,r\}$ with $(m,r) = 1$, $r < m$, a central skewfield over $\hat{K}$ of dimension m^2 with the invariant r. Thus, for a fixed field $\hat{K}$, and a distinguished prime element $\hat{\pi} \in \hat{R}$ there is established a one-one correspondence between the set of all central skewfields of finite dimension over $\hat{K}$ and the set of pairs of relatively prime positive integers, $\{m,r\}$, $r < m$, r/m is called the <u>Hasse invariant of $\hat{A}$ over $\hat{K}$</u>.

We proceed now to prove these facts.

6.13 <u>Theorem</u> (Hasse [2]): Let $\hat{A}$ be a central skewfield of dimension m^2 over $\hat{K}$ and let $\omega \in \hat{A}$ be such that $\hat{L} = \hat{K}(\omega)$ is a field of inertia of $\hat{A}$. Then there exists $\gamma \in \hat{\Gamma}$ such that

(i) $\hat{\Gamma}\gamma = \hat{P} = \text{rad } \hat{\Gamma}$

(ii) $\gamma\omega\gamma^{-1} = \omega^{q^r}$, with $(r,m) = 1$, $0<r<m$,

(iii) $\hat{R}\gamma^m = \text{rad } \hat{R}$.

<u>Proof</u>: Since $\omega \longmapsto \omega^q$ induces an automorphism of $\hat{L}$ (cf. (6.10)), and since $\hat{L}$ is a maximal subfield of $\hat{A}$, this automorphism is con-

jugation by an element of $\hat{A}$, (cf. III,(6.8)), i.e., $\omega^q = a\omega a^{-1}$ for some $0 \neq a \in \hat{A}$, and, if necessary by multiplying with an element of $\hat{R}$, we may assume that $a \in \hat{\Gamma}$. Now, by (5.2), the left ideal $\hat{\Gamma}a$ is a power of $\hat{P}$, that is $\hat{\Gamma}a = \hat{P}^s$, for some positive integer s. We show that s and m are relatively prime: if we put $n_1 = m/(s,m) \leq m$, where (s,m) denotes the greatest common divisor of s and m, then $a^{n_1} \in \hat{P}^{s\cdot n_1} \subset \hat{P}^m = \hat{\Gamma}\hat{\pi}$, and thus $a^{n_1} = u\hat{\pi}^t$, for some unit $u \in \hat{\Gamma}$, $t \in \underline{\underline{N}}$. But then

$$\omega^{q^{n_1}} = a^{n_1}\omega a^{-n_1} = u\hat{\pi}^t\omega\hat{\pi}^{-t}u^{-1} = u\omega u^{-1},$$

and, reducing modulo $\hat{P}$, we obtain $\bar{\omega}^{q^{n_1}} = \bar{\omega}$, since $\hat{\Gamma}/\hat{P}$ is commutative. It follows that $n_1 = m$ and therefore $(s,m) = 1$. If $s = 1$ we set $\gamma = a$; otherwise there are positive integers $r < m$ and j such that $1 = sr - mj$, and we set $\gamma = a^r/\hat{\pi}^j$. γ has all the desired properties. For, $\hat{\Gamma}a^r = \hat{P}^{sr} = \hat{P}^{1+mj} = \hat{P}\hat{\pi}^j$, thus $\gamma \in \hat{\Gamma}$ and $\hat{\Gamma}\gamma = \hat{P}$. Moreover, $\gamma\omega\gamma^{-1} = a^r\omega a^{-r} = \omega^{q^r}$, and clearly $(r,m) = 1$. Finally, $\gamma^m\omega\gamma^{-m} = \omega$, since $\gamma^m = \frac{a^{rm}}{\hat{\pi}^{jm}} = u^r\hat{\pi}^{tr-mj}$; therefore γ^m lies in the center of $\hat{A} = \hat{K}(\omega,\gamma)$. This is also true if $s = 1$. Moreover, $\gamma^m \in \hat{R}$ because it is integral over $\hat{R}$, thus $\gamma^m \in \hat{R} \cap \text{rad}\,\hat{\Gamma} = \text{rad}\,\hat{R}$, and, m being the smallest integer for which this holds, we conclude that $\hat{R}\gamma^m = \text{rad}\,\hat{R}$. #.

6.14 __Corollary:__ If $\gamma \in \hat{\Gamma}$, $\hat{\pi} \in \hat{R}$ and $r \in \underline{N}$ are given such that $\hat{\pi}\hat{R} = \text{rad}\,\hat{R}$, $\hat{\Gamma}\gamma = \hat{P}$ and $\gamma\omega\gamma^{-1} = \omega^{q^r}$, then there exists $\alpha \in \hat{\Gamma}$ such that $\alpha^m = \hat{\pi}$, $\hat{\Gamma}\alpha = \hat{P}$ and $\alpha\omega\alpha^{-1} = \omega^{q^r}$.

__Proof:__ Let $\gamma^m = \hat{\pi}_1$ (observe $\gamma^m\omega\gamma^{-m} = \omega$ implies $\gamma^m \in \hat{R}$). Then, by (6.13) $\hat{\pi}_1\hat{R} = \text{rad}\,\hat{R} = \hat{\pi}\hat{R}$, and $\hat{\pi} = u_1\hat{\pi}_1$ for some unit $u_1 \in \hat{R}$. Since $\hat{L} = \hat{K}(\omega)$ is unramified over $\hat{K}$ (cf. (6.9)), we can employ (6.11) to find $u \in \hat{L}$, such that $N_{\hat{L}/\hat{K}}(u) = u_1$. Since u_1 is a unit in $\hat{R}$, u is a unit in $\hat{L}$ (cf. Ex. 6,4). We set $\alpha = u\gamma$. Then

$$\alpha\omega\alpha^{-1} = u\gamma\omega\gamma^{-1}u^{-1} = u\,\omega^{q^r}u^{-1} = \omega^{q^r},$$

since $u \in \hat{K}(\omega)$ commutes with ω. Moreover, if we denote by σ_i the homomorphism $\hat{L} \longrightarrow \hat{L}$ induced by $\omega \longmapsto \gamma^i\omega\gamma^{-i} = \omega^{q^{ri}}$, then $\{\sigma_i\}_{0\le i\le m-1} = \mathrm{Gal}(\hat{L}/\hat{K})$, since $(r,m) = 1$. It follows from (Ex. 6,2) that

$$u_1 = N_{\hat{L}/\hat{K}}(u) = \prod_{i=0}^{m-1} \sigma_i(u) = \prod_{i=0}^{m-1} \gamma^i u\gamma^{-i} = (u\gamma)^m \cdot \gamma^{-m}.$$

Hence $\alpha^m = (u\gamma)^m = u_1\gamma^m = u_1\hat{\pi}_1 = \hat{\pi}$, and α has the desired properties. #

By (6.13) and (6.14) there exists to every field of inertia $\hat{L} = \hat{K}(\omega)$ an element $\alpha \in \hat{A}$ which is a root of the irreducible polynomial $X^m - \hat{\pi} \in \hat{R}[X]$, and which acts by conjugation as a Frobenius automorphism σ_r on $\hat{L}$. Clearly $\hat{K}(\alpha)$ is a totally ramified field of dimension m over $\hat{K}$ and thus a maximal subfield of $\hat{A}$. Moreover, if γ is any element in $\hat{A}$ such that $\gamma^m = \hat{\pi}$, then $\hat{K}(\gamma) \cong \hat{K}(\alpha)$, by $\alpha \longmapsto \gamma$, and it follows that $\gamma = b\alpha b^{-1}$ for some $b \in \hat{\Gamma}$, so that also $\hat{\Gamma}\gamma = \mathrm{rad}\,\hat{\Gamma}$ (observe that $\hat{\Gamma}\alpha b = \hat{\Gamma} b\alpha = \hat{\Gamma}\gamma b$ since $\mathrm{rad}\,\hat{\Gamma} = \hat{\Gamma}\alpha$ is a two-sided ideal) and $b\omega b^{-1} \longmapsto \gamma b\omega b^{-1}\gamma^{-1} = (b\omega b^{-1})^{q^r}$ induces the Frobenius automorphism σ_r on the field of inertia $\hat{K}(b\omega b^{-1})$. It still remains to show that r is an invariant of $\hat{A}$. Assume that $\gamma_1^m = \hat{\pi}$ and $\gamma_1\omega_1\gamma_1^{-1} = \omega_1^{q^s}$ for some primitive (q^m-1)-th root ω_1 of 1. Then $\omega_1 = b\omega b^{-1}$ for some $b \in \hat{\Gamma}$, and if we set $\gamma = b^{-1}\gamma_1 b$, we have $\gamma\omega\gamma^{-1} = \omega^{q^s}$ as well as $\alpha\omega\alpha^{-1} = \omega^{q^r}$. From this we obtain $\gamma^{-1}\alpha\,\omega\,\alpha^{-1}\gamma = \omega^{q^{r-s}}$. Now since $\hat{\Gamma}\alpha = \hat{\Gamma}\gamma$, $\gamma^{-1}\alpha$ is a unit $u \in \hat{\Gamma}$ and thus, reducing modulo rad $\hat{\Gamma}$, we find that $\bar{\omega} = \bar{u}\bar{\omega}\bar{u}^{-1} = \bar{\omega}^{q^{r-s}}$; hence, since $r,s < m$ we have indeed $r = s$.

Now, since $[\hat{\Gamma}/\hat{P} : \hat{R}] = m$, $\hat{\Gamma}/\hat{P}$ is the finite field of order q^m and its multiplicative group is generated by a primitive (q^m-1)-th root of 1. Thus the polynomial $X^{q^m-1} - 1$ has an irreducible monic factor of degree m in $\hat{R}[X]$, which has a primitive (q^m-1)-th root of 1 as a root in $\bar{\Gamma} = \hat{\Gamma}/\hat{P}$. But then all its roots are primitive, since they are conjugate under the Galois group. Moreover, by Hensel's lemma $X^{q^m-1}-1$ has then an irreducible factor of degree m over $\hat{R}$ all of whose roots are primitive (q^m-1)-th roots of 1 belonging to $\hat{\Gamma}$ (cf. (6.9)). Conversely, observe that every irreducible monic polynomial $f_m(X) \in \hat{R}[X]$ remains irreducible under reduction modulo rad $\hat{R}$ (cf. (6.1)). We summarize these results as follows.

6.15 Theorem (Hasse [2]): Let $\hat{A}$ be a central skewfield of dimension m^2 over $\hat{K}$. Let $q = [\hat{R}/\text{rad}\ \hat{R} : 1]$ and fix $\hat{\pi}$ such that rad $\hat{R} = \hat{\pi}\hat{R}$. Then there exists a positive integer $r < m$, uniquely determined by $\hat{A}$, and an irreducible factor $f_m[X] \in \hat{R}[X]$ of $X^{q^m-1}-1$ so that there are α, $\omega \in \hat{A}$ satisfying the conditions

$$\hat{A} = \hat{K}(\omega, \alpha), f_m(\omega) = 0, \alpha^m = \hat{\pi}, \alpha\omega\alpha^{-1} = \omega^{q^r}.$$

Note that $(r,m) = 1$ is an immediate consequence of these conditions. Having thus shown that $\hat{A}$ is completely characterized by the integers m and r, with $(r,m) = 1$ we proceed to show that to every such pair (m,r) there exists a skewfield as in (6.15).

6.16 Theorem (Hasse [2]): To every prime element $\hat{\pi} \in \hat{R}$ and every pair of positive integers r and m, $0<r<m$ with $(r,m) = 1$ there exists a unique central skewfield of dimension m^2 over $\hat{K}$ with the invariant r.

Proof: Once the existence is established, the uniqueness follows immediately. For, if ω, α and ω', α' are as in (6.15), then clearly $\omega \mapsto \omega'$, $\alpha \mapsto \alpha'$ induces an isomorphism $\hat{K}(\omega, \alpha) \xrightarrow{\sim} \hat{K}(\omega', \alpha')$ of skewfields over $\hat{K}$. Thus, we assume that m and r, satisfying our conditions, are given and we proceed to construct a skewfield with the desired properties. We recall that $\bar{R} = \hat{R}/\hat{\pi}\hat{R}$ is a Galois field of order q. Any Galois field of order q^m can be considered as a Galois extension of $\bar{R}$, thus the polynomial

$X^{q^m-1} - 1 \in \bar{R}[X]$ has an irreducible factor of degree m whose roots are primitive (q^m-1)-th roots of 1, and it follows by Hensel's lemma that $X^{q^m-1} - 1 \in \hat{R}[X]$ has an irreducible factor with the same properties. We may therefore choose an irreducible polynomial $f_m(X) \in \hat{R}[X]$, so that $f_m(X)$ is of degree m and that its roots are primitive (q^m-1)-th roots of 1. Let ω be such a root. We recall that $\hat{K}$ can be embedded in an algebraically closed field $\hat{K}^*$, and so we may choose $\omega \in \hat{K}^*$, and put $\hat{W} = \hat{K}(\omega)$. $\hat{W}$ is then an unramified Galois extension of degree m of $\hat{K}$, and the correspondence $\omega \longmapsto \omega^{q^r}$ induces the Frobenius automorphism σ_r that generates the Galois group $Gal(\hat{W}/\hat{K})$ since $(r,m) = 1$. We write $x^{(s)}$ for $\sigma_r^s(x)$, with $x \in \hat{W}$; thus, in particular $\omega^{(s)} = \omega^{q^{rs}}$. Now we construct $\hat{A}$ as a subalgebra of the algebra $(\hat{W})_m$ of m×m-matrices over $\hat{W}$. We put

$$\underline{\omega} = \begin{bmatrix} \omega^{(0)} & 0 & 0 & \dots & 0 \\ 0 & \omega^{(1)} & 0 & \dots & 0 \\ \vdots & \vdots & \vdots & \dots & \vdots \\ 0 & \cdot & \dots & \cdot & 0 \\ 0 & 0 & \dots & 0 & \omega^{(m-1)} \end{bmatrix} \text{ and } \underline{\alpha} = \begin{bmatrix} 0 & 1 & 0 & \dots & 0 & 0 \\ 0 & 0 & 1 & \dots & 0 & 0 \\ \vdots & \vdots & \vdots & \dots & \vdots & \vdots \\ 0 & 0 & 0 & \dots & 0 & 1 \\ \hat{\pi} & 0 & 0 & \dots & 0 & 0 \end{bmatrix}$$

and set $\hat{A} = \hat{K}(\underline{\omega}, \underline{\alpha})$. $\underline{\alpha}$ is the <u>companion matrix</u> of the polynomial $X^m - \hat{\pi}$, and so $\underline{\alpha}^m = \hat{\pi}\underline{E}_m$, where $\underline{E}_m$ stands for the m × m identity matrix, and $X^m - \hat{\pi} = \min_{\hat{A}/\hat{K}}(\underline{\alpha})$. Clearly $\underline{\omega}$, being a diagonal matrix with primitive (q^m-1)-th roots of 1 as entries, is itself such a root in $\hat{K}(\underline{\omega})$, and we have an isomorphism $\sigma : \hat{W} \longrightarrow \hat{K}(\underline{\omega})$, induced by $\omega \longmapsto \underline{\omega}$, which maps $x \in \hat{W}$ to the diagonal matrix $\underline{x} = \mathrm{diag}(x^{(0)}, x^{(1)}, \dots, x^{(m-1)})$. In $\hat{K}(\underline{\omega})$ again $\underline{\omega} \longmapsto \underline{\omega}^{q^r}$ induces the r-th Frobenius automorphism which we shall also denote by σ_r. We write again $\underline{w}^{(s)}$ for $\sigma_r^s(\underline{w})$, with $\underline{w} \in \hat{K}(\underline{\omega})$, and observe that $\sigma_r(\underline{w}^{(i)}) = \underline{w}^{(i+1)}$ and that σ_r permutes the entries in the matrix $\underline{w}$

cyclically along the diagonal. Moreover, we have

$$\underline{\alpha}\,\underline{\omega} = \begin{bmatrix} 0 & \omega^{(1)} & 0 & \cdots & 0 & 0 \\ 0 & 0 & \omega^{(2)} & \cdots & 0 & 0 \\ \vdots & \vdots & \vdots & \cdots & \vdots & \vdots \\ 0 & 0 & 0 & \cdots & 0 & \omega^{(m-1)} \\ \hat{\pi}\omega & 0 & 0 & \cdots & 0 & 0 \end{bmatrix} = \mathrm{diag}(\omega^{(1)}, \omega^{(2)}, \ldots, \omega^{(m-1)}, \omega)\,\underline{\alpha} = \underline{\omega}^{(1)}\,\underline{\alpha}.$$

For the sake of completeness we make the obvious observation that

$$\underline{\alpha}^{-1} = \begin{bmatrix} 0 & 0 & \cdots & 0 & \hat{\pi}^{-1} \\ 1 & 0 & \cdots & 0 & 0 \\ \vdots & \vdots & \cdots & \vdots & \vdots \\ 0 & 0 & \cdots & 1 & 0 \end{bmatrix}.$$

We thus have established that for $\hat{A} = \hat{K}(\underline{\omega}, \underline{\alpha})$,
$\underline{\omega}$ is a primitive (q^m-1)-th root of 1, $\underline{\alpha}^m = \hat{\pi}$ and $\underline{\alpha}\,\underline{\omega}\,\underline{\alpha}^{-1} = \underline{\omega}^{q^r}$.
It remains to show that $\hat{A}$ is a central m^2-dimensional skewfield over $\hat{K}$. As to the dimension, it is easily seen that the set $\{\underline{\omega}^{(i)}\underline{\alpha}^j\}_{0\le i,j\le m-1}$ forms a basis for $\hat{A}$ over $\hat{K}$. Let $\underline{a} = \sum_{i=0,\,j=0}^{m-1,\,m-1} k_{ij}\,\underline{\omega}^i\,\underline{\alpha}^j \in \hat{A}$; then $b_j = \sum_{i=0}^{m-1} k_{ij}\,\omega^i \in \hat{K}(\omega)$, and

$$\underline{a} = \begin{bmatrix} b_0 & b_1 & \cdots & b_{m-2} & b_{m-1} \\ \hat{\pi} b_{m-1}^{(1)} & b_0^{(1)} & \cdots & b_{m-3}^{(1)} & b_{m-2}^{(1)} \\ \vdots & \vdots & \cdots & \vdots & \vdots \\ \hat{\pi} b_1^{(m-1)} & \hat{\pi} b_2^{(m-1)} & \cdots & \hat{\pi} b_{m-1}^{(m-1)} & b_0^{(m-1)} \end{bmatrix}$$

where $b_j^{(s)} = \sum_{i=0}^{m-1} k_{ij}\,\omega^{(s)i}$. Thus, if $\underline{a} = 0$, so are b_j, $0\le j\le m-1$, and consequently $k_{ij} = 0$, $0\le i,j\le m-1$; i.e., $[\hat{A} : \hat{K}] = m^2$.
To show that $\hat{K}$, or more precisely, $\hat{K}E_m$ is the center of $\hat{A}$, let $\underline{a} = \sum_{j=0}^{m-1} \underline{w}_j\,\underline{\alpha}^j \in \hat{A}$, $\underline{w}_j \in \hat{K}(\underline{\omega})$, be in the center of $\hat{A}$. Then in par-

ticular $\mathbf{a}\boldsymbol{\alpha} = \boldsymbol{\alpha}\mathbf{a}$, and we have

$$\sum_{j=0}^{m-1} \mathbf{w}_j \boldsymbol{\alpha}^{j+1} = \sum_{j=0}^{m-1} \boldsymbol{\alpha}\mathbf{w}_j \boldsymbol{\alpha}^{j} = \sum_{j=0}^{m-1} \mathbf{w}_j^{(1)} \boldsymbol{\alpha}^{j+1}.$$

But then, by linear independence, $\mathbf{w}_j = \mathbf{w}_j^{(1)}$, for $j = 0,1,\dots,m-1$; i.e., the coefficients $\mathbf{w}_j$ are invariant under the Frobenius automorphism and hence belong to $\hat{K}$. Thus we may assume that $\mathbf{a} = \sum_{j=0}^{m-1} k_j \boldsymbol{\alpha}^j$, with $k_j \in \hat{K}$. Now, since $\boldsymbol{\omega}\mathbf{a} = \mathbf{a}\boldsymbol{\omega}$,

$$\sum_{j=0}^{m-1} k_j \boldsymbol{\omega}\boldsymbol{\alpha}^j = \sum_{j=0}^{m-1} k_j \boldsymbol{\alpha}^j \boldsymbol{\omega} = \sum_{j=0}^{m-1} k_j \boldsymbol{\omega}^{(j)} \boldsymbol{\alpha}^j,$$

and, again by linear independence, we conclude that $k_j\boldsymbol{\omega} = k_j \boldsymbol{\omega}^{(j)}$ for all j, but this is only possible if $k_j = 0$, unless $j = 0$. Then $\mathbf{a} = k_0 \in \hat{K}$, as claimed.

It remains to show that $\hat{A} = \hat{K}(\boldsymbol{\omega}, \boldsymbol{\alpha})$ is a skewfield; i.e., that every non-zero $\mathbf{a} \in \hat{A}$ is invertible. For this it clearly suffices to prove that every non-zero element of $\hat{\Omega}(\boldsymbol{\alpha}) \setminus \hat{\pi}\hat{\Omega}(\boldsymbol{\alpha})$ has a left inverse in $\hat{A}$, where $\hat{\Omega}$ is the maximal $\hat{R}$-order in $\hat{L} = \hat{K}(\boldsymbol{\omega})$. For, given $0 \neq \mathbf{x} \in \hat{A}$, then - since $\hat{\Omega}(\boldsymbol{\alpha})$ is clearly an $\hat{R}$-order in $\hat{A}$ -, there exists $r \in \hat{K}$ such that $r\mathbf{x} \in \hat{\Omega}(\boldsymbol{\alpha}) \setminus \hat{\pi}\hat{\Omega}(\boldsymbol{\alpha})$; and if $\mathbf{y} \in \hat{A}$ is a left inverse of $r\mathbf{x}$ then $r\mathbf{y}$ is a left inverse of $\mathbf{x}$. Moreover if every non-zero element in a ring has a left inverse then every non-zero element has a unique left inverse, and this is a two-sided inverse. (Let $yx = 1$, then $xyx - x = (xy-1)x = 0$; if $xy \neq 1$, let $z(xy-1) = 1$ then $z(xy-1)x = x = 0$, a contradiction.) Thus, let $\mathbf{a} = \sum_{j=0}^{m-1} \mathbf{w}_j \boldsymbol{\alpha}^j \in \hat{\Omega}(\boldsymbol{\alpha}) \setminus \hat{\pi}\hat{\Omega}(\boldsymbol{\alpha})$ with $\mathbf{w}_j \in \hat{\Omega}$, for $j = 0,1,\dots,m-1$, and let h be the smallest number j for which $\mathbf{w}_j \notin \hat{\pi}\hat{\Omega}$. Then, since $\hat{L}$ is unramified over $\hat{K}$; i.e., rad $\hat{\Omega} = \hat{\pi}\hat{\Omega}$, $\mathbf{w} = \mathbf{w}_h$ is a unit in $\hat{\Omega}$ and we may - using $\boldsymbol{\alpha}^m = \hat{\pi}$ - write a as follows:

$$\mathbf{a} = \mathbf{w}\Big(\sum_{j<h} \mathbf{w}^{-1}\mathbf{w}_j' \boldsymbol{\alpha}^{m-h+j} + 1 + \sum_{j>h} \mathbf{w}^{-1}\mathbf{w}_j \boldsymbol{\alpha}^{j-h}\Big)\boldsymbol{\alpha}^h = \mathbf{w}(1+\mathbf{b}\boldsymbol{\alpha})\boldsymbol{\alpha}^h =$$
$$= \mathbf{w}\boldsymbol{\alpha}^h(1+\mathbf{b}^{(m-h)}\boldsymbol{\alpha}),$$

where $\hat{\pi}\mathbf{w}_j' = \mathbf{w}_j$, for $j < h$, thus $\mathbf{w}_j' \in \hat{\Omega}$ by choice of h, and $\mathbf{b} \in \hat{\Omega}(\boldsymbol{\alpha})$,

$\underline{\underline{b}}^{(m-h)} = \underline{\underline{\alpha}}^{-h}\underline{\underline{b}}\,\underline{\underline{\alpha}}^{h} \in \hat{\underline{\underline{\Omega}}}(\underline{\underline{\alpha}})$. $\underline{\underline{w}}\,\underline{\underline{\alpha}}^{h}$ is invertible; in fact, its inverse is $\underline{\underline{\alpha}}^{-h}\underline{\underline{w}}^{-1} = (\underline{\underline{w}}^{-1})^{(m-h)}\underline{\underline{\alpha}}^{-h}$. Therefore it suffices to show that every element of the form $1 - \underline{\underline{v}}\,\underline{\underline{\alpha}}$, with $\underline{\underline{v}} \in \hat{\underline{\underline{\Omega}}}(\underline{\underline{\alpha}})$, has an inverse. Since $(\hat{\underline{\underline{\Omega}}}(\underline{\underline{\alpha}})\underline{\underline{\alpha}}) = (\underline{\underline{\alpha}}\hat{\underline{\underline{\Omega}}}(\underline{\underline{\alpha}}))$ and since $(\hat{\underline{\underline{\Omega}}}(\underline{\underline{\alpha}})\underline{\underline{\alpha}})^m = \hat{\pi}\hat{\underline{\underline{\Omega}}}(\underline{\underline{\alpha}})$, $\hat{\underline{\underline{\Omega}}}(\underline{\underline{\alpha}})$ is complete under the $\hat{\underline{\underline{\Omega}}}(\underline{\underline{\alpha}})\underline{\underline{\alpha}}$-adic topology. Thus $\underline{\underline{u}} = \sum_{n=0}^{\infty} (\underline{\underline{v}}\,\underline{\underline{\alpha}})^n$ exists in $\hat{\underline{\underline{\Omega}}}(\underline{\underline{\alpha}})$ and $\underline{\underline{u}}(1-\underline{\underline{v}}\,\underline{\underline{\alpha}}) = \varprojlim(1-(\underline{\underline{v}}\,\underline{\underline{\alpha}})^n) = 1$. #

6.17 Theorem (Hasse [3]): Let $\hat{A}$ be a central skewfield of dimension m^2 over $\hat{K}$ and $\hat{E}$ a finite extension field of $\hat{K}$. If m divides the degree of $\hat{E}$ over $\hat{K}$ then $\hat{E}$ is a splitting field of $\hat{A}$.

Proof: By assumption we have $[\hat{A} : \hat{K}] = m^2$, and if we let $\hat{\Gamma}$ stand for the maximal $\hat{R}$-order in $\hat{A}$, there exist $r \in \underline{\underline{N}}$, and ω, $\alpha \in \hat{A}$ such that

$$\hat{A} = \hat{K}(\omega, \alpha),\ \text{rad}\ \hat{\Gamma} = \hat{\Gamma}\alpha,\ \omega^{q^m-1} = 1,$$
$$\alpha^m = \hat{\pi},\ \alpha\omega\alpha^{-1} = \omega^{q^r}\ \text{and}\ (r,m) = 1.$$

We prove the theorem first for a field $\hat{E}$ whose ramification order e is a multiple of m. Then, if $\hat{\Gamma}_{\hat{E}}$ denotes the maximal $\hat{R}$-order in $\hat{E}$, we have $e' \in \underline{\underline{N}}$ and $\omega', \varepsilon \in \hat{E}$, such that

$$\hat{E} = \hat{K}(\omega', \varepsilon),\ \text{rad}\ \hat{\Gamma}_{\hat{E}} = \hat{\Gamma}_{\hat{E}}\varepsilon,\ \hat{\Gamma}_{\hat{E}}\varepsilon^e = \hat{\Gamma}_{\hat{E}}\hat{\pi}\ \text{and}\ e = m\cdot e'.$$

By III(6.5) there exists a central skewfield $\hat{D}$ over $\hat{E}$ such that

$$\hat{E} \otimes_{\hat{K}} \hat{A} = (\hat{D})_s \text{ for some } s \mid m.$$

Let $\hat{\Gamma}_1$ be the maximal $\hat{R}$-order in $\hat{D}$, $t^2 = [\hat{D} : \hat{E}]$, then $m = st$ and we set

$$\omega_1 = \omega^{(q^m-1)/(q^t-1)} = \prod_{i=0}^{s-1} \omega^{q^{ti}}.$$

Clearly, ω_1 is a primitive (q^t-1)-th root of 1, and belongs to $\hat{\Gamma}_1$. Now let $\hat{\Gamma}'$ be a maximal $\hat{R}$-order in $(\hat{D})_s$ containing $\hat{\Gamma}_{\hat{E}} \otimes_{\hat{K}} \hat{\Gamma}$. Then $\hat{\Gamma}'$ embeds $\hat{\Gamma}_{\hat{E}}$ as well as $\hat{\Gamma}$ and $\hat{\Gamma}_1$, $\hat{\Gamma}_1$ being the unique maximal $\hat{R}$-order in $\hat{D}$. Hence ε, α, $\omega_1 \in \hat{\Gamma}'$, and we find that $\alpha = \varepsilon^{e'}u$ for some unit $u \in \hat{\Gamma}'$. For, by the above identities $\alpha^m = \hat{\pi} = \varepsilon^{me'}u'$ for some unit

$u' \in \hat{\Gamma}_{\hat{E}}$, whence $(\varepsilon^{-e'}\alpha)^m = u'$ is a unit in $\hat{\Gamma}'$, but then so is $u = \varepsilon^{-e'}\alpha$. Thus

$$\omega_1^{q^r} = \alpha\omega_1\alpha^{-1} = \varepsilon^{e'} u\omega_1 u^{-1}\varepsilon^{-e'} = u\omega_1 u^{-1},$$

and reducing modulo rad $\hat{\Gamma}'$ we obtain $\bar{\omega}_1 = \bar{\omega}_1^{q^r} = \bar{\omega}_1^{q^t}$. But $\bar{\omega}_1$ is a primitive (q^t-1)-th root of 1 and, r being prime to $m = s \cdot t$, we have $(r,t) = 1$ and we conclude that $t = 1$. Hence the desired result, $\hat{E} \otimes_{\hat{K}} \hat{A} = (\hat{E})_m$, is established for the case $m \mid e$.

The theorem follows now simply from the transitivity of the tensor product. Namely, let $\hat{E}$ have ramification order ee' and residue class degree ff' with $ef = m$. Then $\hat{E} = \hat{K}(\omega_0, \varepsilon)$, where ω_0 is a primitive $(q^{ff'}-1)$-th root of 1 and rad $\hat{\Gamma}_{\hat{E}} = \hat{\Gamma}_{\hat{E}}\varepsilon$. If we set $\omega' = \omega_0(q^{ff'}-1)/(q^f-1)$, $\hat{L} = \hat{K}(\omega')$, then $\hat{L}$ is isomorphic to a subfield of $\hat{A}$ of dimension f over $\hat{K}$, since ω' is a primitive (q^f-1)-th root of 1. Now

$$\hat{E} \otimes_{\hat{K}} \hat{A} \cong (\hat{E} \otimes_{\hat{L}} \hat{L}) \otimes_{\hat{K}} \hat{A} \cong \hat{E} \otimes_{\hat{L}} (\hat{L} \otimes_{\hat{K}} \hat{A}) \cong \hat{E} \otimes_{\hat{L}} (\hat{A}')_f \cong (\hat{E} \otimes_{\hat{L}} \hat{A}/)_f ,$$

(cf. III,(6.5)), where $\hat{A}'$ is a central skewfield of dimension e^2 over $\hat{L}$. But $\hat{E}$ has ramification order $e \cdot e'$ over $\hat{L}$, since $\hat{L}$ is unramified over $\hat{K}$, and so $\hat{E}$ splits $\hat{A}'$ over $\hat{L}$, by our previous result and we obtain

$$\hat{E} \otimes_{\hat{K}} \hat{A} \cong (\hat{E})_{fe} = (\hat{E})_m ,$$

and $\hat{E}$ does indeed split $\hat{A}$. #

Exercises §6:

We keep the notation of the previous section.

1.) Show that every element $k \in \hat{K}$ has a unique expression $k = u\hat{\pi}^s$, where u is a unit in $\hat{R}$ and $s \in \mathbb{Z}$.

2.) Let K' be a Galois extension of K, G = Gal(K'/K) and $k \in K'$. Show that

$$Pc_{K'/K}(k,X) = \prod_{\sigma \in G}(X - \sigma(k)).$$

3.) Let $\hat{L}$ be a Galois extension of $\hat{K}$; show that every $\sigma \in \mathrm{Gal}(\hat{L}/\hat{K})$ is a continuous function $\sigma : \hat{L} \longrightarrow \hat{L}$.

4.) Let $\hat{A}$ be a separable finite dimensional skewfield over $\hat{K}$, with maximal $\hat{R}$-order $\hat{\Gamma}$. Give a direct proof of the fact that $\gamma \in \hat{\Gamma}$ is a unit if and only if $N_{\hat{A}/\hat{K}}(\gamma)$ is a unit in $\hat{R}$.

5.) For a principal ideal domain R let $f(X) \in R[X]$ be of the form $X^m + C_{m-1}X^{m-1} + \dots + C_o$ with $C_{m-1}, \dots, C_o \in \pi R$ but $C_o \notin \pi^2 R$, π a prime element in R. Show that f(X) is irreducible over R.

6.) Show that not every factor of degree m of $X^{q^m-1}-1$ that is irreducible over $\bar{H}$, has primitive (q^m-1)-th roots of 1 as roots.

7.) Modify the last two step s in the proof of theorem (6.16) as follows:

(a) To show that $\hat{A}$ is a skewfield, it suffices to show that no $\underline{a} \in \hat{\Omega}(\underline{\alpha})$ is a zero divisor, where $\hat{\Omega}$ denotes the maximal $\hat{R}$-order in $\hat{K}(\underline{\omega})$ (as well as that in $\hat{W}$). Assume that $\underline{a}$ is a zero divisor and show that $\underline{a} \in \pi^n \hat{\Omega}(\underline{\alpha})$, for all positive integers n. Set $\underline{a} = \sum_{j=0}^{m-1} \underline{w}_j \underline{\alpha}^j$, and let $\underline{w}_j = \sigma a_j$, where $\sigma : \hat{W} \longrightarrow \hat{K}(\underline{\omega})$ is as in the proof of (6.16). Now show that $0 = \det \underline{a} \equiv N_{\hat{W}/\hat{K}}(a_o) (\mathrm{mod}\ \hat{\pi}\hat{\Omega})$, and hence $a_o \in \hat{\pi}\hat{\Omega}$ since $\hat{W}$ is unramified over $\hat{K}$ (cf. (6.12),(6.11)). Now this yields $\underline{a}\underline{\alpha}^{m-1} \in \hat{\pi}\hat{\Omega}(\underline{\alpha})$, hence $\underline{a}\underline{\alpha}^{-1} = \underline{b} \in \hat{\Omega}(\underline{\alpha})$, and $\underline{b}$ is again a zero divisor. But $b_o = a_1$ and it follows as above that $a_1 \in \hat{\pi}\hat{\Omega}$. Iterating this procedure it follows that $a_i \in \hat{\pi}\hat{\Omega}$ for all $i = 0,1,\dots,m-1$, and thus $\underline{a} = \hat{\pi}\underline{a}'$ for some $\underline{a}' \in \hat{\Omega}(\underline{\alpha})$, which again is a zero divisor. The result now follows by induction.

(b) Show that $\hat{W}$ splits $\hat{A}$, and use this to show that $\hat{A}$ is central over $\hat{K}$.

8.) Show that the tensor product of a totally ramified and an unramified extension of $\hat{K}$ is a field. More specifically, show that if $\hat{E}$ is totally ramified of ramification order e over $\hat{K}$ and $\hat{L}$ is an unramified extension of $\hat{K}$ with residue class degree f, then $\hat{E} \otimes_{\hat{K}} \hat{L}$ is unramified and has residue class degree f over $\hat{E}$.

CHAPTER V

THE HIGMAN IDEAL AND EXTENSIONS OF LATTICES

§1. The different and the inverse different

The different and the inverse different of an order Λ in A over K relative to a non-degenerate bilinear form are defined. They are shown to "commute" with localization and are computed for maximal orders over $\underline{p}$-adically complete rings.

Notation: R = Dedekind domain with quotient field K ,

A = finite dimensional separable K-algebra,

Λ = R-order in A .

1.1 Definitions: Let $f : A \times A \longrightarrow K$ be a non-degenerate bilinear form (cf. III,(3.6),(3.7)). Then the inverse different $\underline{\underline{d}}_f^{-1}(\Lambda)$ and the different $\underline{\underline{d}}_f(\Lambda)$ of Λ over R relative to f are defined as follows:

$$\underline{\underline{d}}_f^{-1}(\Lambda) = \{a \in A : f(\Lambda, a) \subset R\},$$

$$\underline{\underline{d}}_f(\Lambda) = \{a \in A : \underline{\underline{d}}_f^{-1}(\Lambda)a \subset \Lambda\}.$$

1.2 Lemma: Assume that Λ has an R-basis $\{\omega_i\}_{1\leq i\leq n}$. If $\{\omega_i^*\}_{1\leq i\leq n}$ is the dual of this as K-basis relative to f, (cf. III,(3.7)), then

$$\underline{\underline{d}}_f^{-1}(\Lambda) = \oplus_{i=1}^n R\,\omega_i^* \quad \text{and}$$

$$\underline{\underline{d}}_f(\Lambda) = \{a \in A : \omega_i^* a \in \Lambda,\ 1\leq i\leq n\}.$$

Proof: Since $\{\omega_i^*\}_{1\leq i\leq n}$ is also a K-basis for A, we have

$$\underline{\underline{d}}_f^{-1}(\Lambda) = \left\{\sum_{i-1}^n k_i\omega_i^* : f(\omega_j, \sum_{i=1}^n k_i\omega_i^*) = k_j \in R\right\} = \oplus_{i=1}^n R\,\omega_i^* .$$

The rest follows right from the definition. #

1.3 Lemma: For every prime ideal $\underline{p}$ in R:

$$\underline{\underline{d}}_f^{-1}(\Lambda)_{\underline{p}} = \underline{\underline{d}}_f^{-1}(\Lambda_{\underline{p}}), \text{ and } \underline{\underline{d}}_f(\Lambda)_{\underline{p}} = \underline{\underline{d}}_f(\Lambda_{\underline{p}}).$$

Proof: The inclusion $\underline{\underline{d}}_f^{-1}(\Lambda)_{\underline{p}} \subset \underline{\underline{d}}_f^{-1}(\Lambda_{\underline{p}})$ is obvious. To establish the converse inclusion, let $\lambda_1,\ldots,\lambda_n$ be a system of generators of Λ over R and assume that $f(\Lambda_{\underline{\underline{p}}},a) \subset R_{\underline{\underline{p}}}$, then $f(\lambda_i,a) = r_i/t_i$, $r_i,t_i \in R$ and $t_i \notin \underline{\underline{p}}$, $i = 1,2,\ldots,n$. Now set $a' = a\prod_{i=1}^{n} t_i$, then $a' \in \underline{\underline{d}}_f^{-1}(\Lambda)$ and $a \in \underline{\underline{d}}_f^{-1}(\Lambda)_{\underline{\underline{p}}}$. The second inclusion $\subset$ follows from this by $\underline{\underline{d}}_f^{-1}(\Lambda_{\underline{\underline{p}}})\underline{\underline{d}}_f(\Lambda)_{\underline{\underline{p}}} = \underline{\underline{d}}_f^{-1}(\Lambda)_{\underline{\underline{p}}}\underline{\underline{d}}_f(\Lambda)_{\underline{\underline{p}}} = (\underline{\underline{d}}_f^{-1}(\Lambda)\underline{\underline{d}}_f(\Lambda))_{\underline{\underline{p}}} \subset \Lambda_{\underline{\underline{p}}}$. Finally let $a \in \underline{\underline{d}}_f(\Lambda_{\underline{\underline{p}}})$, then $\underline{\underline{d}}_f^{-1}(\Lambda)a \subset \underline{\underline{d}}_f^{-1}(\Lambda_{\underline{\underline{p}}})a \subset \Lambda_{\underline{\underline{p}}}$, and taking a system of generators for $\underline{\underline{d}}_f^{-1}(\Lambda)$, the same argument as above shows that $a \in \underline{\underline{d}}_f(\Lambda)_{\underline{\underline{p}}}$. #

1.4 Remark: From (1.2) we conclude that $\underline{\underline{d}}_f^{-1}(\Lambda)$ and $\underline{\underline{d}}_f(\Lambda)$ are R-lattices in A. Moreover, $\underline{\underline{d}}_f(\Lambda)$ is not only a right Λ-module, which is obvious, but it is a (possibly fractional) two-sided Λ-ideal. It follows from (IV,(4.14)), that if Γ is a maximal order in A, $\underline{\underline{d}}_f^{-1}(\Gamma)$ is indeed the inverse of $\underline{\underline{d}}_f(\Gamma)$ as defined in (IV,(4.12)) and that $\underline{\underline{d}}_f^{-1}(\Gamma)$ is also a two-sided fractional Γ-ideal in A.

We shall next compute the different and the inverse different for a maximal order in a simple algebra A relative to the reduced trace.

1.5 Lemma: Let A be a simple separable K-algebra with center L and Λ an R-order in A. If $\Delta = \Lambda \cap L$ is the central R-order in Λ, then

$$\underline{\underline{d}}_{\mathrm{Trd}_{A/K}}^{-1}(\Lambda) = \underline{\underline{d}}_{\mathrm{Trd}_{A/L}}^{-1}(\Lambda)\,\underline{\underline{d}}_{\mathrm{Tr}_{L/K}}^{-1}(\Delta),$$

where $\underline{\underline{d}}_{\mathrm{Trd}_{A/L}}^{-1}(\Lambda) = \{a \in A : \mathrm{Trd}_{A/L}(\Lambda a) \subset \Delta\}$. (For the notation cf. III,(3.1),(6.11).)

Proof: From (III,6.15) and the fact that $\mathrm{Trd}_{A/L}$ is L-linear and since Λ is a Δ-module we obtain

$\underline{\underline{d}}_{\mathrm{Trd}_{A/K}}^{-1}(\Lambda) = \{a \in A : \mathrm{Trd}_{A/K}(\Lambda a) \subset R\} =$

$= \{a \in A : Tr_{L/K}(Trd_{A/L}(\Lambda a)) \subset R\} = \{a \in A : Tr_{L/K}(\Delta\, Trd_{A/L}(\Lambda a)) \subset R\}$

$= \{a \in A : Trd_{A/L}(\Lambda a) \subset \underline{\underline{d}}^{-1}_{Tr_{L/K}}(\Delta)\}$. From this it follows at once that

$$\underline{\underline{d}}^{-1}_{Trd_{A/L}}(\Lambda)\underline{\underline{d}}^{-1}_{Tr_{L/K}}(\Delta) = \underline{\underline{d}}^{-1}_{Trd_{A/K}}(\Lambda). \quad \#$$

1.6 Remark: (i) Since the trace function is symmetric (cf. III, (6.16)), $\underline{\underline{d}}^{-1}_{Trd_{A/K}}(\Lambda)$ is a two-sided Λ-ideal.

(ii) If Γ is a maximal R-order in the separable K-algebra A, then $\underline{\underline{d}}^{-1}_{Trd_{A/K}}(\Gamma)\underline{\underline{d}}_{Trd_{A/K}}(\Gamma) = \Gamma$, (cf. IV, (4.15)). Moreover, since $\underline{\underline{d}}^{-1}_{Trd_{A/K}}(\Gamma) \supset \Gamma$, this justifies the names different and inverse different.

(iii) If Λ is an R-order in the simple separable K-algebra A, $[A : center(A)] = m^2$, and if $m \neq 0$ in K, then the trace function $A \times A \longrightarrow K$ is non-degenerate, and it follows easily from (1.5) that

$$\underline{\underline{d}}^{-1}_{Tr_{A/K}}(\Lambda) = 1/m\, \underline{\underline{d}}^{-1}_{Trd_{A/K}}(\Lambda) \text{ (cf. III,(6.16)).}$$

1.7 Theorem (Hasse [2]): Let $\hat{K}$ be the $\underline{\underline{p}}$-adic completion of an algebraic number field and $\hat{R}$ its ring of integers, with rad $\hat{R} = \hat{\pi}\hat{R}$. If $\hat{D}$ is a central skewfield of dimension n^2 over $\hat{K}$, with maximal $\hat{R}$-order $\hat{\Gamma}$ then

$$\underline{\underline{d}}_{Trd_{\hat{D}/\hat{K}}}(\hat{\Gamma}) = (\text{rad } \hat{\Gamma})^{n-1} .$$

Proof: By (1.6) it suffices to show that $(\text{rad } \hat{\Gamma})^{1-n} = (\text{rad } \hat{\Gamma})^{-1(n-1)} \subset \underline{\underline{d}}^{-1}_{Trd_{\hat{D}/\hat{K}}}(\hat{\Gamma})$ and $(1/\hat{\pi})\hat{\Gamma} \not\subset \underline{\underline{d}}^{-1}_{Trd_{\hat{D}/\hat{K}}}(\hat{\Gamma})$ (observe $\hat{\pi}\hat{\Gamma} = (\text{rad } \hat{\Gamma})^n$ (cf. IV, (6.7))). We use the notation of IV (6.16) and recall that $\hat{\Gamma}$ has the $\hat{R}$-basis $\{\underline{\underline{\omega}}^{(i)}\underline{\underline{\alpha}}^j\}_{0 \leq i,j \leq n-1}$ where rad $\hat{\Gamma} = \hat{\Gamma}\alpha$, $\underline{\underline{\alpha}}^n = \hat{\pi}$ and $\underline{\underline{\omega}}^{(i)}$ is the diagonal $n \times n$-matrix over $\hat{W} = \hat{K}(\omega)$ whose entries are the n conjugates $\omega^{(i+h)}$,

$h = 0,1,\dots,n-1$, of the primitive (q^n-1)-th root ω of 1, $\omega^{(i)} = \omega^{q^{ri}}$. Moreover, α acts as Frobenius automorphism σ_r on the field $\hat{K}(\underline{\omega})$, $\underline{\omega} = \underline{\omega}^{(0)}$, and $\underline{\alpha}, \underline{\alpha}^2, \dots, \underline{\alpha}^{n-1}$ all have 0 diagonals. It follows that

$$\begin{aligned} & \mathrm{Trd}_{\hat{D}/\hat{K}}(\underline{\omega}^{(i)}\underline{\alpha}^j(1/\hat{\pi})\underline{\omega}^{(h)}\underline{\alpha}^{s+1}) \\ &= \mathrm{Trd}_{\hat{D}/\hat{K}}(\underline{\omega}^{(i+jh)}\underline{\alpha}^{j+s+1}/\hat{\pi}) \\ &= \begin{cases} \mathrm{Tr}_{\hat{W}/\hat{K}}(\omega^{(i+jh)}) \in \hat{R}, & \text{if } j+s+1 = m \\ 0 & \text{otherwise.} \end{cases} \end{aligned}$$

Hence $(1/\hat{\pi})\hat{\Gamma}\alpha = (\mathrm{rad}\ \hat{\Gamma})^{1-n} \subset \underline{d}^{-1}_{\mathrm{Trd}_{\hat{D}/\hat{K}}}(\hat{\Gamma})$. On the other hand the residue class field of $\hat{W}$ is a finite field and thus separable over $\bar{R}$. Thus its trace function is non-degenerate and it follows that there are units in $\hat{\Gamma}$ whose reduced trace does not belong to $\mathrm{rad}\ \hat{R} = \hat{\pi}\hat{R}$. Consequently $(1/\hat{\pi})\hat{\Gamma} \not\subset \underline{d}^{-1}_{\mathrm{Trd}_{\hat{D}/\hat{K}}}(\hat{\Gamma})$, and we conclude that

$$\underline{d}^{-1}_{\mathrm{Trd}_{\hat{D}/\hat{K}}}(\hat{\Gamma}) = (\mathrm{rad}\ \hat{\Gamma})^{1-n}. \qquad \#$$

1.8 <u>Corollary</u>: Let $\hat{A}$ be a central simple $\hat{K}$-algebra and $\hat{\Gamma}$ a maximal $\hat{R}$-order in $\hat{A}$. Let $n^2 = [\hat{D} : \hat{K}]$, where $\hat{A} = (\hat{D})_m$ for some m. Then

$$\underline{d}_{\mathrm{Trd}_{\hat{A}/\hat{K}}}(\hat{\Gamma}) = (\mathrm{rad}\ \hat{\Gamma})^{n-1}.$$

<u>Proof</u>: We show this first for the particular maximal $\hat{R}$-order $\hat{\Gamma}_1 = (\hat{\Omega})_m$, where $\hat{\Omega}$ is the maximal $\hat{R}$-order in $\hat{D}$. But for $\hat{\Gamma}_1$ the result follows immediately from the proof of (1.7), since an $\hat{R}$-basis for $\hat{\Gamma}_1$ is given by $\{\underline{\omega}^{(i)}\underline{\alpha}^j\underline{E}_{\gamma\mu}\}_{0\le i,j\le n-1, 1\le\gamma,\mu\le m}$, where $\underline{E}_{\gamma\mu}$ is the $m \times m$ matrix over D with 1 at position (γ,μ) and zeros elsewhere. If now $\hat{\Gamma}$ is any maximal $\hat{R}$-order in $\hat{A}$, then $\hat{\Gamma} = a\hat{\Gamma}_1 a^{-1}$ for some invertible element $a \in \hat{A}$ (cf. IV,(5.8)). Since the trace

function is symmetric we conclude that $\underline{\underline{d}}_{Trd_{\hat{A}/\hat{K}}}(\hat{\Gamma}) = (\mathrm{rad}\ \hat{\Gamma})^{n-1}$. #

<u>Exercises § 1</u>:

1.) Let Γ be a maximal R-order in the separable K-algebra A. Find formulae for $\underline{\underline{d}}^{-1}_{Trd_{A/K}}(\Gamma)$ and $\underline{\underline{d}}^{-1}_{Tr_{A/K}}(\Gamma)$.

2.) Compute the inverse different relative to the reduced trace for the following orders

(i) $\Lambda = \left\{ \begin{pmatrix} a & p^{-1}b \\ c & d \end{pmatrix} : a, b, c, d \in \underline{\underline{Z}} \right\}$, p a rational prime number,

(ii) $\Lambda = \left\{ \begin{pmatrix} a & b \\ c & a+pd \end{pmatrix} : a, b, c, d \in \underline{\underline{Z}} \right\}$, p a rational prime number .

§2. Projective homomorphisms

Let S be a left noetherian ring. Then $M \in {}_S\underline{\underline{M}}^f$ is projective if and only if $\mathrm{Ext}^1_S(M,X) = 0$, for every $X \in {}_S\underline{\underline{M}}^f$ (cf. II, Ex. 4,2). We shall introduce the concept of a projective homomorphism and establish its properties, which are similar to those of a projective module. We use the notation and concepts introduced in II §5.

2.1 Definition: Let $M,N \in {}_S\underline{\underline{M}}^f$. $\alpha \in \mathrm{Hom}_S(N,M)$ is called a projective homomorphism, if $\alpha\,\mathrm{Ext}^1_S(M,-) = 0$, or more precisely, if $\mathrm{ext}^1_S(\alpha, X) = 0$, for every $X \in {}_S\underline{\underline{M}}^f$ (cf. II, (5.5) and (5.9)).

2.2 Remarks: (i) For $\alpha \in \mathrm{Hom}_S(N,M)$, we have $\alpha\,\mathrm{Ext}^1_S(M,X) \subset \mathrm{Ext}^1_S(N,X)$ (cf. II, (5.7) and (5.8)).

(ii) $M \in {}_S\underline{\underline{P}}^f$ if and only if $1_M \in \mathrm{Hom}_S(M,M)$ is projective.

2.3 Lemma: Let $M,N \in {}_S\underline{\underline{M}}^f$. $\alpha \in \mathrm{Hom}_S(N,M)$ is projective if and only if every diagram with exact row

$$D \qquad \begin{array}{ccccc} & & N & & \\ & \overset{\gamma}{\swarrow} & \downarrow{\scriptstyle\alpha} & & \\ X & \xrightarrow{\beta} & M & \longrightarrow & 0 \end{array}$$

can be completed commutatively.

Proof: (i) Let α be projective. Given D, we put $Y = \mathrm{Ker}\,\beta$ and let $\delta : Y \longrightarrow X$ be the canonical injection. Then we obtain the commutative diagram with exact rows

$$\begin{array}{rcccccccccc} E: & 0 & \longrightarrow & Y & \xrightarrow{\delta} & X & \xrightarrow{\beta} & M & \longrightarrow & 0 \\ & & & \uparrow{\scriptstyle 1_Y} & & \uparrow{\scriptstyle\sigma} & & \uparrow{\scriptstyle\alpha} & & \\ \alpha E: & 0 & \longrightarrow & Y & \longrightarrow & Z & \xrightarrow{\beta'} & N & \longrightarrow & 0 \end{array}$$

(cf. II, (5.4)). Since α is projective, αE splits (cf. II, (5.9)); i.e., there exists $\tau' \in \mathrm{Hom}_S(N,Z)$ such that $\tau'\beta' = 1_N$ (cf. I, (2.2)). Then $\gamma = \tau'\sigma$ completes the diagram D.

(ii) Conversely, assume that every diagram D can be completed. Let E be an exact sequence in $\tilde{E}_S(M,Y)$ (cf. II,(5.1)),
$E : 0 \longrightarrow Y \xrightarrow{\delta} X \xrightarrow{\beta} M \longrightarrow 0$. Then there exists $\gamma \in \mathrm{Hom}_S(N,X)$ such that $\gamma\beta = \alpha$, and we find that the following diagram commutes

$$\begin{array}{lccccccccc}
E : & 0 & \longrightarrow & Y & \xrightarrow{\delta} & X & \xrightarrow{\beta} & M & \longrightarrow & 0 \\
& & & \uparrow 1_Y & & \uparrow \delta \oplus \gamma & & \uparrow \alpha & & \\
E' : & 0 & \longrightarrow & Y & \xrightarrow{\iota_1} & Y \oplus N & \xrightarrow{\pi_2} & N & \longrightarrow & 0 ,
\end{array}$$

where ι_1 and π_2 are the respective canonical injections and projections. Thus $E' \in [\alpha E]$ and since E' is split exact $[\alpha E] = 0$ and $\alpha\,\mathrm{Ext}^1_S(M,Y) = 0$ for all $Y \in {}_S\underline{\underline{M}}^f$ by the natural equivalence between $\mathrm{Ext}^1_S(-,Y)$ and $E_S(-,Y)$ (cf. II,(5.9)). #

2.4 <u>Lemma</u>: Let $M,N \in {}_S\underline{\underline{M}}^f$. Then $\alpha \in \mathrm{Hom}_S(N,M)$ is projective if and only if $\alpha \in \mathrm{Im}\,\mu_{N,M}$, where $\mu_{N,M} : N^* \otimes_S M \longrightarrow \mathrm{Hom}_S(N,M)$.

<u>Proof</u>: We recall the definition of $\mu_{N,M}$ (cf. III,(1.4))

$$\mu_{N,M} : \mathrm{Hom}_S(N,S) \otimes_S M \longrightarrow \mathrm{Hom}_S(N,M)$$
$$\varphi \otimes m \longmapsto (\varphi \otimes m)^{\mu_{N,M}},$$

where $n(\varphi \otimes m)^{\mu_{N,M}} = (n\varphi)m$, for every $n \in N$.

(i) Let α be projective, and choose a free S-module F on symbols $\{f_i\}_{1\le i\le t}$, which maps onto M; then we have the commutative diagram

$$\begin{array}{ccccc}
F & \xrightarrow{\beta} & M & \longrightarrow & 0 \\
& \overset{\gamma}{\nwarrow} & \uparrow \alpha & & \\
& & N & &
\end{array}$$

We write, for every $n \in N$, $n\gamma = \sum_{i=1}^{t} s_i(n)f_i$ with $s_i(n) \in S$. It is easily checked that

$$\varphi_i : N \longrightarrow S; \quad \varphi_i : n \longmapsto s_i(n)$$

is in $N^* = \mathrm{Hom}_S(N,S)$ such that

$$n\alpha = \sum_{i=1}^{t} s_i(n)(f_i)\beta ;$$

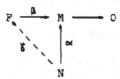

i.e.,

$$\alpha = \left(\sum_{i=1}^{t} \varphi_i \otimes f_i\beta\right)^{\mu_{M,N}},$$

and $\alpha \in \operatorname{Im} \mu_{N,M}$.

(ii) If, conversely, $\alpha \in \operatorname{Im} \mu_{N,M}$, say

$$\alpha = \left(\sum_{i=1}^{t} \varphi_i \otimes m_i\right)^{\mu_{N,M}},$$

then for the diagram with exact row

$$\begin{array}{ccccc} X & \xrightarrow{\beta} & M & \longrightarrow & 0 \\ & & \uparrow{\scriptstyle \alpha} & & \\ & & N & & \end{array}$$

we choose elements $\{x_i\}_{1\le i\le t}$ in X with $x_i\beta = m_i$, $1\le i\le t$.

Then

$$\gamma : N \longrightarrow X; \quad \gamma : n \longmapsto \sum_{i=1}^{t} (n\varphi_i)x_i$$

is an S-homomorphism with $\gamma\beta = \alpha$; i.e., α is projective (cf. (1.3)).
#

2.5 Remark: (i) For $M,N \in {}_{\underline{\underline{S}}}\underline{M}^f$, the projective homomorphisms in $\operatorname{Hom}_S(N,M)$ form an $[\operatorname{End}_S(N), \operatorname{End}_S(M)]$-bimodule (cf. III,(1.4)).

(ii) If $M,N \in {}_{\underline{\underline{S}}}\underline{M}^f$, where S is an R-algebra, R a Dedekind ring, then $\varphi \in \operatorname{Hom}_S(N,M)$ is projective if $\varphi\operatorname{Ext}^1_S(M,X)=0$, $\forall X \in {}_{\underline{\underline{S}}}\underline{M}^o$, (proof of 2.4i).

2.6 Schanuel's lemma (Roiter [5]): Given two exact sequences of left S-modules of finite type

$$E_1 : 0 \longrightarrow M' \longrightarrow M \xrightarrow{\alpha} M'' \longrightarrow 0$$

$$E_2 : 0 \longrightarrow N' \longrightarrow N \xrightarrow{\beta} N'' \longrightarrow 0.$$

If $M'' \cong N''$, and if α and β are projective, then $M' \oplus N \cong N' \oplus M$.

Proof: This generalization of Schanuel's lemma is an immediate consequence of the symmetry and the uniqueness-up to isomorphism - of the pullback (cf. II,(1.13)). Indeed, for any two exact sequences

$$E_1 : 0 \longrightarrow M' \longrightarrow M \xrightarrow{\alpha} M'' \longrightarrow 0$$

$$E_2 : 0 \longrightarrow N' \longrightarrow N \xrightarrow{\beta} M'' \longrightarrow 0$$

the middle terms of βE_1 and of αE_2 are both given by the pullback of (α, β),

$$\begin{array}{ccc} & & N \\ & & \big\downarrow{\scriptstyle\beta} \\ M & \xrightarrow{\alpha} & M'' \end{array}$$

(cf. II,(5.3),(5.4)), and thus we have the commutative diagram with exact rows and exact columns:

$$\begin{array}{ccccccccc}
 & & & & O & & O & & \\
 & & & & \uparrow & & \uparrow & & \\
E_1 : & O & \longrightarrow & M' & \longrightarrow & M & \xrightarrow{\alpha} & M'' & \longrightarrow & O \\
 & & & \Big\| {\scriptstyle 1_{M'}} & & \uparrow{\scriptstyle\beta'} & & \uparrow{\scriptstyle\beta} & & \\
\beta E_1 : & O & \longrightarrow & M' & \longrightarrow & Y & \xrightarrow{\alpha'} & N & \longrightarrow & O \\
 & & & & & \uparrow & {\scriptstyle 1_{N'}} & \uparrow & & \\
 & & & & & N' & \longrightarrow & N' & & \\
 & & & & & \uparrow & & \uparrow & & \\
 & & & & & O & & O & & \\
 & & & & & \alpha E_2 & & E_2 & &
\end{array}$$

Hence, by (II,(5.1)) the middle terms in any sequences in $[\beta E_1]$ and in $[\alpha E_2]$ respectively are isomorphic. Now if α and β are projective and E_1, E_2 are given as in the lemma, with an isomorphism $\sigma : M'' \longrightarrow N''$, then we may identify M'' and N'' because of (2.5), or more precisely, we can replace α by the projective homomorphism $\alpha\sigma$. Now, since αE_1 and βE_2 are split, the pullback of $(\alpha\sigma, \beta)$ is isomorphic to both $M' \oplus N$ and $N' \oplus M$ and it follows that

$$M' \oplus N \cong N' \oplus M. \quad \#$$

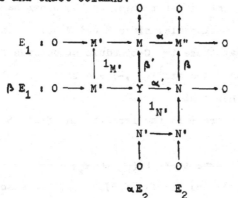

Exercises §2:

1.) Let C be the center of the noetherian ring S. For every $M \in {}_S\underline{M}^f$, we have a homomorphism

$$\Phi_M : C \longrightarrow \mathrm{End}_S(M);$$

$$x \longmapsto \varphi_x,$$

where $m\varphi_x = xm$. (Give an example, where Φ is not monic!) Via Φ_M, Φ_N we can make $\mathrm{Ext}_S^1(M,N)$ into a C-bimodule. On the other hand, $\mathrm{Ext}_S^1(M,N)$ is naturally a C-bimodule, since M and N are C-bimodules (cf. II,(1.12)). Show that these two C-bimodule structures coincide; i.e., show that $\mathrm{ext}_S^1(-,\varphi_x)$ and $\mathrm{ext}_S^1(\varphi_x,-)$ are both multiplication by x. More explicitly, show that

(i) $\Psi(E) \cdot x = [E\varphi_x]$, where Ψ is the isomorphism of II,(5.9), and that

(ii) $[\varphi_x E] = [E\varphi_x]$. (Observe that $[E][\varphi_x,\varphi_x,\varphi_x] = [E]$, for every $E \in \tilde{E}_S(M,N)$ (cf. II,(5.6)), and use II,(5.6). But it is also instructive to work this identity out explicitly.)

§3. The Higman ideal of an order

The Higman ideal of Λ in Δ, consisting of the projective Λ^e-endomorphisms of Λ, is investigated and explicitely calculated in terms of the Gaschütz-Casimir operator. In particular, this result is applied to group rings. Moreover $\mathrm{Ext}^1_\Lambda(M,N)$ is shown to be an R-torsion module with trivial $\underline{\underline{p}}$-components for all $\underline{\underline{p}} \in \underline{\underline{S}}$ that do not divide the Higman ideal.

We shall return now to the study of lattices over orders.

Notation: R = a Dedekind domain with quotient field K,

A = a finite dimensional separable K-algebra,

Λ = an R-order in A,

$\underline{\underline{S}}$ = the set of prime ideals in R,

Δ = the center of Λ,

$\Lambda^e = \Lambda \otimes_R \Lambda^{op}$, the enveloping algebra of Λ (cf. III,(4.1)).

3.1 Definition: The Higman ideal $\underline{\underline{H}}(\Lambda)$ of Λ in R is the R-annihilator of $\mathrm{Ext}^1_{\Lambda^e}(\Lambda,-)$; i.e.,

$$\underline{\underline{H}}(\Lambda) = \{r \in R : r\,\mathrm{Ext}^1_{\Lambda^e}(\Lambda,V) = 0, \text{ for every } V \in {}_{\Lambda^e}\underline{\underline{M}}^f\}.$$

Thus, the Higman ideal consists of all elements r of R that induce projective homomorphisms $\varphi_r \in \mathrm{End}_{\Lambda^e}(\Lambda)$ (cf. Ex. 2,1). We recall that $\mathrm{Ext}^1_{\Lambda^e}(\Lambda,V)$ is an R-module of finite type and that a homomorphism $\alpha\varphi$ is projective whenever φ is projective (cf. (2.3)). Thus, $\underline{\underline{H}}(\Lambda)$ is indeed an ideal in R. In order to arrive at an explicit description of $\underline{\underline{H}}(\Lambda)$, and in view of §2, we shall investigate the ideal of $\mathrm{End}_{\Lambda^e}(\Lambda)$ consisting of all projective homomorphisms. Since $\mathrm{End}_{\Lambda^e}(\Lambda) = \Delta$; this leads to the following definition.

3.2 Definition: The Higman ideal of Λ in Δ is defined as

$$\underline{\underline{H}}_\Delta(\Lambda) = \{\varphi \in \mathrm{End}_{\Lambda^e}(\Lambda) : \varphi\,\mathrm{Ext}^1_{\Lambda^e}(\Lambda,V) = 0, \text{ for every } V \in {}_{\Lambda^e}\underline{\underline{M}}^f\}$$
$$= \{\delta \in \Delta : \delta\,\mathrm{Ext}^1_{\Lambda^e}(\Lambda,V) = 0, \text{ for every } V \in {}_{\Lambda^e}\underline{\underline{M}}^f\}$$

and clearly $\underline{\underline{H}}(\Lambda) = \underline{\underline{H}}_\Delta(\Lambda) \cap R$.

3.3 <u>Lemma</u>: Let $\varepsilon : \Lambda^e \longmapsto \Lambda;\ \lambda' \otimes \lambda^{op} \longmapsto \lambda'\lambda$, be the augmentation map, and $\varphi_o : \Lambda \longrightarrow \operatorname{Ker}\varepsilon$, the derivation defined by $\varphi_o : \lambda \longmapsto \lambda \otimes 1^{op} - 1 \otimes \lambda^{op}$. Denote by $\operatorname{InDer}(\Lambda, M)$ the set of all morphisms in $\operatorname{Hom}_R(\Lambda, M)$ that are inner derivations (cf. III, 4). Then

$$\underline{\underline{H}}_{\Delta}(\Lambda) = \{\delta \in \Delta : \varphi_o\delta \in \operatorname{InDer}(\Lambda, \operatorname{Ker}\varepsilon)\}.$$

<u>Proof</u>: This is an immediate consequence of III,(4.5),(4.6) and (4.9), together with Ex. 2,1. To be more explicit, we recall that every sequence $E \in \tilde{\underline{E}}_{\Lambda^e}(\Lambda, M)$ (cf. II,(5.11)) is the image of the augmentation sequence $E_\varepsilon : 0 \longrightarrow \operatorname{Ker}\varepsilon \xrightarrow{\iota} \Lambda^e \xrightarrow{\varepsilon} \Lambda \longrightarrow 0$ under some map $\alpha \in \operatorname{Hom}_{\Lambda^e}(\operatorname{Ker}\varepsilon, M)$, and that we have the commutative diagram with exact rows of Λ^e-modules.

$$\begin{array}{lcccccccc}
 & & & \Lambda & & & & & \\
 & & & \downarrow{\scriptstyle \varphi_o} & & & & & \\
E_\varepsilon & : & 0 \longrightarrow & \operatorname{Ker}\varepsilon & \xrightarrow{\iota} & \Lambda^e & \longrightarrow & \Lambda & \longrightarrow 0 \\
 & & & {\scriptstyle \alpha}\downarrow & & \downarrow & & \downarrow{\scriptstyle 1_\Lambda} & \\
E = E_\varepsilon\alpha & : & 0 \longrightarrow & M & \longrightarrow & X & \longrightarrow & \Lambda & \longrightarrow 0
\end{array}$$

Now, $\delta \in \Delta$ induces on any left Λ^e-module the same action as $\delta \otimes 1$, which belongs to the center of Λ^e, and thus we have, according to Ex. 2,1: $\delta E = \delta E_\varepsilon \alpha = E_\varepsilon \delta\alpha$. By III,(4.6) and (4.9) then $[\delta E] = 0$ if and only if $\varphi_o\delta\alpha$ is an inner derivation. Now, if $\varphi_o\delta$ is inner, then so is $\varphi_o\delta\alpha$ for all homomorphisms α with domain $\operatorname{Ker}\varepsilon$, and thus $\delta \in \underline{\underline{H}}_{\Delta}(\Lambda)$. Conversely, if $\delta \in \underline{\underline{H}}_{\Delta}(\Lambda)$, then $[\delta E_\varepsilon] = 0$ and $\varphi_o\delta$ must be an inner derivation. We observe, as a consequence of this proof, that $\delta \in \underline{\underline{H}}_{\Delta}(\Lambda)$ if and only if δE_ε splits. #

3.4 <u>Remark</u>: By definition Λ is a separable R-order if and only if Λ is Λ^e-projective, and this in turn means that $1 \in \underline{\underline{H}}_{\Delta}(\Lambda)$. Thus, since $\underline{\underline{H}}_{\Delta}(\Lambda)$ is a two-sided Δ-ideal (cf. (2.5)), Λ is separable if and only if $\underline{\underline{H}}_{\Delta}(\Lambda) = \Delta$. In this sense the Higman ideal $\underline{\underline{H}}_{\Delta}(\Lambda)$ gives a measure for the deviation of Λ from separability. Note also that

$\underline{\underline{H}}_\Delta(\Lambda) = \Delta$ if and only if $\underline{\underline{H}}(\Lambda) = R$.

3.5 Theorem: $\underline{\underline{H}}(\Lambda) \neq 0$ and thus $\mathrm{Ext}^1_{\Lambda^e}(\Lambda, V)$ is an R-torsion module for all $V \in {}_{\Lambda^e}\underline{\underline{M}}^f$.

Proof: By (3.3) it suffices to find $r \in R$ such that $\varphi_o r$ is an inner derivation: $\Lambda \longrightarrow \mathrm{Ker}\,\varepsilon$. But, since A is a separable K-algebra,

$$1 \otimes \varphi_o : A = K \otimes_R \Lambda \longrightarrow K \otimes_R \mathrm{Ker}\,\varepsilon$$

is an inner derivation from A to the kernel of the augmentation map of A, (cf. (3.4) and (3.3)), and thus there exists $b = \sum_{i=1}^{t} k_i \otimes x_i \in K \otimes_R \mathrm{Ker}\,\varepsilon$, such that $1 \otimes \varphi_o : 1 \otimes \lambda \longmapsto (1 \otimes \lambda)b - b(1 \otimes \lambda)$. But then $r\,b \in \mathrm{Ker}\,\varepsilon$, for some $0 \neq r \in R$, and $\varphi_o r : \Lambda \longrightarrow \mathrm{Ker}\,\varepsilon$ is an inner derivation, i.e., $r \in \underline{\underline{H}}(\Lambda)$. #

3.6 Theorem (Reiner [3]): Let $V \in {}_{\Lambda^e}\underline{\underline{M}}^f$. Then

$$\mathrm{Ext}^1_{\Lambda^e}(\Lambda, V) \overset{nat}{\cong} \bigoplus_{\underline{\underline{p}} | \underline{\underline{H}}(\Lambda)} \mathrm{Ext}^1_{\Lambda^e_{\underline{\underline{p}}}}(\Lambda_{\underline{\underline{p}}}, V_{\underline{\underline{p}}}).$$

Proof: By (3.5) $\mathrm{Ext}^1_{\Lambda^e}(\Lambda, V)$ is an R-torsion module of finite type, and we have (cf. I,(8.9) and III,(1.2)),

$$\mathrm{Ext}^1_{\Lambda^e}(\Lambda, V) \cong \bigoplus_{\underline{\underline{p}} \in \underline{\underline{S}}} R_{\underline{\underline{p}}} \otimes_R \mathrm{Ext}^1_{\Lambda^e}(\Lambda, V) \cong \bigoplus_{\underline{\underline{p}} \in \underline{\underline{S}}} \mathrm{Ext}^1_{\Lambda^e_{\underline{\underline{p}}}}(\Lambda_{\underline{\underline{p}}}, V_{\underline{\underline{p}}}).$$

But $\underline{\underline{H}}_{\Delta_{\underline{\underline{p}}}}(\Lambda_{\underline{\underline{p}}}) = R_{\underline{\underline{p}}} \otimes_R \underline{\underline{H}}_\Delta(\Lambda)$ (cf. Ex. 3,1), and thus

$$\mathrm{Ext}^1_{\Lambda^e_{\underline{\underline{p}}}}(\Lambda_{\underline{\underline{p}}}, V_{\underline{\underline{p}}}) = 0, \text{ for every } \underline{\underline{p}} \nmid \underline{\underline{H}}(\Lambda). \quad \#$$

3.7 Corollaries: (i) If $\underline{\underline{p}} \nmid \underline{\underline{H}}(\Lambda)$, then $\Lambda_{\underline{\underline{p}}}$ is a separable $R_{\underline{\underline{p}}}$-order.

(ii) For $M, N \in {}_\Lambda\underline{\underline{M}}^o$, $\mathrm{Ext}^1_\Lambda(N, M) \overset{nat}{\cong} \bigoplus_{\underline{\underline{p}} | \underline{\underline{H}}(\Lambda)} \mathrm{Ext}^1_{\Lambda_{\underline{\underline{p}}}}(N_{\underline{\underline{p}}}, M_{\underline{\underline{p}}})$.

Proof: (i) is an immediate consequence of (3.4) and Ex. 3,1.
(ii) follows readily from III,(4.11) and from (3.6). #

3.8 Remark: From (i) it follows that $\Lambda_{\underline{\underline{p}}}$ is separable for almost all prime ideals $\underline{p}$ of R, and (ii) implies that $\Lambda_{\underline{p}}$ is hereditary for almost all $\underline{p} \in \underline{S}$.

3.9 Theorem (Higman [2]): Let $\{\omega_i\}_{1 \le i \le n}$ and $\{\omega_i^*\}_{1 \le i \le n}$ be a pair of dual bases of A with respect to some non-degenerate bilinear form f; and let $\gamma_f : a \longmapsto \sum_{i=1}^{n} \omega_i^* a \omega_i$, be the associated Gaschütz-Casimir operator, then

$$\underline{H}_\Delta(\Lambda) = \gamma_f(\underline{d}_f(\Lambda)).$$

Proof: Since the formation of the Higman ideal, as well as that of the different and the Gaschütz-Casimir operator, commute with localization (cf. (1.2) and Ex. 3,1 and 3,2), it suffices to prove the theorem locally. Thus we may assume that R is a principal ideal domain. We recall that by definition $\underline{H}_\Delta(\Lambda)$ consists of all projective Λ^e-endomorphisms of Λ. On the other hand we have the exact sequence of Λ^e-modules

$$E_\varepsilon : \quad 0 \longrightarrow \operatorname{Ker}\varepsilon \xrightarrow{\iota} \Lambda^e \xrightarrow{\varepsilon} \Lambda \longrightarrow 0$$

which gives rise to the exact homology sequence

$$0 \longrightarrow \operatorname{Hom}_{\Lambda^e}(\Lambda, \operatorname{Ker}\varepsilon) \xrightarrow{\iota_*} \operatorname{Hom}_{\Lambda^e}(\Lambda, \Lambda^e) \xrightarrow{\varepsilon_*} \operatorname{Hom}_{\Lambda^e}(\Lambda, \Lambda) \longrightarrow$$
$$\xrightarrow{\delta_E} \operatorname{Ext}^1_{\Lambda^e}(\Lambda, \operatorname{Ker}\varepsilon) \longrightarrow \dots .$$

The kernel of the connecting homomorphism δ_E is

$$\operatorname{Im}\varepsilon_* = \{\varphi\varepsilon : \varphi \in \operatorname{Hom}_{\Lambda^e}(\Lambda, \Lambda^e)\}.$$

We show that

$$(3.9') \qquad \underline{H}_\Delta(\Lambda) = \operatorname{Im}\varepsilon_*.$$

By (2.4) we have

$$\underline{H}_\Delta(\Lambda) = \{\alpha \in \operatorname{End}_{\Lambda^e}(\Lambda) : \alpha \text{ is projective}\} = \operatorname{Im}\mu_\Lambda,$$

where

$$\mu_\Lambda : \operatorname{Hom}_{\Lambda^e}(\Lambda, \Lambda^e) \otimes_{\Lambda^e} \Lambda \longrightarrow \operatorname{End}_{\Lambda^e}(\Lambda);\ \varphi \otimes 1 \longmapsto (\varphi \otimes 1) = \varphi\varepsilon,$$

since $\lambda(\varphi \otimes 1)^{\mu_\Lambda} = (\lambda\varphi)1$ and $(x \otimes y)1 = xy = (x \otimes y)\varepsilon$. Moreover, every element of the tensor product $\operatorname{Hom}_{\Lambda^e}(\Lambda, \Lambda^e) \otimes_{\Lambda^e} \Lambda$ is of the form $\varphi \otimes 1$ with $\varphi \in \operatorname{Hom}_{\Lambda^e}(\Lambda, \Lambda^e)$, and thus (3.9') is established.

Now, since A is separable it follows immediately from (3.4) that

$$1_K \otimes \varepsilon_* : \mathrm{Hom}_{A^e}(A,A^e) \longrightarrow \mathrm{End}_{A^e}(A)$$

is epic. But in (III,(6.20)) it was shown that

$$\mathrm{Im}(1_K \otimes \varepsilon_*) = \{\textstyle\sum_{i=1}^n \omega_i^* a \omega_i : a \in A\},$$

and, since we are assuming R to be a principal ideal domain, we may choose the K-basis $\{\omega_i\}_{1\leq i\leq n}$ so that it is also an R-basis for Λ. From (III,(6.19)) we have

$$\mathrm{Hom}_{\Lambda^e}(\Lambda, \Lambda^e) = \{\varphi_b : b \in A, \lambda\varphi_b = \textstyle\sum_{i=1}^n \lambda\omega_i^* b \otimes \omega_i^{op} \in \Lambda^e\}$$

and it follows that

$$\underline{\underline{H}}_\Delta(\Lambda) = \mathrm{Im}\,\varepsilon_* = \{\textstyle\sum_{i=1}^n \omega_i^* b\omega_i : b \in A, \omega_i^* b \in \Lambda, 1\leq i\leq n\}$$
$$= \gamma_f(\underline{\underline{d}}_f(\Lambda)),$$

since (cf. (1.3)) $\underline{\underline{d}}_f(\Lambda) = \{a \in A : \omega_i^* a \in \Lambda, 1\leq i\leq n\}$. #

3.10 <u>Corollary</u>: Let G be a finite group whose order n is relatively prime to the characteristic of the field K, and let $\Lambda = RG$ be the group ring of G over R. Then

$$\underline{\underline{H}}_\Delta(\Lambda) = \{\textstyle\sum_{g\in G} g\lambda g^{-1} : \lambda \in \Lambda\}, \text{ and } \underline{\underline{H}}(\Lambda) = nR.$$

<u>Proof</u>: Since char K $\nmid$ n, the group algebra KG is separable (cf. III, (3.8)). Now, $f : KG \times KG \longrightarrow K$ defined by

$$f(g,g') = \begin{cases} 0 & \text{if } g' \neq g^{-1} \\ 1 & \text{if } g' = g^{-1}, \end{cases}$$

is a non-degenerate bilinear form (cf. III,(3.8)), and the dual basis to $\{g\}_{g\in G}$ relative to f is $\{g^{-1}\}_{g\in G}$. Thus

$$\underline{\underline{d}}_f^{-1}(\Lambda) = \underline{\underline{d}}_f(\Lambda) = \Lambda,$$

and the first statement follows from (3.9). Now,

$$\underline{\underline{H}}(\Lambda) = \underline{\underline{H}}_\Delta(\Lambda) \cap R = \{\textstyle\sum_{g\in G} g\lambda g^{-1} \in R : \lambda \in \Lambda\}.$$

Thus, at any rate $\underline{\underline{H}}(\Lambda) \supset nR$. For the converse inclusion we observe that $\lambda = r\, g_1$, for some $r \in R$, where g_1 is the unit of G, whenever

$\sum_{g \in G} g \lambda g^{-1} \in R$, and thus indeed $\underline{\underline{H}}(\Lambda) = n R$. #

Exercises § 3:

1.) Show that for every prime ideal $\underline{\underline{p}} \in \underline{\underline{S}}$, $\underline{\underline{H}}_{\Delta_{\underline{\underline{p}}}}(\Lambda_{\underline{\underline{p}}}) = (\underline{\underline{H}}_{\Delta}(\Lambda))_{\underline{\underline{p}}}$.

2.) Show that $\gamma_f(\underline{\underline{d}}_f(\Lambda_{\underline{\underline{p}}})) = (\gamma_f(\underline{\underline{d}}_f(\Lambda)))_{\underline{\underline{p}}}$.

3.) Let $\hat{K}$ be the $\underline{\underline{p}}$-adic completion of the field K, and let $\hat{D}$ be a central skewfield over $\hat{K}$ with maximal order $\hat{\Gamma}$. Compute the Higman ideal $\underline{\underline{H}}(\hat{\Gamma})$.

§4 Extensions of lattices

In §3 we gave an explicit description of the Higman ideal, which measures the deviation of an R-order from separability. However, from a module theoretic point of view, projective lattices play an outstanding rôle and thus we shall define an ideal $\underline{\underline{J}}(\Lambda)$ which gives a measure for how far removed Λ is from being hereditary (cf. IV,(4.1)). But unfortunately no explicit description of $\underline{\underline{J}}(\Lambda)$ seems to be available in general.

We keep the notation of the previous sections.

4.1 <u>Definition</u>: For an R-order with center Δ, we define the <u>$\underline{\underline{J}}$-ideal of</u> Λ by

$$\underline{\underline{J}}_\Delta(\Lambda) = \bigcap_{M,N \in {}_\Lambda\underline{\underline{M}}^o} \operatorname{ann}_\Delta(\operatorname{Ext}^1_\Lambda(M,N))$$

$$= \{\delta \in \Delta : \delta \operatorname{Ext}^1_\Lambda(M,N) = 0, \text{ for all } M,N \in {}_\Lambda\underline{\underline{M}}^o\}$$

$$= \{\delta \in \Delta : \text{multiplication by } \delta \text{ is a projective } \Lambda\text{-endomorphism for all } M \in {}_\Lambda\underline{\underline{M}}^o\}.$$

With some abuse of notation we shall consider $\Delta \subset \operatorname{End}_\Lambda(M)$ (observe that this embedding need not be monic (cf. Ex. 2,1)) and thus, because of (2.4), we have

$$\underline{\underline{J}}_\Delta(\Lambda) = \Delta \cap (\bigcap_{M \in {}_\Lambda\underline{\underline{M}}^o} \operatorname{Im} \mu_M).$$

Moreover we set $\underline{\underline{J}}(\Lambda) = \underline{\underline{J}}_\Delta(\Lambda) \cap R$.

4.2 <u>Remark</u>: (i) Λ is hereditary if and only if $1 \in \underline{\underline{J}}_\Delta(\Lambda)$; since $\underline{\underline{J}}_\Delta(\Lambda)$ is an ideal in Δ, this means that $\underline{\underline{J}}_\Delta(\Lambda) = \Delta$.

(ii) $\underline{\underline{H}}_\Delta(\Lambda) \subset \underline{\underline{J}}_\Delta(\Lambda)$, since

$$\operatorname{Ext}^1_{\Lambda^e}(\Lambda, \operatorname{Hom}_R(M,N)) \overset{nat}{\cong} \operatorname{Ext}^1_\Lambda(M,N), \text{ for every } M,N \in {}_\Lambda\underline{\underline{M}}^o \text{ (cf. III, (4.11))}.$$

In particular $\underline{\underline{J}}_\Delta(\Lambda) \neq 0$ (cf. (3.5)). In general, this inclusion is proper, as will be shown in Ex. 3.3.

(iii) We observe that to every $M \in {}_{\Lambda}\underline{\underline{M}}^{f}$ there exists a morphism $\nu_M \in \mathrm{Hom}_{\Delta}(\Delta, \mathrm{End}_{\Lambda}(M))$ that maps $\delta \in \Delta$ onto multiplication by δ. Clearly ν_M is in general not monic. $J(\Delta)$ is the intersection of the preimages of the sets of projective homomorphisms in $\mathrm{Im}\,\nu_M$, $M \in {}_{\Lambda}\underline{\underline{M}}^{o}$.

4.3 <u>Lemma</u> (Jacobinski [1]): Let Γ be a hereditary R-order in A containing Λ. The <u>central conductor</u> of Γ in Λ is defined as $(\Lambda : \Gamma)_{\Delta} = \{\delta \in \Delta : \Gamma\delta \subset \Lambda\}$, and we have $(\Lambda : \Gamma)_{\Delta} \subset \underline{\underline{J}}_{\Delta}(\Lambda)$.

<u>Proof</u>: It clearly suffices to show that every element $\delta \in (\Lambda : \Gamma)_{\Delta}$ is projective as a Λ-endomorphism for every Λ-lattice. We use (2.3) to show this. Let

(1)
$$\begin{array}{ccccc} & & M & & \\ & & \downarrow{\scriptstyle\delta} & & \\ X & \xrightarrow{\ \beta\ } & M & \longrightarrow & 0 \end{array}$$

be a diagram of Λ-maps with an exact row, $M \in {}_{\Lambda}\underline{\underline{M}}^{o}$ and $\delta \in (\Lambda : \Gamma)_{\Delta}$. Since $M \subset K \otimes_R M = KM = AM$, we have $\Gamma M \in {}_{\Gamma}\underline{\underline{M}}^{o}$ and $M \subset \Gamma M$. Moreover $\Gamma M \in {}_{\Gamma}\underline{\underline{P}}^{f}$, Γ being hereditary. β induces an epimorphism $\beta_{\Gamma} : \Gamma X \longrightarrow \Gamma M$, since M is an R-lattice. Thus, we can complete the diagram

$$\begin{array}{ccccc} & & \Gamma M & & \\ & \swarrow{\scriptstyle\alpha} & \downarrow{\scriptstyle 1_{\Gamma M}} & & \\ \Gamma X & \xrightarrow{\ \beta_{\Gamma}\ } & \Gamma M & \longrightarrow & 0 \end{array} \quad ;$$

but then also the diagram

$$\begin{array}{ccccc} & & M & & \\ & \swarrow{\scriptstyle\alpha\delta} & \downarrow{\scriptstyle\delta} & & \\ X & \xrightarrow{\ \beta\ } & M & \longrightarrow & 0 \end{array}$$

is commutative, if $\delta \in (\Lambda : \Gamma)_{\Delta}$; and δ is projective. #

4.4 <u>Corollary</u>: If Λ and Γ are as in (4.3), and if e is a central idempotent in A and $(\Lambda : \Gamma e)_{\Delta} = \{\delta \in \Delta : \Gamma e \delta \subset \Lambda\}$, then we have for all $M \in {}_{\Lambda}\underline{\underline{M}}^{o}$, for which $eM = M$, $(\Lambda : \Gamma e)_{\Delta} \subset \mathrm{Im}\,\mu_M$.

<u>Proof</u>: It should be observed, that if suffices to show, that every diagram of the form

$$\begin{array}{ccccc} & & M & & \\ & & \Big\downarrow{\scriptstyle\delta} & & \\ X & \longrightarrow & M & \longrightarrow & 0 \end{array}$$

can be completed, where $eX = X$. Now, the result follows as in the proof of (4.3), since for $\delta \in (\Lambda : \Gamma e)_{\Delta}$, $\delta \Gamma X = \Gamma \delta e X \subset \Lambda X = X$. #

4.5 <u>Theorem</u> (Reiner [9]): Let G be a finite group of order n, such that char $K \nmid n$, set $A = KG$, $\Lambda = RG$ and let $M \in {}_{\Lambda}\underline{\underline{M}}^{o}$. Then $\varphi \in \mathrm{End}_{\Lambda}(M)$ is projective if and only if there exists $\psi \in \mathrm{End}_R(M)$ such that $\varphi = \psi^G$, where $m\,\psi^G = \sum_{g \in G} g^{-1}((gm)^{\psi})$.

<u>Proof</u>: If we are willing to violate our convention to write R-endomorphisms as exponents, we may express this theorem in the following form: $\mathrm{Im}\, \mu_M = \{ \sum_{g \in G} g \psi g^{-1} : \psi \in \mathrm{End}_R(M)\}$, where the endomorphisms are written on the left.

We have the R-homomorphisms

$$\Phi : \mathrm{End}_R(M) \longrightarrow \mathrm{End}_{\Lambda}(M), \quad \psi \longmapsto \psi^G, \text{ and}$$

$$\Psi_o : \mathrm{Hom}_R(M,R) \longrightarrow \mathrm{Hom}_{\Lambda}(M,\Lambda), \quad \psi \longmapsto \psi^G.$$

Moreover, the R-balanced map $\mathrm{Hom}_{\Lambda}(M,\Lambda) \times M \longrightarrow \mathrm{Hom}_{\Lambda}(M,\Lambda) \otimes_{\Lambda} M$ induces an R-epimorphism $\Psi' : \mathrm{Hom}_{\Lambda}(M,\Lambda) \otimes_R M \longrightarrow \mathrm{Hom}_{\Lambda}(M,\Lambda) \otimes_{\Lambda} M$, and putting $\Psi = (\Psi_o \otimes 1_M)\Psi'$ it is easily verified that the following diagram of R-homomorphisms commutes:

$$\begin{array}{ccc} \mathrm{Hom}_R(M,R) \otimes_R M & \xrightarrow{\ \mu_M^R\ } & \mathrm{End}_R(M) \\ \Psi \Big\downarrow & & \Big\downarrow \Phi \\ \mathrm{Hom}_{\Lambda}(M,\Lambda) \otimes_{\Lambda} M & \xrightarrow{\ \mu_M^{\Lambda}\ } & \mathrm{End}_{\Lambda}(M) \end{array}.$$

Clearly μ_M^R is epic since M is R-projective. But Ψ is also an epimorphism. To show this, it suffices to note that Ψ_o is epic. Thus, let $\varphi \in \mathrm{Hom}_{\Lambda}(M,\Lambda)$ and $m \in M$. Then for some $\{r_g(m)\}_{g \in G}$,

$m\varphi = \sum_{g \in G} r_g(m)g^{-1}$, since $\{g^{-1}\}_{g \in G}$ is an R-basis for Λ. Moreover, since φ is a Λ-homomorphism, we have $r_g(m) = r_1(gm)$, and $\psi : m \longmapsto r_1(m)$ is a map in $\mathrm{Hom}_R(M,R)$, so that $\varphi = \psi^G$, and Ψ_o is epic. It follows now from the commutativity of the above diagram that $\mathrm{Im}\,\mu_M^\Lambda = \mathrm{Im}\,\Phi$. #

4.6 <u>Corollary</u>: $\underline{\underline{J}}_\Delta(RG) \cap R = n \cdot R$, if n is the order of the group G.

<u>Proof</u>: By (3.10) and (4.2) we have

$$n \cdot R = \underline{\underline{H}}_\Delta(\Lambda) \cap R \subset \underline{\underline{J}}_\Delta(\Lambda) \cap R.$$

To prove the converse inclusion we choose R for M in (4.5) with the trivial G-module structure; i.e., $R \in {}_\Lambda\underline{\underline{M}}^o$, by $gr = r$, $g \in G$. Then $\mathrm{Im}\,\mu_R \cap R = nR$, since $\mathrm{End}_R(R) = R$, because every R-endomorphism of R as trivial G-module is determined by its action on $1 \in R$, and multiplication by any element of R is such an endomorphism. Since the action of G on R is trivial, $r\psi^G = r \cdot \sum_{g \in G} 1\psi = nr\psi$. But by definition (cf. (4.1),(4.2)) $\vartheta_R : \underline{\underline{J}}_\Delta(\Lambda) \longrightarrow \mathrm{Im}\,\mu_R$, and thus $\underline{\underline{J}}_\Delta(\Lambda) \cap R \subset \mathrm{Im}\,\mu_R \cap R = nR$. #

4.7 <u>Theorem</u> (Roggenkamp [10]): Let Γ be an R-order in A containing the R-order Λ, and let

$$(\Lambda : \Gamma)_l = \{x \in A : \Gamma x \subset \Lambda\}$$

be the <u>left conductor of</u> Γ <u>in</u> Λ, then, for Γ, considered as a left Λ-lattice, we have

$$\mathrm{Im}\,\mu_{{}_\Lambda\Gamma} = (\Lambda : \Gamma)_l \cdot \Gamma.$$

<u>Proof</u>: By (IV,(1.14)) we have $\mathrm{End}_\Lambda({}_\Lambda\Gamma) = \mathrm{End}_\Gamma({}_\Gamma\Gamma) \cong \Gamma$. Under this isomorphism $\mathrm{Hom}_\Lambda(\Gamma, \Lambda) \cong (\Lambda : \Gamma)_l$ and, observing that $\mu_{{}_\Lambda\Gamma} : \mathrm{Hom}_\Lambda(\Gamma, \Lambda) \otimes_\Lambda \Gamma \longrightarrow \mathrm{End}_\Lambda(\Gamma)$, maps $x \otimes y$, with $x \in (\Lambda : \Gamma)_l$ and $y \in \Gamma$ onto multiplication by xy, we find that indeed $\mathrm{Im}\,\mu_{{}_\Lambda\Gamma} = (\Lambda : \Gamma)_l \cdot \Gamma$. #

4.8 <u>Corollary</u>: If Γ is a hereditary R-order containing Λ, and if the left conductor $(\Lambda : \Gamma)_l$ is a two-sided ideal in Γ, then

$$\underline{\underline{J}}_{\Delta}(\Lambda) = (\Lambda : \Gamma)_{\Delta} = (\Lambda : \Gamma)_1 \cap \Delta .$$

<u>Proof</u>: The inclusion $\supset$ was established in (4.3), while the converse inclusion follows from (4.7). #

4.9 <u>Lemma</u>: If an R-order Γ' in A contains a hereditary R-order Γ in A then

$$(\Gamma : \Gamma')_1 \cdot \Gamma' = \Gamma'.$$

<u>Proof</u>: Since Γ is hereditary $1_{\Gamma'} \in \text{Im}\, \mu_{\Gamma}{}_{\Gamma'} = (\Gamma : \Gamma')_1 \cdot \Gamma'$. But then $(\Gamma : \Gamma')_1 \cdot \Gamma' = \Gamma'$. #

4.10 <u>Lemma</u>: Assume that A has a unique maximal R-order Γ. Then

(i) $(\Lambda : \Gamma)_1$ is a two-sided Γ-ideal for all R-orders Λ in A,

(ii) Γ is the only hereditary R-order in A.

(iii) $\underline{\underline{J}}_{\Delta}(\Lambda) = (\Lambda : \Gamma)_1 \cap \Delta$, for all R-orders Λ in A.

<u>Proof</u>: It is clear that (ii) and (iii) follow from (i) by (4.9) and (4.8) respectively. To prove (i) we have to show that $\Lambda_r((\Lambda : \Gamma)_1) = \text{End}_{\Gamma}((\Lambda : \Gamma)_1) = \Gamma$. By (IV,(5.5)) $\text{End}_{\Gamma}((\Lambda : \Gamma)_1)$ is a maximal R-order in A because Γ is maximal and $(\Lambda : \Gamma)_1 \in {}_{\Gamma}\underline{\underline{M}}^{o}$; but then, since there is only one maximal R-order in A, the desired result follows. #

4.11 <u>Corollary</u>: The conclusions of (4.10) hold if

(i) A is commutative, or

(ii) A is a direct sum of skewfields over a complete field $\hat{K}$.

<u>Proof</u>: From IV,(5.8) and IV,(4.8) it follows that a commutative algebra has exactly one maximal R-order, and (ii) follows from (IV,(5.2)). #

4.12 <u>Theorem</u> (Jacobinski [1]; Roggenkamp [10]): Let $\Lambda = RG$ be the group ring of a finite group G of order n, such that char K $\nmid$ n. Then

$$\underline{\underline{J}}_{\Delta}(\Lambda) = (\Lambda : \Gamma)_{\Delta} \text{ and } (\Lambda : \Gamma)_1 = \{ nx \in A : Tr_{A/K}(x\Gamma) \subset R \}$$

for every hereditary R-order Γ in A that contains Λ.

Proof: For any R-lattice M in A = KG we define the dual M^* with respect to the trace function from A to K as follows:

$$M^* = M^*_{Tr_{A/K}} = \{x \in A : Tr_{A/K}(xM) \subset R\}.$$

The correspondence $M \longmapsto M^*$ is strictly inclusion reversing; i.e.,

$$M_1 \subsetneqq M_2 \text{ implies } M_2^* \subsetneqq M_1^*.$$

That inclusions are reversed is an obvious consequence of the definition. To establish the preservation of strict inclusions it suffices to prove this locally, since localization and dualization commute; i.e., $R_{\underline{p}} \otimes_R M^* = (M_{\underline{p}})^*$ for every prime ideal $\underline{p} \in \underline{S}$, and we have $M^* = \bigcap_{\underline{p} \in \underline{S}} M^*_{\underline{p}}$, (cf. I,(8.6)), because M^* is also an R-lattice (cf. Ex. 4,1). Let us therefore assume that R is a principal ideal domain, and let $\{\omega_i\}_{1\le i\le n}$ be an R-basis for M_1. Then $\{\omega_i\}_{1\le i\le n}$ is also a K-basis for A and, since A = KG is separable, there exists a dual basis $\{\omega_i^*\}_{1\le i\le n}$ relative to the trace function $Tr_{A/K}$ (cf. III,(3.6)). (We note that the trace function $Tr_{KG/K} : KG \times KG \longrightarrow K$ is non-degenerate, since $\{1/n\ g^{-1}\}_{g \in G}$ is a dual basis of KG with respect to the trace function.) Thus, observing that $Tr_{A/K}$ is symmetric, we obtain, (cf. Ex. 1,1),

$$M_1^* = \oplus_{i=1}^n R\,\omega_i^*.$$

But, if $M_1 \subset M_2$ and $M_1 \ne M_2$, then there exists $m = \sum_{i=1}^n k_i\,\omega_i \in M_2$ with, at least for some i, $k_i \notin R$. But then $Tr_{A/K}(M_1^* m) \not\subset R$ and $M_1^* \not\subset M_2^*$. From the definition and the symmetry of the trace it follows at once that $M \subset M^{**}$, thus also $M^* \subset M^{***}$ as well as $M^{***} \supset M^*$ since inclusions are reversed. But then, because of the strictness of the inclusion reversal, we obtain

$$M = M^{**}.$$

(Observe the rôle that the symmetry of our bilinear form is playing in these arguments.) Moreover, for any R-order Ω in A,

$$M \in {}_{\underline{\underline{\Omega}}}\underline{M}^{o} \text{ if and only if } M^* \in \underline{\underline{M}}^{o}_{\Omega}.$$

If suffices to prove one implication since the other one follows from the fact that $M = M^{**}$. But clearly, if $M \in {}_{\underline{\underline{\Omega}}}\underline{M}^{o}$ and $Tr_{A/K}(xM) \subset R$ then $Tr_{A/K}((x\Omega)M) \subset R$ and thus $M^* \in \underline{\underline{M}}^{o}_{\Omega}$. Returning now to the proof of (4.12) we observe that

$$\Lambda^* = \bigoplus_{g \in G} (1/n)R\, g^{-1} = (1/n)\Lambda$$

since $\{(1/n)g^{-1}\}_{g \in G}$ is a dual basis to $\{g\}_{g \in G}$ relative to the trace function, (cf. III,(3.8)), and since Λ is R-free. If now Γ is any R-order in A containing Λ, then $\Lambda^*\Gamma$ is the smallest right Γ-lattice containing Λ^*. Hence $(\Lambda^*\Gamma)^*$ is the largest left Γ-lattice contained in Λ, i.e.,

$$(\Lambda^*\Gamma)^* = (\Lambda : \Gamma)_1.$$

But also,

$$(\Lambda^*\Gamma)^* = ((1/n)\Lambda\Gamma)^* = n(\Gamma)^*$$

is a right Γ-module, so that the left conductor $(\Lambda : \Gamma)_1$ is a two-sided Γ-ideal. This proves the second equality and the first equality of (4.12) follows from (4.8) in case Γ is hereditary. #

We remark that this proof can be carried over to the case of any R-order in A provided that the dual of Λ with respect to some non-degenerate bilinear form is of the form $\delta\Lambda$ with δ in the center of A.

<u>Exercises §4</u>:

1.) Let f be a non-degenerate bilinear form, $f : A \times A \longrightarrow K$. Define the dual M^* of an R-lattice M with respect to f, show that for every $\underline{\underline{p}} \in \underline{\underline{S}}$, $(M^*)_{\underline{\underline{p}}} = (M_{\underline{\underline{p}}})^*$, and show that M^* is again an R-lattice in A. (Distinguish between right and left!)

2.) Show that the maps Φ and Ψ_o used in the proof of (4.5) are R-homomorphisms.

3.) Let $\Lambda = RG$ be the group ring of a finite group G such that char $K \nmid |G|$. Let the bilinear form $f : KG \times KG \longrightarrow K$ be defined as follows:

$$f(g,g') = \begin{cases} 0 & \text{if } g' \neq g^{-1} \\ 1 & \text{if } g' = g^{-1}. \end{cases}$$

Compute $\Lambda_f^* = \{x \in A ; f(x, \Lambda) \subset R\}$; use this to prove (4.12) and compare your result with the proof of the text.

4.) Show that $(\underline{\underline{J}}_{\Delta}(\Lambda))_{\underline{\underline{p}}} = \underline{\underline{J}}_{\Delta_{\underline{\underline{p}}}}(\Lambda_{\underline{\underline{p}}})$, for all primes $\underline{\underline{p}}$ of R.

§5. Annihilators of some special classes of Λ-lattices.

We define $\underline{\underline{J}}_\Delta(L)$ for $L \in {}_{\Lambda}\underline{\underline{M}}^f$. $\underline{\underline{J}}_\Delta(L)$ and $\underline{\underline{J}}_\Delta(\Lambda)$ are computed in case $\Lambda = RG$ is a group ring.

We retain the notation of the previous sections.

5.1 Lemma: Let Γ be a hereditary R-order in A containing Λ and e a central idempotent in A. If $M \in {}_{\Gamma e}\underline{\underline{M}}^o$ is a progenerator, and if Δ is the center of Λ, then

$\operatorname{Im} \mu_{\Lambda M} \cap \Delta e = (\Lambda : \Gamma e)_1 \cdot \Gamma e \cap \Delta e$, where $(\Lambda : \Gamma e)_1 =$

$= \{a \in A : \Gamma e a \subset \Lambda\}$.

Proof: By (4.4) and a slight modification of (4.7) we have

$$\operatorname{Im} \mu_{\Lambda \Gamma e} = (\Lambda : \Gamma e)_1 \cdot \Gamma e,$$

and it remains to show that

$$\operatorname{Im} \mu_{\Lambda \Gamma e} \cap \Delta e = \operatorname{Im} \mu_{\Lambda M} \cap \Delta e.$$

By (2.4) this amounts to showing that multiplication with $\sigma \in \Delta e$ is a projective Λ-endomorphism on Γe if and only if it is a projective Λ-endomorphism on M. Now, since M is a Γe-progenerator, there are Γe-lattices X and Y such that:

$X \oplus M \cong (\Gamma e)^{(s)}$ and $Y \oplus \Gamma e \cong M^{(t)}$, for some $s, t \in \underline{\underline{N}}$.

For $\sigma \in \Delta e$, and any $N \in {}_{\Lambda}\underline{\underline{M}}^f$, we have

$$\begin{aligned}(\sigma \operatorname{Ext}^1_\Lambda(\Gamma e, N))^{(s)} &\cong \sigma \operatorname{Ext}^1_\Lambda((\Gamma e)^{(s)}, N) = \sigma \operatorname{Ext}^1_\Lambda(X \oplus M, N)\\ &\cong \sigma \operatorname{Ext}^1_\Lambda(X, N) \oplus \sigma \operatorname{Ext}^1_\Lambda(M, N), \text{ and}\\ (\sigma \operatorname{Ext}^1_\Lambda(M, N))^{(t)} &\cong \sigma \operatorname{Ext}^1_\Lambda(M^{(t)}, N) \cong \sigma \operatorname{Ext}^1_\Lambda(Y \oplus \Gamma e, N)\\ &\cong \sigma \operatorname{Ext}^1_\Lambda(Y, N) \oplus \sigma \operatorname{Ext}^1_\Lambda(\Gamma e, N).\end{aligned}$$

From these natural isomorphisms it is immediately apparent that σ annihilates $\operatorname{Ext}^1_\Lambda(\Gamma e, -)$ if and only if it annihilates $\operatorname{Ext}^1_\Lambda(M, -)$. #

5.2 Remark: If Γ is a maximal R-order in A and $M \in {}_{\Gamma}\underline{\underline{M}}^o$, then $\operatorname{ann}_\Gamma(M) = \Gamma(1-e)$ for some central idempotent e in A, and $M \in {}_{\Gamma e}\underline{\underline{M}}^o$ is a progenerator (cf. IV,(5.5)), so that in this case (5.1) is applicable.

5.3 Theorem (Roggenkamp[10]): Let $A = \oplus_{i=1}^{n} (K_i)_{n_i}$, where the K_i are finite separable extensions of K, and let $\{e_i\}_{1 \le i \le n}$ be a complete set of orthogonal primitive central idempotents of A. Let Λ be an R-order in A whose center Δ is a subdirect sum of the maximal R-orders Ω_i of the K_i; i.e., $\Delta e_i = \Omega_i$, $1 \le i \le n$, and let M be an irreducible Λ-lattice. Then there is a maximal R-order Γ_M in A, containing Λ, such that

$$\operatorname{Im} \mu_M \cap \Delta = (\Lambda : \Gamma_M e)_1 \cdot \Gamma_M e,$$

where $e \in \{e_i\}$ is the idempotent for which $e M \neq 0$.

Proof: Once it is shown that under the above hypotheses there exists to every irreducible Λ-lattice M a maximal R-order Γ_M in A containing Λ, such that M is a Γ_M-lattice, the result follows from (5.2). For, then M is also an irreducible $\Gamma_M e$-lattice and $\Gamma_M e$ is a maximal R-order. The existence of Γ_M will be established in Ch. VI,(5.12). #

5.4 Lemma: Let Λ be an R-order in the separable K-algebra A and define

$$\underline{\underline{J}}_{\Delta}(L) = \bigcap_{\{M \in {}_{\Lambda}\underline{\underline{M}}^{o} : KM = L\}} \operatorname{Im} \mu_M \cap \Delta, \text{ for } L \in {}_{A}\underline{\underline{M}}^{f}.$$

Then,

(i) $$\bigcap_{\{i : e_i L \neq 0\}} (\Lambda : \Gamma e_i)_1 \cap \Delta \subset \underline{\underline{J}}_{\Delta}(L) \subset \bigcap_{\{i : e_i L \neq 0\}} (\Lambda : \Gamma e_i)_1 \Gamma e_i \cap \Delta,$$

whenever Γ is a maximal R-order in A containing Λ and $\{e_i\}_{1 \le i \le n}$ is a complete set of orthogonal primitive central idempotents in A.

(ii) If for a maximal R-order Γ in A, containing Λ, and for every central primitive idempotent e with $eL \neq 0$, $(\Lambda : \Gamma e)_1 \cdot \Gamma e \subset (\Lambda : \Gamma e)_1$, then $\underline{\underline{J}}_{\Delta}(L) = \bigcap_{\{i : e_i L \neq 0\}} (\Lambda : \Gamma e_i)_1 \cap \Delta$.

Proof: (ii) clearly follows from (i). As to (i), it should be observed, that it suffices to prove the theorem locally (cf. Ex. 4,5 and 5,1). Thus we may assume that $R = R_{\underline{\underline{p}}}$ for some prime ideal $\underline{\underline{p}}$ in

R and let $M \in {}_{\Lambda}\underline{\underline{M}}^{o}$ be such that $KM \cong L$, for fixed $L \in {}_{A}\underline{\underline{M}}^{f}$. If Γ is a maximal R-order in A then, by IV,(5.7), the Krull-Schmidt theorem is valid for Γ-lattices. Hence we may write

$$\Gamma M = \bigoplus_{\{i : e_i L \neq 0\}} M_i^{(n_i)}$$

where, for each i, $1 \leq i \leq n$, M_i is the - up to isomorphism unique - indecomposable Γe_i-lattice, and the integers n_i are unique. Thus

$$\operatorname{Im} \mu_{{}_{\Lambda}\Gamma M} \cap \Delta = \bigcap_{\{i : e_i L \neq 0\}} \operatorname{Im} \mu_{{}_{\Lambda}M_i} \cap \Delta .$$

The first inclusion now follows from (4.4) and the second is established by the method used in the proof of (5.1). #

5.5 <u>Theorem</u> (Jacobinski [1]; Reiner [9]; Roggenkamp[10]): Let G be a finite group whose order n is relatively prime to the characteristic of the field K, and set $A = KG$, $\Lambda = RG$ and $\Delta = \text{center}\ (\Lambda)$. If $A = \bigoplus_{i=1}^{s} Ae_i$ is the decomposition of A into simple algebras Ae_i, we set $A_i = Ae_i$, denote the center of A_i by K_i, and set $r_i^2 = [A_i : K_i]$, $1 \leq i \leq s$. If Γ is a maximal R-order in A, let Γ decompose into the maximal R-orders Γ_i in A_i; i.e., $\Gamma = \bigoplus_{i=1}^{s} \Gamma_i$, and for each i, $1 \leq i \leq s$, let R_i be the center of Γ_i; i.e., the integral closure of R in K_i. Then

(i) $\underline{\underline{J}}_{\Delta}(\Lambda) = \bigoplus_{i=1}^{s} (n/r_i)\, \underline{\underline{d}}^{-1}_{\mathrm{Tr}_{K_i/K}}(R_i) = (\Lambda : \Gamma)_l \cap \bigoplus_{i=1}^{s} R_i$;

(ii) $\underline{\underline{J}}_{\Delta}(L) = \{ \delta \in \Delta : \delta e_i \in (n/r_i)\, \underline{\underline{d}}^{-1}_{\mathrm{Tr}_{K_i/K}}(R_i)$, for all i with $Le_i \neq 0\}$,

and

$$\underline{\underline{J}}_{\Delta}(L) \cap R = \bigcap_{\{i : Le_i \neq 0\}} (n/r_i)(\underline{\underline{d}}^{-1}_{\mathrm{Tr}_{K_i/K}}(R_i) \cap Ke_i), \text{ for } L \in {}_{A}\underline{\underline{M}}^{f};$$

(iii) if L is absolutely irreducible; i.e., if $\operatorname{End}_A(L) \cong K$, then $\underline{\underline{J}}_{\Delta}(L) \cap R = (n/r_i)R = (n/\dim_K(L))R = (\Lambda : \Gamma e_i)_l \cap R$, if $Le_i \neq 0$, in particular, for every idempotent e of Γ:

$$(\Lambda : \Gamma e)_l\, \Gamma e = (\Lambda : \Gamma e)_l .$$

<u>Proof</u>: (i) In (4.12) it was shown that $\underline{\underline{J}}_{\Delta}(\Lambda) = (\Lambda : \Gamma)_l \cap \Delta$ and $(\Lambda : \Gamma)_l = n \cdot \Gamma^*$. But from (1.5) we obtain

$$\Gamma^* = \oplus_{i=1}^{s} \Gamma_i^* = \oplus_{i=1}^{s} (1/r_i)\, \underline{\underline{d}}^{-1}_{Trd_{A_i/K_i}}(\Gamma_i) \cdot \underline{\underline{d}}^{-1}_{Tr_{K_i/K}}(R_i).$$

Hence

$$\underline{\underline{J}}_{\Delta}(\Lambda) = \oplus_{i=1}^{s} (n/r_i)\, \underline{\underline{d}}^{-1}_{Trd_{A_i/K_i}}(\Gamma_i) \cdot \underline{\underline{d}}^{-1}_{Tr_{K_i/K}}(R_i).$$

Now we claim that

5.5' $(n/r_i)\underline{\underline{d}}^{-1}_{Trd_{A_i/K_i}}(\Gamma_i) \cdot \underline{\underline{d}}^{-1}_{Tr_{K_i/K}}(R_i) \cap R_i = (n/r_i)\underline{\underline{d}}^{-1}_{Tr_{K_i/K}}(R_i) \cap R_i$.

To show this it clearly suffices to prove that

$$\underline{\underline{d}}^{-1}_{Trd_{A_i/K_i}}(\Gamma_i) \cap K_i \subset R_i ,$$

and since the inverse different localizes properly, we need only prove this locally. Thus, let $\underline{\underline{p}}$ be a prime ideal in R_i, then, by (1.7),

$$\underline{\underline{d}}^{-1}_{Trd_{A_i/K_i}}((\Gamma_i)_{\underline{\underline{p}}}) = (\mathrm{rad}(\Gamma_i)_{\underline{\underline{p}}})^{1-r_i} .$$

On the other hand

$$\underline{\underline{d}}^{-1}_{Trd_{A_i/K_i}}((\Gamma_i)_{\underline{\underline{p}}}) \cap K_i \subset (\mathrm{rad}(R_i)_{\underline{\underline{p}}})^{t}, \text{ for some } t \in \underline{\underline{Z}}.$$

Thus $(\mathrm{rad}(R_i)_{\underline{\underline{p}}})^t \cdot (\Gamma_i)_{\underline{\underline{p}}} \subset (\mathrm{rad}(\Gamma_i)_{\underline{\underline{p}}})^{1-r_i}$; i.e.,

$(\mathrm{rad}(\Gamma_i)_{\underline{\underline{p}}})^{tr_i} \subset (\mathrm{rad}(\Gamma_i)_{\underline{\underline{p}}})^{1-r_i}$ (cf. IV,(6.13)). But this clearly implies that $t \geq 0$, whence (5.5') is established. From this we obtain (i), since obviously $(\Lambda : \Gamma)_l \cap \oplus_{i=1}^{s} R_i = (\Lambda : \Gamma)_l \cap \Delta$.

To prove (ii), we recall that, by (5.4),

$$\Delta \cap \bigcap_{\{i : Le_i \neq 0\}} (\Lambda : \Gamma_i)_l \subset \underline{\underline{J}}_{\Delta}(L) \subset \bigcap_{\{i : Le_i \neq 0\}} (\Lambda : \Gamma_i)_l\, \Gamma_i \cap \Delta .$$

But $(\Lambda : \Gamma e_i) = \{x \in A : \Gamma e_i x \subset \Lambda\} = \{x \in A : e_i x \in (\Lambda : \Gamma)_l\}$
$= \{x \in A : e_i x \in \oplus_{i=1}^{s} (n/r_i)\, \underline{\underline{d}}^{-1}_{Trd_{A_i/K_i}}(\Gamma_i) \cdot \underline{\underline{d}}^{-1}_{Tr_{K_i/K}}(R_i)\}$ which

is a two-sided Γ_1-ideal, since $\underline{\underline{d}}^{-1}_{Trd_{A_1/K_1}}(\Gamma_1)$ is one.

Now (ii) follows by (5.5').

(iii) finally follows from (ii) and (5.3). For, if L is absolutely irreducible corresponding to the central idempotent e_1, then $R_1 = R$, $r_1 = \dim_K(L)$ and $\underline{\underline{d}}^{-1}_{Tr_{R_1/R}}(R_1) = R$. #

<u>Exercise §5</u>:

1.) Show that conductors localize properly; i.e., show that for any prime ideal $\underline{\underline{p}}$ in R $((\Lambda : \Gamma)_1)_{\underline{\underline{p}}} = (\Lambda_{\underline{\underline{p}}} : \Gamma_{\underline{\underline{p}}})_1$.

BIBLIOGRAPHY

ALBERT, A.A.

1. Structure of Algebras. AMS Colloquium publications XXIV, 1961.

ARTIN, E.

1. Zur Arithmetik hypercomplexer Zahlen. Hamb. Abh. 5 (1927), 261-289.

2. Theory of algebraic numbers. Göttingen, 1959.

ARTIN, E. - C. NESBITT - R.M. THRALL

1. Rings witn minimum condition. Univ. of Michigan, Ann Arbor, 1944.

AUSLANDER, M.

1. On the dimension of modules and algebras (III). Nagoya Math. J. 9 (1955), 67-77.

2. Rings, Modules and Homology. Lecture notes, Brandeis Univ. 1959/60.

AUSLANDER, M. - O. GOLDMAN

1. Maximal orders. Trans. Am. Math. Soc. 97 (1960), 1-24.

2. The Brauer group of a commutative ring. Trans. Am. Math. Soc. 97 (1960), 367-409.

ASANO, S.

1. On the automorphism ring of a division algebra. Kodai Math. Sem. Rep. 18 (1966), 368-372.

AZUMAYA, G.

1. Corrections and supplementaries to my paper concerning Krull-Remak-Schmidt's theorem. Nagoya Math. J. 1 (1950), 117-124.

2. Completely faithful modules and self-injective rings. Nagoya Math. J. 27 (1966), 697.

BACHMUTH.S. - H.Y. MOCHIZUKI

1. Cyclotomic ideals in group rings. Bull. AMS 72 (1966), 1018.

BANASCHEWSKI, B.

1. On the character ring of finite groups. Can. J. Math. 15 (1963), 605-612.

BASS, H.

1. Projective modules over algebras. Ann. of Math. 73 (1961), 532-542.

2. The Morita Theorems. Mim. notes, Oregon, 1962.

3. Torsion free and projective modules. Trans. Am. Math. Soc. 102 (1962), 319-327.

4. On the ubiquity of Gorenstein rings. Math. Zeit. 82 (1963), 8-28.

5. K-theory and stable algebra. Math. I.H.E.S. 22 (1964), 489-544.

6. The Dirichlet unit theorem, induced characters, and Whitehead groups. Topology 4 (1966), 391-410.

BASS, H.

7. Finistic dimension and homological generalization of semi-primary rings. Trans. Am. Math. Soc. 95 (1960), 466-488.

8. Algebraic K-theory. Benjamin, New York-Amsterdam, 1968.

BASS, H. - A. HELLER - R. SWAN

1. The Whitehead group of a polynomial extension. I.H.E.S. 22 (1964), 61-79.

BASS, H. - M.P. MURTHY

1. Grothendieck groups and Picard groups of abelian group rings. Ann. Math. 86 (1967), 16-73.

BERMAN, S.D.

1. On certain properties of integral group rings. Dokl. Akad. Nauk S.S.S.R. 91 (1953), 7-9.

2. Integral representations of finite groups. Dokl. Akad. Nauk S.S.S.R. 152 (1963), 1286-1287.

3. A contribution to the theory of integral representations of finite groups. Dokl. Akad. Nauk S.S.S.R. 157 (1964), 506-508.

4. Representations of group rings over arbitrary fields and over rings of integral numbers. Izv. Akad. Nauk S.S.S.R. 30 (1966), 69-132.

5. On isomorphism of centers of group rings of p-groups. Dokl. Akad. Nauk 91 (1953), 185-187.

6. On a necessary condition for isomorphism of integral group rings. Dopovidi Akad. Nauk Ukrain. RSR (1953), 313-316.

7. On the equation $x^m = 1$ in an integral group ring. Ukrain. Math. Ẑ. 7 (1955), 253-261.

BERMAN, S.D. - P.M. GUDIVOK

1. On the integral representations of finite groups. Dokl. Akad. Nauk S.S.S.R. 145 (1962), 1199-1201.

2. Indecomposable representations of finite groups. Izvestia Akad. Nauk S.S.S.R. 28 (1964), 875-910.

3. On integral representations of finite groups. Dokl. Uzhgorod Univ. Nauk 5 (1962), 74-76.

BOERNER, H.

1. Darstellungen von Gruppen. Springer, Berlin-Heidelberg-New York 1967.

2. Darstellungstheorie der endlichen Gruppen. Enzykl. Math. Wiss., Bd. I,1, Heft 6, II, no. 15, Teubner, Stuttgart, 1967.

BOREVICH, Z.I. - D.K. FADDEEV

1. Theory of homology in groups I, II. Proc. Leningrad Univ. 4 (1956), 3-39; 7 (1959), 72-87.

2. Integral representations in quadratic rings. Proc. Leningrad Univ. 8 (1960), 52-64.

BOREVICH, Z.I. - D.K. FADDEEV

3. Representations of orders of cyclic index. Trudy Mat. Inst. Steklov Akad. Nauk S.S.S.R. 80 (1965), 51-65.

BOREVICH, Z.I. - I.R. SAFAREVICH

1. Number Theory. Academic Press, New York-London, 1966.

BOURBAKI

1. Algèbre. Ch. 1-8, Hermann, Paris, 1961.

2. Algèbre commutative. Ch. 1,2,3,...,7, Hermann, Paris, 1961.

BRUMER, A.

1. Structure of hereditary orders. Bull. Am. Math. Soc. 69 (1963), 721-729.

2. Addendum, "Structure of hereditary orders". Ibid. 70 (1964), 185.

BRANDT, H.

1. Zur Komposition der quaternären, quadratischen Formen. J. für reine angew. Math. 143 (1913), 106-129.

2. Der Kompositionsbegriff bei den quaternären, quadratischen Formen. Math. Ann. 91 (1924), 300-315.

3. Über das associative Gesetz bei der Komposition der quaternären, quadratischen Formen. Math. Ann. 96 (1927), 360-366.

4. Ideal Theorie in Quaternionenalgebren. Math. Ann. 99 (1928), 1-29.

BRAUER, R.

1. Über Darstellungen von Gruppen in Galoischen Feldern. Act. Sci. Ind. 195, Paris, 1935.

2. On modular and p-adic representations of algebras. Proc. Nat. Acad. Sci. 25 (1939), 252-258.

3. On the Cartan invariants of groups of finite order. Ann. of Math. 42 (1941), 53-61.

4. On the connection between ordinary and modular characters of groups of finite order. Ann. of Math. 42 (1941), 926-935.

5. On the representation of a group of order g in the field of the g-th roots of unity. Am. J. Math. 67 (1945), 461-471.

6. On blocks of characters of groups of finite order I, II. Proc. Nat. Acad. Sci. 32 (1946), 182-186; 215-219.

7. Zur Darstellungstheorie der Gruppen endlicher Ordnung I, II. Math. Zeit. 63 (1956), 406-444; 72 (1959), 25-46.

8. Some application of the theory of blocks of characters of finite groups, I, II. J. of Algebra 1 (1964), 152-167; 307-334.

BRAUER, R. - H. HASSE - E. NOETHER

1. Beweis eines Hauptsatzes in der Theorie der Algebren. J. für Math. 167 (1931), 399-404.

BRAUER, R. - C. NESBITT

1. On the modular representations of finite groups. Univ. of Toronto Studies, Math. Ser. 4 (1937).

BURNSIDE, W.

1. On an arithmetical theorem connected with roots of unity, and its application to group characteristics. Proc. London Math. Soc. (2), 1 (1904), 112-116.

2. On the complete reduction of any transitive permutation group and on the arithmetic nature of the coefficients in its irreducible components. Proc. London Math. Soc. (2), 3 (1905), 239-252.

3. On the arithmetical nature of the coefficients in a group of linear substitutions of finite order. Proc. London Math. Soc. (2), 4 (1906), 1-9.

4. Theory of groups of finite order. Second edition, Cambridge Univ. Press, Cambridge, 1911.

CARTAN, H. - S. EILENBERG

1. Homological algebra. Princeton, Princeton Univ. Press, 1956.

CHEVALLEY, C.

1. L'arithmétique dans les algèbres de matrices. Act. Sci. Ind. 323 (1936), 1-35.

2. On the theory of local rings. Ann. Math. 44 (1943), 690-708.

3. Fundamental concepts of algebra. Academic Press, New York, 1956.

COHN, P.M.

1. Morita Equivalence and Duality. Queen Mary College, Mathematics Notes.

CONLON, S.B.

1. Twisted group algebras and their representations. J. Austral. Math. Soc. 4 (1964), 152-173.

2. Certain representation algebras. J. Austral. Math. Soc. 5 (1965), 83-99.

CURTIS, C.W.

1. Quasi-Frobenius rings and Galois theory. Ill. J. Math. 3 (1959), 134-144.

CURTIS, C.W. - J.P. JANS

1. On algebras with a finite number of indecomposable modules. Trans. Am. Math. Soc. 114 (1965), 122-132.

CURTIS, C.W. - I. REINER

1. Representation theory of finite groups and associative algebras. Interscience, New York, 1962.

DADE, E.C.

1. Some indecomposable group representations. Ann. of Math. 77 (1963), 406-412.

DADE, E.C. - O. TAUSSKY - H. ZASSENHAUS

1. On the theory of orders, in particular on the semi-group of ideal classes and genera of an order in an algebraic number field. Math. Ann. 148 (1962), 31-64.

DEDEKIND, R.
1. Gesammelte Mathematische Werke. Chelsea Publ. Co., 1968.

DE MEYER, F.R.
1. The trace map and separable algebras. Osaka J. Math. 3 (1966), 7-11.

DEURING, M.
1. Algebren. Springer, Berlin, 1935.

2. Galoische Theorie und Darstellungstheorie. Math. Ann. 107 (1932), 140-144.

DIEUDONNÉ, J.
1. Les déterminants sur un corps non-commutatif. Bull. Soc. Math. France 71 (1943), 27-45.

2. Remarks on quasi-Frobenius rings. Ill. J. Math. 2 (1958), 346-354.

DIEDERICHSEN, F.E.
1. Über die Ausreduktion ganzzahliger Gruppendarstellungen bei arithmetischer Äquivalenz. Hamb. Abh. 14 (1940), 357-412.

DROZD, JU.A.
1. On the representations of cubic Z-rings. Dokl. Akad. Nauk S.S.S.R. 174 (1967), 16-18.

DROZD, JU.A. - V.V. KIRICHENKO
1. On representations of rings, lying in matrix algebras of the second kind. Ukrain. Math. Ž. 19 (1967), 107-112.

DROZD, JU.A. - V.V. KIRICHENKO - A.V. ROITER
1. On hereditary and Bass orders. Izv. Akad. Nauk S.S.S.R. 31 (1967), 1415-1436.

DROZD, JU.A. - A.V. ROITER
1. Commutative rings with a finite number of integral indecomposable representations. Izv. Akad. Nauk S.S.S.R. 31 (1967), 783-798.

DROZD, JU.A. - V.M. TURCHIN
1. On the number of representation modules in the genus for integral matrix rings of 2nd order. Mat. Zametki Akad. Nauk Sojirza S.S.S.R. 2 (1967).

ECKMANN, B. - A. SCHOPF
1. Über injektive Moduln. Archiv der Math. 4 (1953), 75-78.

EICHLER, M.
1. Bestimmung der Idealklassenzahl in gewissen normalen einfachen Algebren. J. reine angew. Math. 176 (1937), 192.

2. Über die Idealklassenzahl hyperkomplexer Systeme. Math. Zeit. 43 (1937), 481-494.

3. Über die Idealklassenzahl total definiter Quaternionenalgebren. Math. Zeit. 43 (1938), 102-109.

4. Allgemeine Kongruenzklasseneinteilung der Ideale einfacher Algebren über algebraischen Zahlkörpern und ihre L-Reihen. J. reine angew. Math. 179 (1938), 227-251.

EICHLER, M.
1. Zur Zahlentheorie der Quaternionenalgebren. J. reine angew. Math. 195 (1955), 127-151.

ENDO, S.
1. Completely faithful modules and quasi-Frobenius algebras. J. Math. Soc. Japan 19 (1967), 437-456.

ENDO, S. - Y. WATANABE
1. The centers of semi-simple algebras over a commutative ring. Nagoya Math. J. 30 (1967), 285-293.

2. On separable Algebras over a Commutative Ring. Osaka, J. Math. 4 (1967), 233-242.

FADDEEV, D.K.
1. On the semi-group of genera in the theory of integral representations. Izv. Akad. Nauk S.S.S.R. 28 (1964), 475-478.

2. An introduction to the multiplicative theory of integral representation modules. Trudy Mat. Inst. Steklov 80 (1965), 145-182.

FEIT, W.J.
1. Characters of finite groups. Benjamin, New York, 1967.

FEIT, W. J. - J. THOMPSON
1. Solvability of groups of odd order. Pacific J. Math. 13 (1963) 775-1029.

FOSSUM, R.
1. Maximal orders over Krull domains. J. of Algebra 10 (1968), 321-332.

FROBENIUS, G.
1. Über Gruppencharaktere. Sitzber. preuss. Akad. Wiss. 1896, 985-1021.

2. Über die Darstellung der endlichen Gruppen durch lineare Substitutionen I, II. Sitzber. preuss. Akad. Wiss. 1897, 994-1015; 1899, 482-500.

3. Über Relationen zwischen den Charakteren einer Gruppe und derer ihrer Untergruppen. Sitzber. preuss. Akad. Wiss. 1898, 501-515.

4. Über die Composition der Charaktere einer Gruppe. Sitzber. preuss. Akad. Wiss. 1899, 330-339.

5. Über die Charaktere der symmetrischen Gruppe. Sitzber. preuss. Akad. Wiss. 1900, 516-534.

6. Über die Charaktere der alternierenden Gruppe. Sitzber. preuss. Akad. Wiss. 1901, 516-534.

7. Über die Äquivalenz der Gruppen linearer Substitutionen. Sitzber. preuss. Akad. Wiss. 1906 b, 209-217.

FROBENIUS, G. - I. SCHUR
1. Über die reellen Darstellungen der endlichen Gruppen. Sitzber. preuss. Akad. Wiss. 1906, 186-208.

FROBENIUS, G. - I. SCHUR
2. Über die Äquivalenz der Gruppen linearer Substitutionen. Sitzber. preuss. Akad. Wiss. 1906, 209-217.

FRÖHLICH, A.
1. Invariants for modules over commutative separable orders. Quart. J. Math., Oxford 2nd series 16 (1965), 193-232.

2. Radical modules over a Dedekind domain. Nagoya Math. J. 27 (1966), 643-662.

3. Resolvents, discriminants and trace invariants. J. of Algebra 4 (1966), 173-198.

FRÖHLICH, A. - A.M. MC EVETT
1. Forms over rings with involutions. J. of Algebra 12 (1969), 79-104.

2. Representations of groups by automorphisms of forms. J. of Algebra 12 (1969), 114-133.

GABRIEL, P.
1. Des catégories abeliennes. Bull. Soc. Math. France 90 (1962), 323-448.

GAUSS, C.F.

1. Werke, 12 vol., Göttingen, 1870-1927.

GREEN, J.A.
1. The indecomposable representations of a finite group. Math. Zeit. 70 (1959), 430-445.

2. A lifting theorem for modular representations. Proc. Roy. Soc. London, 252 A (1959), 135-142.

3. Blocks of modular representations. Math. Zeit. 79 (1962), 100-115.

4. The modular representation algebra of a finite group. Ill. J. Math. 6 (1962), 607-619.

5. A transfer theorem for modular representations. J. of Algebra 1 (1964), 73-84.

GUDIVOK, P.M.
1. Integral representations of groups of type (p,p). Dokl. Uzhgored Univ., ser. Phys.-Mat. Nauk 5 (1962), 73.

2. On p-adic integral representations of finite groups. Dokl. Uzhgorod Univ. ser. Phys.-Mat. Nauk 5 (1962), 81-82.

3. Representations of finite groups over certain local rings. Dopovidi Akad. Nauk Ukrain. RSR (1964), 173-176.

4. Representations of finite groups over quadratic rings. Dokl. Akad. Nauk S.S.S.R. 159 (1964), 1210-1213.

5. Representations of finite groups over number rings. Izv. Akad. Nauk S.S.S.R. 31 (1967), 799-834.

HANNULA, T.A.
1. The integral representation ring a (R_kG). Trans. Am. Math. Soc. 133 (1968), 553.

HANNULA, T. - T. RALLEY - I. REINER
1. Modular representation algebras. Bull. Am. Math. Soc. 73 (1967), 100-101.

HARADA, M.
1. Hereditary orders. Trans. Am. Math. Soc. 107 (1963), 273-290.

2. Structure of hereditary orders over local rings. J. of Math., Osaka City Univ. 14 (1963), 1-22.

3. Multiplicative ideals theory in hereditary orders. J. of Math., Osaka City Univ. 14 (1963), 83-106.

4. On special type of hereditary abelian categories. Osaka J. Math. 4 (1967), 243.

5. Note on orders over which a hereditary order is projective. Osaka J. Math. 4 (1967), 151-156.

HASSE, H.
1. Bericht über neuere Untersuchungen und Probleme aus der Theorie der algebraischen Zahlkörper. Jahresber. dtsch. Math. Verein. 35 (1926); 36 (1927); 39 (1930).

2. Über p-adische Schiefkörper und ihre Bedeutung für die Arithmetik hyperkomplexer Zahlsysteme. Math. Ann. 104 (1931), 495-534.

3. The theory of cyclic algebras over an algebraic number field. Trans. Am. Math. Soc. 34 (1932), 171-214.

HATTORI, A.
1. Rank element of a projective module. Nagoya Math. J. 25 (1965), 113-120.

HECKE, E.
1. Vorlesungen über die Theorie der algebraischen Zahlen. Chelsea, New York, 1948.

HELLER, A.
1. On group representations over a valuation ring. Proc. Nat. Acad. Sci. 47 (1961), 1194-1197.

HELLER, A. - I. REINER
1. Indecomposable representations. Ill. J. Math. 5 (1961), 314-323.

2. Indecomposable representations of cyclic groups. Bull. Am. Math. Soc. 68 (1962), 220-222.

3. Representations of cyclic groups in rings of integers I, II. Ann. of Math. 76 (1962), 73-92; 77 (1963), 318-328.

4. Grothendieck groups of orders in semi-simple algebras. Trans. Am. Math. Soc. 112 (1964), 344-355.

5. Grothendieck groups of integral group rings. Ill. J. Math. 9 (1965), 349-360.

HERSTEIN, N.I.
1. Theory of rings. Lecture notes, Univ. of Chicago, 1961.

2. Topics in ring theory. Lecture notes, Univ. of Chicago, 1965.

HIGMAN, D.G.
1. Indecomposable representations at characteristic p. Duke Math. J. 21 (1954), 377-381.

2. On orders in separable algebras. Can. J. Math. 7 (1955), 509-515.

3. Induced and produced modules. Can. J. Math. 7 (1955), 490-508.

4. Relative cohomology. Can. J. Math. 9 (1957), 19-34.

5. On isomorphisms of orders. Mich. J. Math. 6 (1959), 255-257.

6. Representations of orders over Dedekind domains. Can. J. Math. 12 (1960), 107-125.

HIGMAN, D.G. - J.E. MAC LAUGHLIN
1. Finiteness of the class numbers of representations of algebras over function fields. Mich. Math. J. 6 (1959), 401-404.

HOCHSCHILD, G.
1. Relative homological algebra. Trans. Am. Math. Soc. 82 (1956), 246-269.

HOEHNKE, H.J.
1. Über Beziehungen zwischen Problemen von H. Brandt aus der Theorie der Algebren und den Automorphismen der Normalform. Math. Nach. 34 (1967), 229-255.

ISAACS, I.M. - T. NAKAYAMA
1. Groups with representations of bounded degree. Can. J. Math. 16 (1964), 299-309.

ITO, N.
1. On the degrees of irreducible representations of a finite group. Nagoya Math. J. 3 (1951), 5-6.

JACOBSON, N.
1. Theory of Rings. Math. Surveys II, Am. Math. Soc., 1943.

2. The structure of rings. Am. Math. Soc. Providence, 1956.

JACOBINSKI, H.
1. On extensions of lattices. Mich. Math. J. 13 (1966), 471-475.

2. Sur les ordres commutatifs avec un nombre fini de réseaux indécomposables. Acta Mathematica 118 (1967), 1-31.

3. Über Geschlechter von Ordnungen. J. reine angew. Math. 230 (1968), 29-39.

4. Genera and direct decomposition of lattices over orders. Acta Mathematica 121 (1968), 1-29.

5. On embeddings of lattices belonging to the same genus. To appear, Proc. Am. Math. Soc.

JANS, J.P.

1. On the indecomposable representations of algebras. Ann. of Math. 66 (1957), 418-429.

2. On orders in quasi-Frobenius rings. J. of Algebra 7 (1967), 35-43.

JENNER, W.E.

1. On the class number of non-maximal orders in p-adic division algebras. Math. Scand. 4 (1956), 125-128.

JONES, A.

1. Groups with a finite number of indecomposable integral representations. Mich. J. Math. 10 (1963), 257-261.

2. On representations of finite groups over valuation rings. Ill. J. Math. 9 (1965), 297-303.

KAPLANSKY, I.

1. Modules over Dedekind rings and valuation rings. Trans. Am. Math. Soc. 72 (1952), 327-340.

2. Submodules of Quaternion algebras. Proc. London Math. Soc. 19 (1969), 219-232.

3. Composition of binary quadratic forms. Studia Math. 31 (1968), 523-530.

KASCH, F. - M. KNESER - H. KUPISCH

1. Unzerlegbare modulare Darstellungen endlicher Gruppen mit zyklischer p-Sylow-Gruppe. Archiv der Math. 8 (1957), 320-321.

KATO, T.

1. Characterizations of self-injective rings. Proc. Japan Acad. 44 (1968), 294-297.

KIRICHENKO, V.V.

1. Orders all of whose representations are completely resolvable. Mat. Zametki 2 (1967), 139-144.

KNEBUSCH, M.

1. Elementarteilertheorie über Maximalordnungen. J. reine angew. Math. 226 (1967), 175-183.

KNEE, D.I.

1. The indecomposable integral representations of finite cyclic groups. Ph.D. thesis, M.I.T., 1962.

KNESER, M.

1. Einige Bemerkungen über ganzzahlige Darstellungen endlicher Gruppen. Arch. der Math. 17 (1966), 377-380.

LAM, T.Y.

1. Induction theorems for Grothendieck groups and Whitehead groups of finite groups. Ann. Sci. École Norm. Sup. (4) 1 (1968), 91-148.

2. A theorem on Green's modular representation ring. To appear, J. of Algebra.

3. Artin exponent of finite groups. To appear, J. of Algebra.

LAM, T.Y. - I. REINER
1. Relative Grothendieck rings. Bull. Am. Math. Soc. 75 (1969), 496-498.

2. Relative Grothendieck groups. J. of Algebra 11 (1969), 213-242.

3. Reduction theorems for relative Grothendieck rings. To appear, Trans. Am. Math. Soc.

LANG, S.
1. Algebra. Addison Wesley, 1965.

LARSON, R.G.
1. Group rings over Dedekind domains II. J. of Algebra 7 (1967), 278.

LICHTMAN, A.I.
1. On group rings of p-groups. Isvestia Akad. Nauk S.S.S.R. 27 (1963), 795-800.

MAC LANE, S.
1. Homology. Springer, Berlin, 1963.

2. Categorical algebra. Bull. Am. Math. Soc. 71 (1965), 40-106.

MARANDA, J.M.
1. On P-adic integral representations of finite groups. Can. J. Math. 5 (1953), 344-355.

2. On the equivalence of representations of finite groups by groups of automorphisms of modules over Dedekind domains. Can. J. Math. 7 (1955), 516-526.

MASCHKE, H.
1. Über den arithmetischen Charakter der Coeffizienten der Substitutionen endlicher linearer Substitutionsgruppen. Math. Ann. 50 (1898), 482-498.

MICHLER, G.
1. Halberbliche, fastlokale Ordnungen in einfachen Artinringen. Archiv. Math. 28 (1967), 456.

2. Asano orders. Proc. London Math. Soc. (3) 19 (1969), 421-443.

3. Idempotent ideals in perfect rings. Can. J. Math. 21 (1969), 301-309.

MILNOR, J.W.
1. On the Whitehead homomorphism. Bull. Am. Math. Soc. 64 (1958), 444-449.

2. Whitehead torsion. Bull. Am. Math. Soc. 72 (1966), 358-426.

MORITA, K.
1. On group rings over a modular field which possess radicals expressible as principal ideals. Sci. Rep. Toyo Bemzika Daigaku 4 (1951), 177-194.

2. Duality of modules. Science reports of the Tokyo Kyoiku Daigaku, Sec. A, 6, no. 150 (1958), 83-142.

3. The endomorphism ring theorem for Frobenius extension. Math. Zeit. 102 (1967), 385.

MURTHY, M.P.
1. Modules over regular local rings. Ill. J. Math. 7 (1964), 558-565.

NAKAYAMA, T.
1. On Frobenius algebras I, II. Ann. of Math. 40 (1939), 611-633; 42 (1941), 1-21.

NAGAO, H.
1. A proof of Brauer's theorem on generalized decomposition numbers. Nagoya Math. J. 22 (1963), 73-77.

NAGATA, M.
1. Local rings. Interscience, New York-London, 1962.

NAZAROVA, L.A. - A.V. ROITER
1. Correction of one theorem of Bass. Dokl. Akad. Nauk S.S.S.R. 176 (1967), 266-268.

NEHRKORN, H.
1. Über absolute Idealklassengruppen und Einheiten in algebraischen Zahlkörpern. Abh. math. Sem. Hamburg Univ. 9 (1933), 318-334.

NOETHER, E.
1. Hyperkomplexe Grössen und Darstellungstheorie. Math. Zeit. 30 (1929), 641-692.

2. Nichtkommutative Algebra. Math. Zeit. 37 (1933), 513-541.

NORTHCOTT, D.G.
1. Introduction to Homological Algebra. Cambridge, 1962.

NUNKE, R.J.
1. Modules of extensions of Dedekind rings. Ill. J. Math. 3 (1959), 222-241.

OBAYASHI, T.
1. On the induction theorem in Whitehead torsions of finite groups. Sci. Reports Tokyo Kyoiku Daigaku 9 (1968), 177-183.

O'REILLY, M.F.
1. On the semi-simplicity of the modular representation algebra of a finite group. Ill. J. Math. 9 (1965), 261-276.

PLOTKIN, B.J.
1. Some questions in the general theory of group representations. Isv. Akad. Nauk S.S.S.R. 27 (1963), 855-882.

QUILLEN, D.G.
1. On the associated graded ring of a group ring. J. of Algebra 10 (1968), 411.

REINER, I.
1. Maschke modules over Dedekind rings. Can. J. Math. 8 (1956), 329-334.

2. Integral representations of cyclic groups of prime order. Proc. Am. Math. Soc. 8 (1957), 142-146.

3. On the class number of representations of an order. Can. J. Math. 11 (1959), 660-672.

REINER, I.

4. The non-uniqueness of irreducible constituents of integral group representations. Proc. Am. Math. Soc. 11 (1960), 655-657.

5. The behavior of integral group representations under ground ring extension. Ill. J. Math. 4 (1960), 640-651.

6. The Krull-Schmidt theorem for integral representations. Bull. Am. Math. Soc. 67 (1961), 365-367.

7. Indecomposable representations of non-cyclic groups. Michigan Math. J. 9 (1962), 187-191.

8. Failure of the Krull-Schmidt theorem for integral representations. Michigan Math. J. 9 (1962), 225-232.

9. Extensions of irreducible modules. Michigan Math. J. 10 (1963), 273-276.

10. The integral representation of a finite group. Michigan Math. J. 12 (1965), 11-12.

11. Nilpotent elements in rings of integral representations. Proc. Am. Math. Soc. 17 (1966), 270-274.

12. Integral representation algebras. Trans. Am. Math. Soc. 124 (1966), 111-121.

13. Relations between integral and modular representations. Michigan Math. J. 13 (1966), 357-372.

14. Module extensions and blocks. J. of Algebra 5 (1967), 157-163.

15. Representation rings. Michigan Math. J. 14 (1967), 385-391.

16. An involution in $K^0(ZG)$. Mat. Zametki 3 (1968), 532-527.

REINER, I. - H. ZASSENHAUS

1. Equivalence of representation under extensions of local ground rings. Ill. J. Math. 5 (1961), 409-411.

RILEY, J.A.

1. Reflexive ideals in maximal orders. J. of Algebra 2 (1965), 451-465.

RIM, D.S.

1. Modules over finite groups. Ann. of Math. 69 (1959), 700-712.

2. On projective class groups. Trans. Am. Math. Soc. 98 (1961), 459-467.

ROBERTS, L.G.

1. K_1 of some abelian categories. Trans. Am. Math. Soc. 138 (1969), 377-382.

ROBSON, J.C.

1. Non-commutative Dedekind rings. J. of Algebra 9 (1968), 249-265.

ROITER, A.V.

1. Categories of representations. Ukrain. Mat. Zh. 15 (1963), 448-452.

ROITER, A.V.

2. Categories with division and integral representations. Dokl. Akad. Nauk S.S.S.R. 153 (1963), 46-48.

3. Divisibility in categories of representations over a complete Dedekind domain. Ukrain Mat. Žh. 17 (1965), 124-129.

4. Integer valued representations belonging to one genus. Izv. Akad. Nauk S.S.S.R. 30 (1966), 1315-1324.

5. Analogon to a theorem of Bass for representation modules of non-commutative orders. Dokl. Akad. Nauk S.S.S.R. 168 (1966), 1261-1264.

6. The unboundedness of the dimensions of the indecomposable representations of algebras that have an infinite number of indecomposable representations. Izv. Akad. Nauk S.S.S.R. 32 (1968), 1275-1282.

ROGGENKAMP, K.W.

1. Darstellungen endlicher Gruppen in Polynombereichen. Mitt. Math. Sem. Giessen 71 (1967), 1-72

2. Gruppenringe von unendlichem Darstellungstyp. Math. Zeit. 96 (1967), 393-398.

3. Darstellungen endlicher Gruppen in Polynomringen. Math. Zeit. 96 (1967), 399-407.

4. Das Krull-Schmidt Theorem für projektive Gitter über lokalen Ringen. Mitt. Math. Sem. Giessen 80 (1969), 29-50.

5. Grothendieck groups of hereditary orders. J. reine angew. Math. 235 (1969), 29-40.

6. On irreducible lattices of orders. To appear, Can. J. Math.

7. Projective modules over clean orders. To appear, Comp. Math.

8. Charakterisierung von Ordnungen in einer direkten Summe kompletter Schiefkörper, die nur endlich viele nicht isomorphe unzerfällbare Darstellungen haben. Habilitationsschrift, Giessen, 1969, 1-122.

9. A necessary and sufficient condition for orders in direct sums of complete skewfields to have only finitely many non-isomorphic indecomposable representations. To appear, Bull. Am. Math. Soc.

10. R-orders in a split algebra have finitely many non-isomorphic lattices as soon as R has finite class number. To be published.

ROGGENKAMP, K.W.

11. A remark on separable orders. To appear, Can. Bull. Math.

12. Projective ideals in clean orders. To appear, Comp. Math.

13. Projective homomorphisms and extensions of lattices. To be published.

14. A counterexample to a conjecture of Roiter. Notices Am. Math. Soc. 14 (1967), 530, 67T-372.

15. Swan's version of Jacobinski's cancellation law. Mimeo. notes, Univ. of Illinois, 1969, 1-27.

16. Jacobinski's cancellation law for orders in separable algebras over function fields of a finite field. To be published.

SAKSONOV, A.I.

1. Some integral rings associated with a finite group. Soviet Math. 7 (1966), no. 6, 1315.

2. On group rings of finite p-groups over some integral domains. Dokl. Akad. Nauk S.S.S.R. 11 (1967), 204-207.

3. Group algebras of finite groups over a number field. Dokl. Akad. Nauk S.S.S.R. 11 (1967), 302-305.

SCHUR, I.

1. Über die Darstellung der endlichen Gruppen durch gebrochene lineare Substitutionen. J. für Math. 127 (1904), 20-50.

2. Neue Begründung der Theorie der Gruppencharaktere. Sitzber. preuss. Akad. Wiss. (1905), 406-432.

3. Untersuchungen über die Darstellung endlicher Gruppen durch gebrochene lineare Substitutionen. J. für Math. 132 (1907), 85-137.

4. Über die Darstellung der symmetrischen und der alternierenden Gruppen durch gebrochene lineare Substitutionen. J. für Math. 139 (1911), 155-250.

SERRE, J.P.

1. Modules projectifs et espaces fibrés à fibre vectorielle. Sem. Dubreil, Paris, 1958.

2. Corps Locaux. Herman, Paris, 1962.

3. Sur les modules projectifs. Sem. Dubreil, Paris, 1960.

4. Algèbre Locale;Multiplicités. Lecture notes in Math. Springer, Berlin-Heidelberg-New York, 1965.

5. Représentation linéaires des groupes finis. Herman, Paris, 1968.

SHUKLA, U.

1. On the projective cover of a module and related results. Pacific J. Math. 12 (1962), 709-717.

SPEISER, A.
1. Allgemeine Zahlentheorie. Vierteljahresschrift d. Naturf. Ges. Zürich 71 (1926).

STANCEL, D.L.
1. Multiplication on the Grothendieck ring of an integral group ring. Bull. Am. Math. Soc. 73 (1967), 92-94.

STEINITZ, E.
1. Rechteckige Systeme und Moduln in algebraischen Zahlkörpern I, II. Math. Ann. 71 (1911), 328-354; 72 (1912), 297-345.

STROOKER, J.R.
1. Faithfully projective modules and clean algebras. Proefschrift, Utrect, 1965.

SWAN, R.G.
1. Projective modules over finite groups. Bull. Am. Math. Soc. 65 (1959), 365-367.
2. Induced representations and projective modules. Ann. of Math. 71 (1960), 552-578.
3. Periodic resolution for finite groups. Ann. Math. 72 (1960), 267-291.
4. Projective modules over group rings and maximal orders. Ann. of Math. 76 (1962), 55-61.
5. The Grothendieck ring of a finite group. Topology 2 (1963), 85-110.
6. The numbers of generators of a module. Math. Zeit. 102 (1967), 318-322.
7. Algebraic K-Theory. Lecture-notes in Math. 76 (1968), Springer Verlag, Berlin-Heidelberg-New York.

TAKAHASHI, S.
1. Arithmetic of group representations. Tohoku Math. J. 11 (1959), 216-246.

TSUSHIMA, Y.
1. Radicals of group algebras. Osaka J. Math. 4 (1967), 179.

VASCONCELOS, W.V.
1. Reflexive modules over Gorenstein rings. Proc. Am. Math. Soc. 19 (1968), 1349.

WANG, S.
1. On the commutator group of a simple algebra. Am. J. Math. 72 (1950), 323-334.

WEDDERBURN, J.H.M.
1. Lectures on matrices. Am. Math. Soc., New York, 1934.

WEIL, A.
1. Basic number theory. Springer Verlag, New York, 1967.

WIELAND, H.
1. Permutation representations. Ill. J. Math. 13 (1969), 91-94.

WILLIAMSON, S.
1. Equivalence classes of maximal orders. Nagoya Math. J. 31 (1968), 131-172.

WITT, E.
1. Schiefkörper über diskret bewerteten Körpern. J. reine angew. Math. 176 (1937), 153-156.

WU, L.E.T. - J.P. JANS
1. On quasi-projectives. Ill. J. Math. 11 (1967), 439.

ZARISKI, O. - P. SAMUEL
1. Commutative Algebra I, II. Van Norstrand, 1963.

ZAKS, A.
1. Residue rings of semi-primary hereditary rings. Nagoya Math. J. 30 (1967), 279.

ZASSENHAUS, H.
1. Neuer Beweis der Endlichkeit der Klassenzahl bei unimodularer Äquivalenz ganzzahliger Substitutionsgruppen. Hamb. Abh. 12 (1938), 276-288.

2. Units in orders. Mimeo. notes, Ohio State Univ., 1966.

3. Über die Äquivalenz ganzzahliger Darstellungen. Nachrichten d. Akad. d. Wissenschaften in Göttingen 12 (1967).

INDEX